Professor Richard K Chang

A Lifelong Teacher and Friend

Professor Richard K Chang

A Lifelong Teacher and Friend

Edited by

Ali Serpengüzel
Koç University, Türkiye

Sung-Wen (Sunny) Chang

Grace Metcalfe

PY Nicole Chang

Hui Cao
Yale University, USA

PM Natasha Chang

PH Kevin Chang

KK Phua
World Scientific Publishing Company, Singapore

World Scientific

NEW JERSEY • LONDON • SINGAPORE • BEIJING • SHANGHAI • TAIPEI • CHENNAI

Published by

World Scientific Publishing Co. Pte. Ltd.
5 Toh Tuck Link, Singapore 596224
USA office: 27 Warren Street, Suite 401-402, Hackensack, NJ 07601
UK office: 57 Shelton Street, Covent Garden, London WC2H 9HE

Library of Congress Control Number: 2025936138

British Library Cataloguing-in-Publication Data
A catalogue record for this book is available from the British Library.

PROFESSOR RICHARD K CHANG
A Lifelong Teacher and Friend

ISBN 978-981-98-0324-8 (hardcover)
ISBN 978-981-98-0325-5 (ebook for institutions)
ISBN 978-981-98-0326-2 (ebook for individuals)

For any available supplementary material, please visit
https://www.worldscientific.com/worldscibooks/10.1142/14094#t=suppl

Desk Editor: Carmen Teo Bin Jie

Contents

Foreword

This volume contains articles dedicated to the memory of Professor Richard K. Chang, an outstanding experimental physicist, who was a faculty member of Yale University for more than 40 years. This memorial volume features the recollections of Richard Chang and the contributions by his former colleagues, students, family, and friends. Professor Richard Chang will always be remembered.

K.K. Phua
Professor of Physics
Singapore
Chairman and Editor-in-Chief
World Scientific Publishing
Singapore

Photographs

Chang Family Album

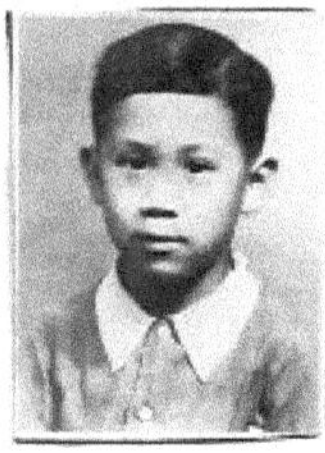

in Chunking

in Tokyo

in Tokyo

1956 in The Stoney Brook School with roommate Larry Foster. Larry's father was professor of economics at Lincoln University in Pennsylvania and was adviser to the first president of Ghana, Kwame Nkrumah. Richard spent Thanksgiving and Christmas with Larry's family.

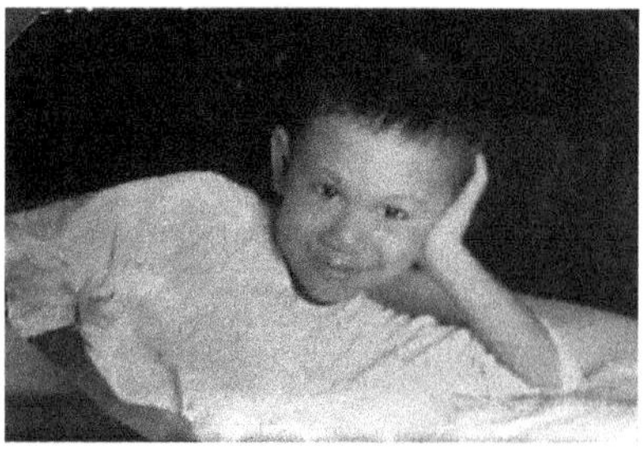

in Chongqing

1960 at Sunny's home in New York City.

Chang family's move from Harvard University, Cambridge MA, to Yale University, New Haven, CT, summer 1966.

Chang family's visit to Sunny's sister in Woodstock, VT in the 1970s.

Nicole and Valdez, 1969.

Visit to the summer home of Yale Colleague Art and Kaye Ross in Maine, 1970's.

Family on Triennial Leave in India, 1983.

Family in Paris, 1983.

Front: Franz Tuteur, Dorothy Greenhouse, Wei-Wen Chang (Sunny's sister).
Back: Sunny, Richard, visiting scientist, Nai-Yong Chang (Sunny's brother-in-law).

Kevin, Mutsuko Kobayashi, Sunny, Susie, and Jean Webber.

Mutsuko Kobayashi, Takao Kobayashi, Richard, Kobayashi's son, Sunny, Trish Niece, Bob Benner, Nicole, Kevin.

Stanley Mroczkowski and Marshall Long.

Memorial for Sergio Porto in Brazil. Richard, Ali Javan, and Eli Bernstein, 1980.

Richard's parents in California, 1980's.

Susie Engelman, 1980's.

Kim Marie Mertz (Richard's goddaughter), daughter of Rose-Marie and James Mertz (Notre Dame and UC Santa Barbara).

In the Caribbean — favorite place. 1984.

One of Richard's favorite pastimes.

Richard shaking hands with Deng Xiaoping at China's first international conference on lasers in Beijing, 1980.

Sunny and Natasha in JinJoo Song's parents' home in Seoul, Korea, 1986.

Nicole, Kevin, and Natasha Chang shaking hands with Deng Xiaoping at China's first international conference on lasers in Beijing, 1980.

Picnic with Prof. Crocker and family in Hawaii, 1986.

JinJoo Song's parents, Mr. and Mrs. In Sang Song, in Cambridge, MA, 1980s.

JinJoo Song's mother in Seoul, Korea, 1986.

Pang-Mei Natasha's Harvard graduation, 1987.

Richard going to the tomb inside the pyramid in Cairo, 1988.

Richard and Kevin in Egypt, 1988.

Richard and Peter Schultheiss planting an azalea in memory of Richard's dad in 1985.

Ulla and Peter Schultheiss planting an azalea in memory of Richard's dad in 1985.

Richard in Helsinki as "Opponent" in J.P. Laine's Ph.D. Defense, 2003.

Prof. Nicolaas Bloembergen, wife Deli and Sunny in Tucson, AZ, 2003.

Home office in Hamden, CT

Visiting the Petting Farm
in VT with grandson
Ming-Wei in 2001.

Richard's 60th birthday party in New Haven, CT.

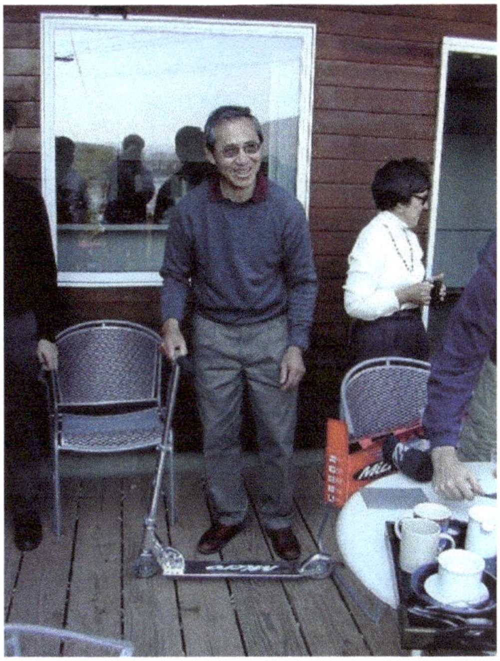

Richard's 60th birthday party in Brattleboro, VT.

The Chang Family in Hamden, CT in the 1980s.

Chang Family in far east, 2012.

Visiting India in 2011.

Richard reading his Festschrift in Sri-Lanka in 2011.

Preface

Shortly after Richard's passing, Professor K.K. Phua, Chairman and Editor-in-Chief of World Scientific Publishing suggested that World Scientific Publishing publish a "memorial book" on him. At the time, I did not really understand what a memorial book was and how it would be different from the scientific books that World Scientific Publishing usually produced. Over a lunch Professor Phua explained the difference between a memorial book and a regular scientific book. A memorial book would be more anecdotal, reflecting personal interactions Richard had had with his students, friends, or colleagues. Naturally, these interactions could include scientific discussions or thoughts around solving scientific problems. Or possibly the impact Richard had on some aspects of his students' lives, perhaps highlighting Richard's character in a way not reflected in his scientific papers. There would be more insights into the relationships and interconnections among fellow scientists.

My first thought was "Who would be interested in such a book?" Would the processes that went into his scientific research be interesting enough? Would his life as a professor be interesting enough? As his wife, of course I thought he was wonderful and unique. However, he was not a celebrity, and his life was not all that different from those of the many other wonderful professors we knew at Yale and at other universities. Would people be interested in getting a glimpse into the relationships between a professor and his students? Or the global relationships formed by scientists from all different parts of the world bound together to unravel the mysteries of nature? Professor Phua knew several of Richard's students and thought that Richard had had a very special relationship with them. He suggested that one of Richard's former students might be able to take on the lead of this project. That was when I remembered a handwritten note from the late Professor Allan Bromley (Yale).

Professor Allan Bromley's note followed a celebration in the autumn of 2000 when Richard's former students, led by Professor Marshall Long (Yale), organized a special day to celebrate Richard's 60th birthday. It was a memorable event where former students, colleagues, friends, and family from all over the world gathered for the celebration. Many former students gave short scientific presentations on "Life after Yale" in the afternoon, before attending a dinner at the Yale Graduate Club. The after-dinner speakers included students, colleagues, and family members. Professor Alan Bromley, then Dean of Engineering and Applied Physics was among the attendees of this birthday celebration. After this event, he sent a note to Richard. A part of that note said: *"In my 75 years I have had the privilege of attending a great many birthday celebrations. But your last evening was unique. I have never witnessed such an outpouring of deeply felt love and admiration both from your children — who would be the envy of any parent — and from your graduate students to whom you have obviously contributed so much!"*

Thinking back to Professor Bromley's note, I felt encouraged to follow up on Professor Phua's suggestion of a memorial book for Richard. Richard's former Ph.D. student Ali Serpengüzel, Professor of Physics at Koç University in Istanbul, Türkiye took the lead on this project. Other editors involved were: Hui Cao, Professor of Applied Physics at Yale, Professor K.K. Phua of World Scientific Publishing, Richard's former Ph.D. student Grace Metcalfe, our children Pang-Hua Kevin Chang, Pang-Mei Natasha Chang and Pang-Yuan Nicole Chang, and myself.

Richard happily retired from Yale in 2008 because he knew that his beloved laboratory in the basement of Becton Center would be taken over by a brilliant scientist and wonderful human being, Professor Hui Cao, then from Northwestern University. When Ali Serpengüzel and I started to reach out to potential contributors for this Memorial Book, many people responded that they could not write anything better than what they had already written in a "memory book" for Richard when he retired from Yale in 2008. Grace Metcalfe and P.Y. Nicole Chang had put together a "memory book" filled with letters and photos from former students, colleagues, family, and friends as a surprise retirement gift for Richard. They presented that book to him at the end of his retirement dinner in 2008.

This Memorial Book contains many contributions from that 2008 surprise memory book, new contributions to this Memorial Book, and various letters to me written upon learning of Richard's passing. Deciding how to organize the contents of this Memorial Book was not an easy task. We ultimately decided to order the contents chronologically, based on the time they first met or worked with Richard. The decision to do it in a chronological order made the most sense. We hope you will agree.

I would like to personally thank everyone who contributed to this book. This book would not be possible if they had not generously shared their memories, experiences, insights, and thoughts. Thank you, contributors, for your generosity!

I would particularly like to thank Professor Ali Serpengüzel who selflessly gave up many hours on Sundays for over a year to devote to this project.

As we read the various contributions from different individuals, a few common threads stood out: the universality of scientists working together without regard for politics; there were no boundaries for discussions of basic science, although strict boundaries were set for never discussing classified matter; how deeply a professor can influence a student's career, or even life in the future; how one can always learn from others regardless of age or stature; and how one's actions matter a great deal as examples of things to avoid or to emulate. I was surprised how many people were encouraged to face their own adversities with a better attitude after witnessing Richard's strength in facing his many health challenges.

As many people referred to Richard's health issues, I have appended a summary of his challenges with his near drowning episode in 2008. This accident was in addition to his ongoing Parkinson's Disease, a traumatic brain injury in 2009 from a general anesthesia for a cataract operation, and a terrible fall in 2014 where he hit his head hard on a marble windowsill in our apartment.

The following are the 2 emails I sent to family and friends updating them on the status of Richard's health after his near drowning episode and a 3rd email announcing his death.

Sung-Wen (Sunny) Pu Chang, Ed.D.

July 6, 2008 — Richard's Near Drowning Episode

Richard had a major accident (a near drowning episode) on day 6 of his retirement in July 2008. He was in our Club swimming pool in Brattleboro, Vermont when he had a very bad leg cramp that made him sink to the bottom of the 10 ft deep pool. I was by the pool with 3 of our very young grandchildren and another couple. Thinking he was practicing for his upcoming scuba diving trip, I did not think too much about his being down 10 feet in the water. However, when he did not come up, our friend dove in to get him up. He was not successful the first time, but on the second attempt, he was able to pull Richard up and his wife did compressions on Richard's chest while they swam towards the edge of the pool. I called 911 and pulled all three of them toward the shallow end of the pool while our three little grandchildren watched in horror.

Richard was not breathing and had no heartbeat when we got him out of the pool. The rescue team stuck something into his throat to get him breathing in a gasping sort of way and gave him atropine and epinephrine to get his heart going. The ambulance brought the comatose Richard to the Brattleboro Memorial Hospital. Some hours later, I learned that they thought the likelihood of his waking up from his comatose state was around 2%. One doctor explained that it was as if he had a massive stroke, even though he did not have one.

I happened to remember that an acupuncturist doctor friend of ours had told me that in China, they give acupuncture to patients immediately after they suffer a stroke. I got permission from the hospital to allow our acupuncturist doctor to treat Richard in the hospital. The accident occurred on Sunday afternoon and the acupuncturist came on Monday evening. More than two hours and about 300 acupuncture needles later, we got no response. On Tuesday morning the acupuncturist doctor came again and another two hours and 300 plus needles later, there was still no response from Richard. Tuesday afternoon, however, Richard woke up and said that he was hungry. The hospital staff referred to Richard as the miracle man. Since then, I have been making a point to share with others the possibility of using acupuncture in cases of strokes.

November 6, 2008 — Update on Richard's Condition

It was four months ago today that Richard had his terrible "near drowning" episode in a swimming pool only six days after his retirement from 42 years of teaching at Yale. As many of you already know, he was not breathing and had no heartbeat when we pulled him out of the pool. It is truly a miracle that he is doing as well as he is, and continues to recover daily, albeit slowly. This miracle can be attributed to the combination of the wonders of modern Western medicine, the marvels of Chinese acupuncture, and the many prayers from all of you around the world.

When I get tired or a little discouraged, I think back to the afternoon he woke up from his 52 hours of comatose state and the immense feeling of gratitude I had that Richard was given a chance to recover at all. Richard was dubbed "The Miracle Man" by the staff at the Brattleboro Memorial Hospital in Vermont. Our acupuncturist's business has increased tremendously from recommendations by the staff at the hospital since Richard's near drowning episode.

After 11 days in the hospital where he became stabilized and out of danger, he was sent to the Cheshire Medical Center (a hospital affiliated with Dartmouth Medical School) in Keene, New Hampshire to receive <u>acute</u> rehabilitation. There he received intensive physical, occupational, recreational and speech therapy 3–6 hours daily, 5 days per week and 1–3 hours per day during the weekends. As soon as he was physically fit enough to walk on his own and transition from walking to a chair or a bed (4 weeks' time), he was discharged to a <u>sub-acute</u> rehab center for further rehabilitation. Whether one remains at an acute rehab center (as opposed to a subacute rehab center) is dependent upon the amount of physical therapy needed and not dependent upon speech therapy where most of the cognitive skills are addressed.

From the Cheshire Medical Center, Richard moved to the Keene Center in Keene, New Hampshire. This is the facility where Richard spent the most time (8 weeks). The rehabilitation section was combined with the nursing care section so there were many patients whom Richard described as being "not with it." On a "special event permission to leave the facility" (my mother's 95th birthday celebration) when Richard was able to spend one night in our Vermont home, I noticed how happy he was to be interacting with people he knew prior to his accident. I thought it would be better for him to be in a rehab center located in Brattleboro, Vermont rather than a 35-minute drive away in Keene, New Hampshire. This is where he is now…Eden Park, in Brattleboro.

The facilities at Eden Park are not as nice as at the Keene Center in New Hampshire, but there is a wonderful speech therapist with whom Richard has bonded very well. Of course, being closer to home makes a world of difference in terms of visits from family and friends. Unfortunately, Richard had a demented roommate who had been injured in a car accident and who called out continually "Help me! Help me!" This was not a good situation because Richard would get up and try to help him. Fortunately, Richard's bed had an alarm so that when he got out of bed, the nurses would be alerted. Last week he spoke up during a "family meeting" with the therapists and nurses about the desperate cries of his roommate and how they were keeping him from sleeping. Luckily, he was moved to a newly available private room this week.

Being a close observer of Richard's recovery process, I am simply amazed at the neuroplasticity of the brain. One can really notice the little improvements with time and the returning of memory as the brain begins its healing process. Some traits/characteristics return immediately, while others require more coaxing and cueing. For example, the first noticeable trait that came back was his sense of humor and wit. During the first few days of his waking up from his comatose state, I would keep asking him "Do you know who I am?" After a few days of this, he answered "No, I don't know who you are, Sung-Wen Pu Chang." Now that he is much more cognizant of his surroundings, he is constantly challenging the nurses, arguing that he should be allowed to go up and down the elevator at will, roam around as he pleases, etc. When the nurses explain the rules and regulations, he asks them, "Are you running an army base here?" Richard continues to behave in his "gentlemanly manner" to the dismay and amusement of the nurses/staff. Because of space problems, the nursing patients are frequently mixed in with the rehab patients on the same floor. Consequently, Richard is often surrounded by patients in wheelchairs who ask him to push them "here or there." Naturally, he begins to do just that. When I am with him, I steer him away from doing such a thing. The nurses are, of course, afraid that he cannot be trusted not to fall himself, let alone push around other patients. So, they quickly rush to get him away from the patients in wheelchairs.

Until he moved to the facilities at Eden Park in Brattleboro, VT, I had been spending most nights and days with him, trying to reinforce what the therapists had taught him. During the few times he had a roommate, and I could not spend the night with him, I usually returned to find him all packed and ready to leave the facility upon my arrival. My mother and my sister-in-law Sen have been simply fantastic about preparing food for Richard and me while we stayed at these medical facilities. Chinese food never tasted so good as after you have been eating hospital food for a while.

It is difficult to express to you how very touched and appreciative we are of your thoughtful acts. From all over the world and all different parts of America, your flowers, plants, cards, foods, DVDs, books, emails, and telephone calls provided us with encouragement and sustainability. Your visits and prayers have all meant so much to us. I especially appreciated them because they gave me hope and strength to cope with Richard's recovery process. They made us realize how very much we appreciated friends and family. Richard would read each card and say to me, "How very thoughtful of him to think of me." or "Tell me, how did I know this person? Where is she now?" I pasted all the cards, notes from plants, flowers, DVDs, etc. into a notebook. Richard reads through these regularly and we talk about the friends who sent them. Often tears well up and he might say, "They are so nice." He carries two books around with him all the time: the "Retirement Memory Book" that Grace Chern Metcalf and P.Y. Nicole Chang put together with over 80 letters from you, and a book of short articles by Richard Feynman sent to him from Paul Fleury.

In the beginning of his recovery, he would sometimes have difficulty putting a name to a face, but that is improving as well. I kept the loaded digital picture frame from Giselle, Maria, and Terri (the Yale Applied Physics Administrative Trio) running so that he could see the faces of the various individuals. I haven't learned how to load extra pictures into them yet. He remembers certain things very well but cannot remember his retirement party or that he is really retired. Fortunately, Dan Prober videotaped the event and sent him a DVD of it. Jayne Miller took Dan's DVD and combined it with Richard's favorite music and the photographs taken by the official Yale photographer to make it into an even more complete rendition of his very special retirement party. Richard watches this DVD regularly so that some of the memory of the event is coming back to him.

On a side note, the few things he talked about when he first woke up were the things that had been on his mind right before the accident. He needed to write thank-you letters to those who had attended his retirement party. He needed to send off the corrections for his OPN article. Check whether the Senior Research Scientist working with him got his U.S. citizenship. It is interesting that now that he is more coherent, he does not really remember his retirement party or the fact that he was adamant about not wanting to be in a teaching hospital in early July when the fresh (and inexperienced) interns arrive.

About one month after the accident, he remembered what had happened to him in the swimming pool. He had a Charley Horse (a very bad leg cramp) in his left leg and all he could think about was the terrible pain in his leg. This made sense because once the doctors had ruled out a stroke or a heart attack, we were not sure what had happened. I had thought that perhaps one of his Parkinson's medications had made him freeze, but the doctor at the rehab center thought that was unlikely.

We are expecting that Richard will be able to come home in about two weeks' time. We will probably stay in Vermont for the Thanksgiving holidays and then decide whether to return to Connecticut in December or wait until after the New Year's. He will still have a great deal of work ahead of him to fully recover, but given how far he has come, we are quite optimistic. As expected, there are times when he is confused about location and time sequence, but usually only when he is tired. His doctor tells us that adequate sleep is critical for the healing process of the brain so it is something I must keep reminding myself when he gets home.

I want to thank each one of you again for your part in helping Richard with this most difficult journey of his life. The fact that he is doing so well is because of the faith, love, and concern you have for him. I do believe God heard our prayers.

Warmly,
Sunny

February 17, 2020 — In Memoriam

Dear Friends,

It is with great sadness that I write to you about Richard's passing on Sunday February 16, 2020, only 4 months before his 80th birthday. He died from sudden cardiac arrest, just two hours after we had been joking with the maids.

Although I am heartbroken, I am immensely grateful for the almost 12 years that he was with me after his near drowning episode. That episode enabled him to visit Heaven briefly where he saw and spoke to deceased family members and friends. The experience was so vivid in his mind that it took me almost 4 years to finally convince him that his sister and Susie Engelman, his devoted secretary at Yale, were truly dead. Ironically, his description of his experience in Heaven is now helping me to cope with his death. I can see him there at the prime of his life happily doing all the things he was not able to do in the last years of his life on earth.

We will be holding his funeral services in Singapore this coming Friday and Saturday. I will be bringing his ashes back to the US sometime in the future.

Please share the news with any of his friends and associates that I might have left out.

Sunny

Gift to Richard and Sunny by well-known painter Professor Xiong Qichuan from Guizhou, China, August 2019.

Tribute to My Father

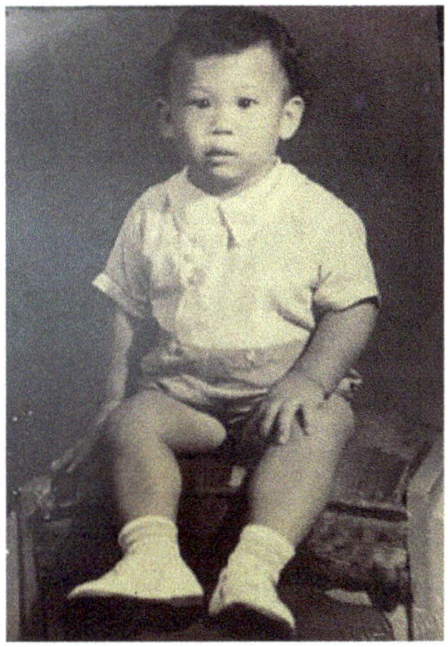

Pang-Hua Kevin Chang, Ph.D.
Singapore
kevin.chang@alum.mit.edu

In the spring of 1972, my 5th grade teacher Miss Davis gave our class an assignment that I'll never forget: each of us had to interview our father and write up a mini-biography, a sort of "Getting to know my father" assignment. So, though Dad's childhood was quite complex, I learned the chronology by heart and remember it still:

- 1940 born in Hong Kong, then returned to the Nationalist base of Chongqing as Shanghai was occupied by the Japanese until the end of WWII.
- 1945–49 back to the family base in Shanghai until the Communist takeover.
- 1949–50 taking stock in Hong Kong following the Nationalist loss to the Communists.
- 1950–54 Tokyo, Japan, which had the promise of growth and reconstruction.
- 1954–56 São Paulo, where his parents next settled (as had other Chinese businessmen).
- Feb. 1956–June 1957, as a boarding student at the Stony Brook School, NY.
- 1957–61 MIT undergraduate studies in Electrical Engineering.
- Sept. 1961 Married my mother Sunny whom he had met at Thanksgiving 1959.
- 1961–62 Harvard M.A. in Applied Physics.
- 1962–65 Harvard Ph.D. in Applied Physics.
- 1965–66 Harvard Postdoctoral year.
- 1966– First job as an Assistant Professor at Yale University, with the family settling in Hamden, Connecticut.

Although intensely dedicated to his scientific research, my father regularly told us stories at bedtime, often including animated synopses of action movies such as "Dirty Harry," "The Mechanic," or "The Magnificent Seven." Dad and Mom's one night out sometimes turned into multiple nights of bedtime narrative for me and my sisters. Dad loved these movies — and found them to be the perfect setting in which to unwind when deeply immersed in a scientific problem. With movies often merely a means to an end, my parents sometimes found themselves inadvertently repeating movies Dad had previously seen while preoccupied with other thoughts and ideas.

A recurrent theme of some of these action films was the unexpected victory of the underdog, the unfavored "long shot" of the horse race, the unsung hero, or the underrated "good guy". After overcoming obstacles and disappointments, and making sacrifices, somehow the protagonist manages to achieve heroic deeds. Of course, the path is never easy, and the character may face humbling setbacks along the way, but ultimately he emerges from the struggle as a winner.

The themes of sacrifice and humility were hallmarks of Dad's real life as well. Even as a chair professor at Yale, and the author of a few hundred academic publications, Dad saw himself modestly as the little "good guy" just trying to do his best in a world where privilege must never be taken for granted. For example, he typically put his name last in papers he co-authored with his students and postdocs. When socializing with work contacts inside or outside Yale, he would regularly cover expenses from his own pocket, rather than use work funds or take from others. I even remember finding some of our prized back-to-school stationery purchases in his office, and (disappointingly from a child's perspective) hardly ever the reverse! When his secretary had trouble with the icy winter walk from her parking space to the office, he offered to trade his nearer professor's spot with hers.

In listening to Dad's bedtime stories, we learned not only about these movie heroes, but also about his own childhood, with anecdotes that became part of our family legend. These vignettes echoed Dad's identification as the humble underdog, a hopeful but vulnerable adventurer in a challenging world.

A few examples include:

- The time he threw a ball over a wall in São Paulo (Brazil) and accidentally hit a passerby, who later stopped by seeking damages for her facial injuries.
- His arriving in New York in February 1956 (summer in Brazil) with no winter coat and having to spend $20 of the $50 from his parents on a winter coat.
- Being assigned to room with classmate Larry Foster, who was black, as the two only nonwhites at Stony Brook School at that time.
- Having to part with his hard-earned cash to repair his only pair of glasses one summer when working as a farm hand. Though the farmer's wife had stepped on them — she blamed Dad for leaving his glasses on the floor.
- An especially difficult exam (class average 29) he'd taken as an MIT undergrad, where luckily the professor's solutions shared after the test validated Dad's answers. Unfortunately, Dad's test paper was lost but the professor offered to award him whatever score he thought he had earned. When Dad said "96," the professor accused Dad of abusing the situation. When later the exam was found, Dad had scored around 96, and the professor immediately offered to write him a graduate school recommendation letter.

But our favorite story was how he met my mother:

- In Thanksgiving of 1959, Richard Chang visited the Flushing, NY home of Sunny Pu, introduced by his relative who had a Kweichow connection. Richard and Sunny were both college students in Boston; he at MIT, and she at Brandeis. As it turns out, Sunny was out on a date that very night, so he met her only briefly. Yet, he was determined to date and, after one date, eventually marry her, even when it was initially against all odds.

As I look back at Dad's life, despite the challenges he faced, I see our "good guy" in the movie smiling contentedly in the final scene, his sacrifices rewarded, his efforts recognized, his good heart appreciated and celebrated. When he arrived alone in America that chilly February of 1956, could anyone ever have imagined that he would one day be delivering physics lectures and keynote seminars around the world as a Yale professor, the mentor of dozens of scientists, Sunny's husband of nearly 60 years, the father of 3 children, with 6 adoring grandchildren? Dad took nothing for granted and always showed great gratitude for the happy moments life afforded him: a night at the symphony, a bike ride in the New England autumn, scuba trips and the underwater world of the Caribbean or Pacific, memorable meals of all cuisines across all continents, a weekend hosting friends in Vermont, holidays with the extended family, or the warm celebrations with his research "family" including his 60th birthday and his retirement parties at Yale in 2000 and 2008, respectively.

So many of Dad's hopes and wishes in life were fulfilled — and sometimes well exceeded, and this filled him with immense pride and joy. I learned only much later that out of the six Assistant Professors who were hired in Dad's department at Yale in 1966, only he was promoted and later tenured, again defying the odds.

In closing, what words can sum up the journey of Dad's life? Perhaps the words of Richard himself can answer that. Two years after my 5th grade "Getting to know my father" assignment, in 1974, I had unit on mythology in 7th grade English class. Dad wrote on the parent-teacher report that he found this topic very important because young people too needed to "learn to dream the impossible dream".

The "impossible dream" could aptly describe Richard's life — and I am so grateful that so many of his "impossible dreams" were fulfilled over his nearly 80 years on earth. I would like to thank everyone in his life for all you did to make some of these dreams come true. I feel that Richard lives on in each of us, and I pray that somewhere, somehow Dad is still smiling and dreaming more impossible dreams.

Letter to My Father on the Occasion of His Retirement in June 2008

For much of my childhood, my father's career remained a mystery to me. I knew that he spent long intense hours in the lab, or many evenings and weekends at home writing and re-writing, with papers and open books spread all over. Sometimes, even during dinner, the phone would ring and my father would disappear for hours, energized and focused, immersed in an unintelligible discussion with jargon-like words such as "Surface-enhanced Raman scattering," scribbling notes on whatever writing surface (or napkin) might be in sight.

I occasionally found myself in his office waiting for a ride home. I must say, I have never been in so productive an environment for my own studies — simply because it was so completely free of tempting distraction. Though the walls were lined with books, and the papers were piled high on the desks, the most interesting reading I could ever find were the New Haven Yellow Pages, and the Yale Directory, where if I was lucky I might recognize the names of a few of my schoolmates' parents. In comparison, the Cross Campus or Sterling Library were infinitely entertaining — with newspapers, magazines, novels.

What exactly did my father do when he met others to "talk science"? One of my sisters thought it was a simple exchange of scientific facts. For example: "Jupiter has 16 moons," one party would say. "–40 degrees is the same in Fahrenheit and Celsius," the other would respond. "Lead is denser than water," the first party might add — and so on, for a few hours. Or, as one student once asked a Columbia professor at a bar on the Upper West Side: "What exactly do you do with these papers you write — do you save them somewhere?" Somehow, this was all a different world — as the family pediatrician discovered for himself when he asked my father for the temperature of my sister suffering from fever, and my father gave the answer in Kelvin. "Could I please speak to Sunny?" was the doctor's unsurprising response.

It all seemed a rather eccentric, sometimes grim and dull life, especially for the poor graduate students who appeared to work 7 days a week, for long hours every day, in windowless offices and labs. Surely, I thought, some other career has got to be more rewarding than this — at least on a per hour basis?

But over the years, my father's career became a bit less mysterious and a few pieces of the puzzle started to fall into place. At the lab parties, dinners, and the department parties (including a few memorable casino nights), we discovered that these people were much more multi-faceted than they at first seemed — with partners, personalities (!), hobbies, and fantastic talents ranging from Frisbee to guitar to biking. I was especially impressed with the depth of connection between people who had worked together — bonding in the lab, sharing their lives so fully.

When I started a Ph.D. in Economics, I came to realize how much discipline and dedication was required to get through a dissertation. As an Assistant Professor, often perishing more than publishing, I came to realize just how much my father and his colleagues had accomplished over the years — the publications, tenure, chair professorships, invitations as keynote speaker, membership in prestigious international panels and academies.

Later, in the private sector, I met many people who tolerated their jobs as a necessary economic evil, dreading Monday mornings and heading straight to the nearest bar with a sigh of relief on Friday afternoons. In contrast, my father and his confreres had the good fortune to find their true passion in life, and to dedicate themselves fully to what they loved to do. For them, the thrill of learning and discovery brought an adrenalin rush and sense of excitement that no ordinary day-job could offer. The spectacular graphs and 3-D color images I saw spoke to the poetry of this science, shedding light where there was darkness, seeing glimmers of the truth fit together for the first time — the "Lux et Veritas" of Yale's motto.

Traveling around the world with my father, I was most impressed to discover that even in some remote corners, his name is known, his work is respected, and people have read the papers that resulted from those late nights in the lab, those dinner-time phone conversations. Far from being lonely and isolated, this network of scientists was united — a community of kindred souls and minds around the world. As children and as adults, we were often the beneficiaries of much warmth, hospitality, and kindness shown by my father's friends and fellow scientists around the world, many of whom are celebrating with us this June.

It has been such a privilege to meet these people with the courage and passion to dedicate themselves wholeheartedly to science, individuals with the strength and perseverance to spend those lonely nights in the lab — to try out new ideas, risking disappointment and frustration, so that over time, the field could be enriched and mankind's pool of common knowledge advanced. I stand beside you with great admiration and respect. May you continue to enjoy this magical process, this passionate pursuit of the truth, and the inspiring human bonds and relationships you have built with each other, and with my father, over the last four-plus decades — and I hope for many decades to come.

In Memory of My Father

Pang-Mei Natasha Chang, J.D.
New York, NY, USA
pmnchang@gmail.com

Tell us about love, Daddy

When you read us Bible stories at bedtime,
taking us to Babel and Bethlehem

And fling open curtains Sunday morns
so we can rise,
to omelettes and a hike on
the stomach trail of a slumbering giant.

Easter morns,
searching for lilies for mom.

And in the books of our lives

The bible of our world
Memories.

Green hills biking,
Diving still waters
Sabbatical forays to new lands.
Peaches, pears, peanuts
And key lime pie.

Driving, high school physics,
Spiral learning.
If we were to find a teacher, what better than you.

Tell us about love, Daddy.

Tell us about love,

so we can bear your death
flooded with sorrow.

Letter to My Father on the Occasion of His Retirement in June 2008

Like the great artist Picasso who was said to make no distinction between his work and play, my father is one of those people who simply loves his work. He used to tell us how lucky he was that every morning he got up and went to a job that "didn't feel like a job." He loved the science — the trail of logic and the leaps of intuition, the cold discipline of experimentation and the hot flashes of breakthrough. He loved the connections he made with other people through intense collaborations and interactions. Because my mother loved my father, she too joined him in his work. His "flow" between work and play became complete.

What I Owe My Father

I owe my father many things, my life, for one, but I would have never written, nor could I have ever written *Bound Feet and Western Dress* without him.

Published by Doubleday in 1996, *Bound Feet and Western Dress* is a memoir of my great-aunt, Zhang Youyi, who was married by arrangement in 1915 and then divorced in 1924 in what is considered the first modern divorce in China, modern because it was based on love, or the lack thereof. Even thirty years later, the book is one of my crowning achievements. It's been translated into over ten languages, and still has readers today for classes and recreation. Recently, a new audio version is out, as well as a new version in China.

It was my father who introduced me to Youyi when she moved to New York from Hong Kong in the 1980s. From his childhood in Shanghai, Dad was close to many of his aunts and their children, holding a special place as little brother, "Di-di", the only boy amid many girl cousins. I learned an entire set of manners, customs and practices watching my father interact with this side of the family. With Youyi, Dad began the tradition of inviting her to our house in Connecticut to stay for a few days each summer. He would drive down (alone) to New York to pick her up and bring her back to us. I loved how deliberately Dad treated elderly Youyi, always thinking about what would make her comfortable and happy. Through his actions, he showed me Chinese values of *zuo ren*, acting with integrity.

My interest in stories of old China originated for me with my father. Dad was the consummate storyteller, with a good memory for people, places and recall of the exact feeling at a certain time. I interviewed Youyi for over five years, and would often interview my father on the side to find out more about Chinese customs, why people acted the way they did, what was the mentality of the Chang family. He had an all-encompassing view of the world and the people in it, and I wanted to bring that kind of humanity to my writing.

My father also brought Youyi to see the historian Jonathan Spence at Yale, and encouraged me to show Professor Spence my own writings from as early as 1980 when I first visited China. Professor Spence was the one who encouraged me to seek a publisher, even introducing me to his own editor at the time, Elisabeth Sifton, at Viking Penguin. Later, Professor Spence joined my thesis board at Harvard.

I wrote the bulk of the manuscript in 1987–1988, the year after I graduated from college. Dad was my companion through this journey. What was he working on that year? It seemed that I was always sitting upstairs in one bedroom writing and he was in his study downstairs also writing. He was familiar with long-term big projects and knew how to follow through. He could see big picture, small picture and process. He who had once said that he wanted us to study music so that we understood the "incremental nature of time." This was how he understood the work of the book.

We shared a printer and hours of revision. He would come up to my room and say to me: we can't be machines — let's go to the movies in a few hours. Or, we should recharge by going for a bicycle ride. At one point that year, I was subsisting on grapefruit and M&M's. He told me that this would not hold me, that I needed to stay fit for the long run.

He was also thoughtful — I could measure it in technology and the amount of Apple products he helped purchase for me each time there was an upgrade beginning with the first Macintosh. We were also among the first to own a laser printer.

Somewhere in the process of writing the book on Youyi's divorce, in a way of life imitating art, I felt angry at men and even at my father. I had learned about my parents' past: that my mother had left her PhD program one year shy of research only for completion because my father got the position at Yale, and they had moved from Cambridge with two young children (I was one and a half) and a third on the way. To my mind at the time, my mother had subjugated herself and her ambitions to my father's, and my father had allowed it.

Worried at my resentment, my mother enrolled at Columbia Teacher's College in 1993 to complete her docurate. The program would be daunting — a full courseload, weekends and summers on campus in New York City, all while working fulltime. My father wholeheartedly supported the move. He said he would take care of all the housework and meal preparation. He had only one stipulation: that my mother work out every day either cycling with him or at the gym for 90 minutes. At the time, my mother was upset at this requirement, but she's subsequently told me that she couldn't have gotten through this stressful period without physical exercise! And that my father had found a way to help her and for them to spend time together. Again with my mother, my father could see the big picture and translate it into small actions.

Pang-Mei Natasha Chang

Letter to Dad

Pang-Yuan Nicole Chang, J.D., MBA
Glendale, CA 91201, USA
pynicole@fasquelle.com

Jan 1, 2022

Dear Dad,

I hope you are getting a chance to enjoy some of the people and things that were precious to you. Auntie Mae Jeane must be thrilled to have her little brother by her side to pamper.

It is New Year's Day and we are in Los Angeles, CA. We are so fortunate to have Mom with us for the holidays; she had not been back to the US for over 6 years and I hadn't seen her since your earthly demise in early 2020. As we transition to a new year and we continue to feel the heavy presence of COVID, it causes one to reflect on what continues to be a constant positive force that one can turn to in times of great uncertainty. YOU and the values you helped instill in us is one of those forces.

I share five (5) lifelong values that you modeled and gifted to me, and I continually attempt to uphold in my life: (1) the motivation, discipline, and confidence to constantly grow and learn; (2) creativity and innovation; (3) the generous and selfless giving of time, attention, energy, resources and ability to help society and others become their best self; (4) the importance of kindness, humor and enjoying life; and (5) the blessing of unconditional love.

1. Motivation, Discipline and Confidence to Constantly Grow and Learn

I will never forget all the love and dedication you showed me, especially in my studies. Like you, I liked the sciences but I am not sure I was as gifted as you so it was perseverance and your support that got me through.

I believe it was my Freshman year at Harvard, when I was studying Physics with Professor Paul and had that graduate student that you had turned away from Yale as a TA. You used to drive all the way from Hamden to Harvard on the weekends to tutor me in Physics, as you were determined to make sure I did well in the class (after I unfairly got a less than stellar midterm grade). We would sequester ourselves in a classroom and you would try to get me to understand concepts that were second nature to you (and quite foreign to me). I still remember when you were trying to explain to me the speed at which a certain sized toilet paper roll would unravel when sitting on a spinning record. And of course, I will never forget when you pulled the first all-nighter of your life to help me type the mathematical equations for my economics thesis when I decided at the last minute to study abroad for my senior year. You said that you had managed to get through your PhD without pulling an all-nighter and now here you were pulling one just to help me get through some of the tedious formatting. I clearly remember our sitting in the Music room/study to work on this, and then you helped drive me back down to Harvard so that I could turn it in on time.

2. Creativity and Innovation

You always demonstrated so much creativity and innovation in your work — personal and professional. Your excitement for seeing and doing things in unconventional ways is contagious. I recall with fondness in 2001 when you likened examining the droplets (shaken off) by a wet dog to detecting pathogens when you were helping develop a system to detect anthrax in the mail system (we could really use your ingenuity to deal with the COVID pandemic), and not long after you woke me up at 2 a.m. once when I was back for the holidays and with great excitement showed me how you managed to catch a droplet exploding in a certain way on camera. And of course, I will never forget how you used to use a stopwatch to time me doing calculus problems (you would say to me to press the button when finished and you would be back in 10 min as needed to finish something up) and taught me tricks on how to do complicated math problems accurately and quickly — I have definitely carried on the tradition with my kids.

3. Generous and Selfless Giving of Time, Attention, Energy, Resources and Ability to Help Society and Others Become Their Best Self

Watching how much you cared about the welfare of your students, not only academically but also in a holistic sense, made a big impression on me. You always were very thoughtful to see the person in front of you and support them to make their lives as fulfilling as possible — be it to make sure their apartments were well-equipped and that their romantic and family issues were resolved. You were the first to implement things to improve their life, such as driving lessons, English lessons, taking them shopping for food and clothes, preparing their apartments or gifting things, and even advising on matchmaking (with Mom's help) and resolving family issues.

4. The Importance of Kindness, Humor, and Enjoying Life

I always associate you with much joy and laughter and wanting to show others a good time. You and Mom generously had an open-door policy to people of all walks of life and nationalities. You were the first to suggest showing others the simple joys of life — whether that was a pick-up game of football in the backyard, a bike ride around the paths of Vermont, or a hike to see the New England foliage. You never let your personal concerns get in the way of offering to help the less fortunate with finances or otherwise.

5. Unconditional Love

No matter how important your work and your students were to you, we always knew how much you loved your family. Mom, PH, PM, and I always knew what a very special place you held for us in your heart and you showed us through your thoughts and actions everyday how important we were to you, and we knew this love was unwavering and unconditional. Thank you for gifting us this privilege to grow up in a nurturing home where we always felt loved and as if we belonged.

Ming-Wei and Ming-An are growing up really nicely and you would be so happy to know about the lasting influence you have had on them as well.

Your Only Grandson: Ming-Wei

Ming-Wei (he goes by Cyprien now that he is in college) is a sophomore at Stanford. I actually call him a "froshmore" as in many ways he is like a freshman since he is new to campus and many of his classmates; due to the COVID pandemic, he did his freshman year completely online from your place in Brattleboro, VT. He formed a pod there with 3 other Stanford freshmen, which we would affectionately refer to as SIB ("Stanford in Brattleboro").

Ming-Wei is so much like you. He loves everything about cars and is most likely destined to study engineering. Ming-Wei is driving your silver 2001 3.25i BMW — like you, he treats it like his baby and loves tinkering with it, and currently has the car on campus in Palo Alto. He says what he loves about driving your car is that the engine obeys him and he feels in such control and adores being closer to the road. Ming-Wei spent time learning how to renovate cars one summer at the Peterson Automobile Museum and is looking forward to spending time working on the car to make sure it stays in great condition.

I so wish you were here to mentor him in his classes like you did me! Ming-Wei is currently taking the prerequisite math and basic engineering courses for product design and computer science, and is trying to figure out whether he will continue studies in electrical engineering (like you) or mechanical. He definitely wishes that he had you like I did when I was taking such classes.

Your Singer Granddaughter: Ming-An

Ming-An (she goes by Cybèle now that she is in college) is a freshman at Harvard, your old haunt where you did your graduate studies. Ming-An lives in Matthews dorm in Harvard Yard across from the John Harvard statue. Mom said that when you guys used to cross the yard, it brought back memories of how exhausted you two were as working graduate students raising a young family.

Of course, Ming-An is at a different time in her life, but she takes after you in many ways. She is single-minded and hyper-focused and knows exactly what she wants. She is fiercely independent and does not like being told what to do.

She is very focused on her work and unabashedly plows through the tourists who block her pathway as they stand outside her dorm on the Harvard Yard tours. Ming-An shares stories about her daily Chinese class which keeps her on her toes with weekly quizzes and I am reminded of that Chinese exam that you had to take for part of your doctoral requirements while at Harvard (with Pu Gong Gong coming to the rescue to help you ace it).

Like you, Ming-An loves music and how it makes one feel. As you recall from all of those made-for-you-and-Mom "concerts" in the living room where *Love is a Many Splendored Things* or *Autumn Leaves* were featured (as well as Ming-An in those Marilyn Monroe-ish dresses), Ming-An is a budding singer/songwriter. She is currently also doing a dual degree with Berklee School of Music where she will be getting her masters in composition for film, television and video games.

We miss you so much Dad, and not a day goes by that we don't think about how much our lives would be enhanced by your physical presence, but know that you live on in us and through us.

Much love,
Pang-Yuan

Letter to My Father on the Occasion of His Retirement in June 2008

Dear Dad,

It is hard to imagine that after so many years of preparation for at MIT and Harvard, countless nights leading to months leading to decades of years of perseverance, infinite months of planning per experiment, and seemingly endless hours of grading papers and lecture preparation time that you will be officially retiring from your position at Yale. I know that for essentially half a century you have devoted so much of your time, energy, and most importantly passion and love to this endeavor. The internal rewards of your clear success mirrored by the many professional accolades, countless prestigious publications and the long trail of supporters, admirers and friends you have made along the way make your success the even more poignant. You have achieved what only a few dare dream about. You should be so proud and self-content with this great achievement of a lifetime.

I know part of you will miss being so implicated in this after your retirement, but what a wonderful thing to look back on and there are so many wonderful things to look ahead to now that you will have more time and energy to devote to other very worthwhile endeavors. All of the wonderful friends and relationships you have formed over the years I am sure will be looking forward to traveling with you, sightseeing, and no doubt talk science.

Your grandson Ming-Wei sized it up so well the other day. I asked him to write you a Happy Retirement Card. He looked at me and said, "That is so boring." He then asked, "Can I write him a Sunny Day and Fun things to Do card?" I asked him what this had to do with your retirement. Without hesitation said to me, "Mom, don't you get it. After he retires he will have time to be with me and Ming-An and there are so many fun things we can do together."

CONGRATULATIONS for an exciting road driven and an adventure-packed beautiful drive ahead!

May you cherish your time at Yale as you enjoy the fruits of your labor in your "retirement."

Congratulations again, and we can't wait to see more of you in China (as do many universities here!).

Much love,
Pang-Yuan, Denis, Ming-Wei (7) and Ming-An (5)

Reminiscing About Our Beloved Grandfather "GongGong"

Ming-Wei Cyprien Fasquelle & Ming-An Cybèle Fasquelle
Glendale, CA 91201, USA
mwcfasquelle@stanford.edu & macfasquelle@harvard.edu

From vacations, to visits in China, to living together in Hamden, and finally to visits in Singapore, we have spent a lot of time reminiscing about all the good moments we were able to share with GongGong. While we are certain everyone has a different perception of him, some of the stories his grandchildren remember the best should show another side.

He was one of the best grandfathers we could ever ask for. Despite his Parkinson's, he always found a way to be there for us and show his love and care. When Ming-An was still in a stroller, and Ming-Wei was erratically pushing her around, with Popo and GongGong following behind, GongGong would always try to keep up the stroller. While some of us were worried his jogging may result in a fall, he never did. One step after the other, he kept his hands hovering above the stroller handles, in case Ming-Wei tried to push too hard or lost control. At the dinner table in Hamden, he would always try to talk to us about his work, or his time in Brazil, or explain to us how any mechanical system worked when we would ask. We could tell it sometimes took effort to speak clearly, especially to young ears with a limited vocabulary, yet he persisted patiently, and that made all the difference to us. More recently in Singapore, when Parkinson's even limited a lot of his speech, we relied on his gestures and facial expressions for communication. We would sit next to him in his wheelchair, looking out over Singapore, and tell him about our lives back in California. His breathing would become steady, and his eyes would often fixate on us, with a smile forming every time we said something funny or reminisced on something good that happened. While some may say we didn't get to speak back and forth all that much as we got older, we came to appreciate his presence, even in silent, mutual understanding. The absence of spoken words, which were instead replaced with eye gestures and extensions of the hand, spoke volumes. It was in those little moments we found GongGong to be the most present with us, more so than anyone who would attempt to break the silence. It was in his eyes we could see the laughter and shared celebrations of our childhood stories. He was a part of it all the way.

Message to Our Grandmother on GongGong's Passing

Dearest Popo, We are so sorry about GongGong's death. You were the light in his life, and it was beautiful to see such pure love between you two. We love you very much and will be here with you in your time of need. You are both in our hearts now and forever. Hope to see you soon. Love you so much! Much love from, Ming-Wei and Ming-An.

Memories of My Grandfather Gong-Gong

Ming-Mei Solenne Wolfe
Junior Student
Dartmouth University
Hanover, New Hampshire 03755 USA

Though he bears the affectionate name Gong-Gong in my head, most of my memories of my mother's father are not in my own head. Instead, they live in the collective memory of the family, the imprint that he made on my mother and her siblings, my grandmother, and even more distant relatives. Though I am closer in blood relation to him than most of my older cousins, they remember my "Gong-Gong" as Uncle Richard, who carried them through AP Physics, distilling key concepts at his dining room table in New Haven, or as the scuba diving enthusiast who was a world renowned scientist. I wish that I had gotten even a slice of the memories that they have, of the athletic and erudite man who possessed a firm command over his work and home life. By the time I was old enough to interact with him, he had been living with Parkinson's disease for almost twenty years.

I remember splashes on the kitchen floor from the drooling that afflicted him in his later years and visiting him in the hospital after unfortunate falls or surgeries. Vividly, I remember being in the swimming pool in Brattleboro, a week after his retirement party, jumping off the diving board. In my peripheral vision, he stood completely still but for the cyan water ripples of my cannonball. That moment seems pivotal, an inflection point I should have seen coming. But at age 7, I was unaware of what "calm before the storm" meant. A bright summer's day full of promise turned in instants to an angst-filled blur. The next few minutes were action-packed years when a neighbor lounging poolside dove under to pull my Gong-Gong out of the water.

He had had a Parkinsonian cramp while practicing breath techniques to get back into scuba diving and hadn't been able to get to the surface on his own. He had lost oxygen to his brain for several minutes, the doctors said. My cousins and I prayed, that he would wake up, that he would be healthy. Ultimately, he woke up, but had lost years of his memory. He called me by my mother's name and asked for his mother and sister who had been dead for many years.

In my grandparent's house now, it is impossible not to see the scars of "the accident." My grandmother made a memory wall for Gong-Gong, to help him place himself in a current temporal frame. The cork board wall is divided into three parts: Past, Present, Future. In the past, she wrote "Maejeane dies…" "Richard graduates from MIT." All three timeframes are full of index cards and my grandmother's distinct, neat, teacher handwriting.

Walking through their house now when I visit with friends, I look at the Future column and see my present: "Ming-Mei attends college." It is bittersweet to think that I was never able to see Gong-Gong in his present, at his prime the way the rest of my family remembers him. At the same time, living with him through the memory of others allows me to see his life as a golden thread, reaching across generations to bring my family together in shared reverence and love. His imprint threads through my life as well, as the image of him in the minds of my relatives leaves massive shoes to fill. When I am figuratively swimming in integrals in college physics and feeling completely lost, or physically swimming through beautiful waters, I imagine Gong-Gong in his present, holding my hand and providing a reassuring stability. Though I never knew my Gong-Gong the way that the rest of my family did, my own memory stitched together through theirs reminds me of how integral a part of my life he was, if only through the way that he lives on in the people that I love.

Note: My grandfather would be delighted to know that I will be graduating as an Electrical Engineering (EE) major in June 2024 and continue my masters in EE at Darthmouth University.

My Grandfather Gong-Gong

Ming-Li Sabina Wolfe
Junior Student
Brooklyn Technical High School
New York, NY 10003 USA

In writing this for my grandfather's memorial, I have gotten to know the stem of an amazing scientist, a father, and all-around person, whom I called Gong-Gong. When my mother was born, he was twenty-five and his Parkinson's started when he was around fifty-five. He lived thirty plus years of adult life without disease and then some twenty-five years with disease, almost half of his adult life was spent with Parkinson's. But the remarkable thing is his story is so much greater than his disease, to me it is riddled with anecdotes of love from his kids, colleagues, students, and wife, a love story with science and the world.

The house in Hamden was always my mystical place with its haunted basement and garden where my sister, Ming-Mei, would scare me with bugs. And then when I was ten, I was able to rediscover it as my mother and I spent hours clearing it out. The trees which I had waltzed by were memories of the past for my mother, where she and her siblings had spent hours watering the trees from the seeds Gong-Gong had planted. The super-8 films showed my aunt, uncle, and mother playing tricks with their new camera, pouring water from a pitcher backward, something that their father had worked to gift them.

I wish I had known him in his first half of life, but I knew him more in his second half. He was the gentle giant to me, bringing me to the playground just to watch me for hours — something my mother never had the patience for. Everything he did, he put his soul into, dedicating every minute into it. He liked to blow bubbles for the little kids and smile at our delight. My memory of him is surrounded by laughing children and pools of light. I wish I had the pleasure of knowing him better before he passed, however the roots Gong-Gong has planted run much deeper than those at the Hamden house, he has left us with a love of science, God, and the desire for devotion.

Note: Because of my grandfather's love for Brazilian Music in general and Astrud Gilberto in particular, I submitted the following article to the NYT 2023 Summer Reading Contest.

The New York Times

The Learning Network
Published August 22, 2023, Updated September 12, 2023

Our annual Summer Reading Contest invites teenagers around the world to tell us what New York Times pieces get their attention and why. This week, the ninth of our 10-week challenge, we received 1,263 entries. Thank you to everyone who participated, and congratulations to our winner, **Ming-Li Sabina Wolfe**, as well as to the runners-up and honorable mentions we honor below.

Winner

Ming-Li Sabina Wolfe, 16, from Brooklyn, N.Y., selected the obituary "Astrud Gilberto, 83, Dies; Shot to Fame With 'The Girl From Ipanema'" and wrote:

In 1951, my grandfather's family moved from China to Brazil. He left after one year to study in the U.S., but his parents stayed until 1975. When "Getz/Gilberto" was released, my great-grandparents mailed the record to my grandfather, which he played while writing love notes to his girlfriend — my future grandmother.

When Gilberto released "The Astrud Gilberto Album," his parents sent it again, but this time to my married grandparents' home in Connecticut. From then on, their Sundays were spent cleaning to "Corcovado" and "Dindi."

This summer, 58 years later and two years since my grandfather's death, my grandmother came to visit my mom and me. Their record player had sat in our closet for decades, but when she arrived, we dusted it off. She sat and played those records on repeat. Like Farber mentioned, Gilberto can "evoke images of summers imagined or lost."

The other night, my grandmother, mother and I were walking in the East Village when we heard the familiar chords of "Agua de Beber." I realized then that music is transient, not singularly tied to lonely moments next to a record player or cool summer nights surrounded by family. Through art, the living and the dead are lost in a dance. My ancestors speak to me through the art of a dead singer, while I repeat their mistakes.

Lately, I've been wondering, "What is a life's work?" It's one that keeps people alive and Gilberto has raised three generations.

Our Grandfather Gong-Gong

Cara Ming-Yan Chang & Tamara Ming-Wen Chang
Students
Tanglin Trust School Ltd.
Singapore

When I think back to times I spent with gong gong, I fondly remember going over to play balloon ball with him and spending time talking to him beside his bed. I remember the loving smile on his face when he hit the ball and how he loved listening to my dad play the piano. I (with my family) often meet some of his ex-students and colleagues and it is wonderful to see how he changed so many of their lives both academically and on a personal level.

It's been 3 years since gong gong left us in February 2020 and I wanted to share some significant things in my life since then. I've gone from playing football four times a week (and playing in cairnhill crest) to having training with my team everyday. I've gone from year 6 in the junior school to year 10 in the senior school and am starting my GCSEs. I hope to play football for Singapore and make you proud.

Cara

When I think of Gong Gong Richard, I remember playing ball with him. Cara and I would visit him a few times a week. Now, 3 years after he passed away, I would like to share some of the things I have been doing recently. I play football/soccer 5 times a week with my club and the National Team. I really enjoy it and it is great as Cara also plays football. Sometimes, I have goalkeeper training which is quite special. As you may know, Cara and I sometimes play at your condo's basketball court because we don't have anywhere to play football in our condo. Other than football, I still have piano lessons, and I like learning to play songs that I recognise and like.

I do wish that we could have spent more time together, and I wish I had gotten to know you a bit better. When I was even younger, I remember that you came to my birthday parties. That was very nice of you and I was extremely happy.

I know that you loved me, and I am sending you my love into heaven.

Tamara

Memories of Richard Chang

Robert M. Pu, J.D.
Attorney and Probate Judge (Ret.)
Brattleboro, VT 05301, USA

It was November 1959 and I was 9 years old. Richard's cousin was staying with our family in Flushing, New York and Richard came to see her over the Thanksgiving holidays. I don't recall if I was introduced to him at the time but years later Sunny told me that I had met Richard before she did. The following summer our family was on home leave in Hong Kong. A few days after we checked into our hotel in Kowloon, Sunny started receiving aerogrammes and letters from Brazil. At first I was puzzled and couldn't understand why Sunny was receiving such frequent correspondence from this unknown person in Brazil. My younger brother and I used to tease our sister about the frequency and volume of mail she was receiving from this unknown person. Only later that summer did I learn from my other sisters that the mysterious sender of all these aerogrammes was the same person who had been at our house the previous Thanksgiving.

Richard was 10 years older than me and in many respects he was more like a brother than a brother-in-law. He encouraged my father to send me off to prep school since he had such favorable memories of his time at The Stony Brook School. We visited the school but I eventually ended up attending another prep school in the northeast.

There was a certain amount of drama associated with my courting of Sen who would eventually become my wife, and Richard assuaged my father's concerns about the manner in which I was wooing Sen. Fortunately things worked out and when Sen and I were married a year later Richard acted as our wedding photographer.

After Sen and I got married and settled in Vermont, and my sister Ching-wen and her husband Carl returned to Vermont, Richard and Sunny became regular visitors to Vermont. When my parents moved to Brattleboro to be closer to us, Sunny and Richard's visits to Vermont increased and they eventually purchased a house right around the corner from my mother's house. For as long as I had known him, Richard had always been attracted to places with scenic landscapes and their vacation home in Brattleboro overlooking the Connecticut River was an idyllic spot where they were able to entertain his Yale colleagues, visiting scholars and graduate students. We introduced Sunny and Richard to some of our local friends and on July 1, 2000, an avid hiker friend led us on a hike up Mount Monadnock in neighboring New Hampshire. Mount Monadnock is reputed to be the second-most climbed mountain in the world after Mount Fuji in Japan and the fact that Richard was able to climb it just as his Parkinson's was starting to manifest itself is a testament to his perseverance and determination. We celebrated our successful ascent of Mount Monadnock that evening with a nice bottle of champagne.

Memorial Recollections of Richard K. Chang

Glen A. Pu, M.D., Ph.D.
Brownwood, TX 76801, USA

Richard has been part of my life almost as long as I can remember. According to Sunny, I was 8 years old when I first met Richard. Truthfully Sunny remembers the meeting better than I.

Richard and I are technically of the same generation. In actuality when we met, I was 8 years old and Richard was about 20. Richard was not simply a "big brother", but almost immediately became a father figure. My dad, having been in America less than 10 years, had innumerable great qualities but he lacked the connection with American culture that I needed as a child.

Richard went to an American prep school, followed by MIT and Harvard. Thus, he had the experience and gravitas to fill a huge void in my upbringing and cultural assimilation. Richard taught me the importance of a great work ethic supplemented with the endless idealism of young age. Hard work can overcome any difficult issue: professional or personal. Richard was idealistic and believed that good character was the key for all accomplishments.

I recall shortly after meeting Richard, he needed my dad's help with translating some Chinese documents. Several late night review sessions were spent with Richard and my dad doing meticulous review and translation of Chinese documents. Richard apparently needed this help to satisfy the foreign language requirement for his PhD.

My recollections of Richard and my dad translating Chinese documents serves as a segue to a Chinese document my dad translated late in his life. The document talks about five pillars (cultivations) to "perfect oneself ... achieving the aims of life."

AN ELDERLY STATESMAN'S COUNSEL TO HIS RELATIVE

The way to perfect oneself consists basically in five aspects of cultivation.

To cultivate good health, one has to observe a regular living, maintain a cheerful state of mind, and practice diligence and industry.

To cultivate sound mind, one should cherish a devout belief, a detached outlook, an indomitable will and untiring zeal.

To cultivate supreme wisdom, one must keep a cool head, an objective attitude, a curious spirit and earnest fulfillment.

To cultivate generosity, one ought to be humble to cope with circumstances, broad-minded to associate with people, patient to restrain oneself and harmonious to tolerate others.

To cultivate stature, one should be fair and sincere to dissipate any anger and resentment, ever responsible to inspire public confidence and willing to sacrifice in order to create.

Always bearing these points in mind and practicing (sic) with sincerity and perseverance, one will succeed in establishing oneself and achieving the aims of life.

To reiterate, the five pillars are Health, Sound Mind, Wisdom, Generosity and Stature. Good Health, first on the list, despite the Chinese teaching is not always controllable. Sadly, Richard and my eldest sister had bad luck with health, through no fault of their own. With “indomitable will and untiring zeal”, amazingly Richard lived about half his adult life with Parkinson Disease. In the later years, I found it very painful to watch Richard suffer, but with Sunny’s love, Richard always did his best to present a pleasant, grateful disposition.

Sound Mind and Wisdom, #2 and #3, for Richard, per se nota. Richard’s students and colleagues undoubtedly will provide many examples with their recollections. I learned much from Richard’s wisdom with often unrequested, but always invaluable, advice.

Richard’s Generosity (Pillar #4) to students, colleagues, friends and family is well known. Add to that, Richard’s immense respect for his entire family. Richard’s regard, for his own parents and my parents, is a textbook example of respecting and honoring ancestors.

Final pillar for the elderly Chinese philosopher, Stature, is probably the hardest, most noble and difficult to achieve. Stature requires fairness and sincerity while making an emotional connection to people (inspiring confidence). Lastly, the willingness “to sacrifice in order to create”. Richard was indeed all about giving of himself to help others.

I would add a final “pillar” which I think for the *Elderly Statesman* was unstated but an understood necessity. Sadly, in today’s world, this pillar needs to be stated: Integrity and Honesty.

Today, too many individuals believe “ends justify the means”. Richard was cut from a different cloth and always did everything by the book.

From my adolescent years, I recall this incident: Richard was still relatively junior at Yale, but he lamented the proceedings at a hiring committee meeting. There were reservations about a prospective professor, academically superb, but potentially shunned because of unsubstantiated gossip about his wife. This seemed a bridge too far for Richard, who wanted to focus on professional qualifications and potential academic contributions.

That in a nutshell was Richard. Honest to the core but also outspoken so that the truth was always there. Consistent, reliable character. A wonderful leader yet kind gentle soul. At the university or with friends and family, we always counted on Richard. I am sure his family and students think about Richard all the time and miss him dearly. I certainly do.

Richard Chang

Cynthia Price
Staff Director of International Law and Practice (Ret.)
American Bar Association
Bethesda, MD 20816, USA

Richard Chang was a brilliant scientist and an equally brilliant human being. I met Richard through his wife, Sung-Wen, who was my roommate in boarding school. After college graduation, Sung-Wen and I renewed our friendship when I moved to Boston. I remember inviting them over to see my apartment which was a 5th floor walk-up. We climbed all those stairs that evening and the next day Sung-Wen and Richard were the parents of a baby boy!

Our early friendship endured as we built our lives and I was fortunate to see Richard flourish both professionally and personally, and to meet Richard's parents, Richard's sister Mae Jeanne, and their three children and now their children. Over the years Richard always exemplified kindness, humility, and an abiding curiosity about others, their families, different cultures and the places where he lived and traveled.

In the final words of his retirement speech concluding a forty-two year career at Yale, Richard wrote... "Others have lit the way for me, so I now hope that I may shine some light on the path of others." Richard measured his successes in the successes of others, his family, students, friends, and all who came in contact with him. As I reread his retirement speech, I was struck by his generosity, humbleness, wise counsel and desire to inspire.

I was in attendance when Kevin, Pan-Mei Natasha and Pan-Yuan Nicole organized a luncheon in NYC to celebrate Richard and Sung-Wen's 50th wedding anniversary. Family, former students, colleagues, and friends gathered to honor them with tributes, recollections, and photos of their many rich years together. The children even made a CD of their favorite "love" songs and as we said our good-byes, they gave each guest a CD to take home to remember the occasion.

I am fortunate to have been able to witness Sung-Wen and Richard surrounded by the loving family and community that they loved so dearly. Sung-Wen and Richard were great partners in life. She was his steadfast and caring companion and together they persistently persevered through his early illness and later a near drowning accident. He never gave up on life and continued to live and enjoy every day in Singapore with his family and to travel the world continuing to explore and inspire others. He will be remembered and held dear in the hearts and minds of all of us who were fortunate to have had him as part of our lives.

Remembering Richard Chang

Madeline Levine[1] & Joan Scott[2]
Professors Emerita
[1]University of North Carolina, Chapel Hill, NC 27514, USA
[2]School of Social Science at the Institute for Advanced Study, Princeton, NJ 08540, USA

Richard and Sunny's Wedding Reception at the MIT Faculty Club in 1962 with some of Sunny's Brandeis classmates. Joan Scott is on Richard's immediate right and Madeline Levine is the 6th person on Sunny's left.

We met Richard Chang when he was courting Sung-Wen Pu (Sunny), who soon became his wife. Sung-Wen was our classmate at Brandeis University (we were the class of 1962); she was Madeline's roommate and Joan's friend. It was Kevin, Richard and Sunny's son, who, having heard his mother's fond reminiscences of our friendship, tracked us down in 2021 and reunited the three of us as a loving gift to his mother. After many years of lost contact, our Zoom reunions have been a chance to reconnect, as we share wonderful excursions down memory lane.

Lots of time has passed since our college years and our memories have become hazy, but we have managed to come up with a few sharp images. The images always include Sung-Wen, who introduced us to the intimacies of Chinese culture, and especially its food. Both of us still own the copies of the Joyce Chen cookbook Sunny gave us and we remember delicious and — for us — unusual meals shared with her and Richard. (Sung-Wen insists that Madeline and another of our friends urged her to take up with Richard, who was quietly and persistently pressing his case.) It was probably at one of those meals that Richard, looking ahead to marriage and family, commented that, he would hope to have smiling fat Chinese babies. Madeline realized only much later (having spent time in Taiwan and learned some of the language) that the reference to fat babies wasn't just about size, but about health and well-being. Joan can still see Richard's smiling face as he contemplated the healthy family that he and Sung-Wen would produce together. Kevin, Natasha, and Nicole would come into the world and fulfill that dream.

It was also undoubtedly at one of those meals, while we were discussing their impending marriage, that Richard announced, to our surprise, that he would never wear a wedding-ring while working in the lab. It had nothing to do with contemplating the potential for marital (in)fidelity, he explained, but the opposite — keeping himself healthy for his wife and family. He knew of a colleague who had lost a finger during an experiment with electricity because the ring set off a fiery reaction. Richard would not want to endanger the well-being of his family by incurring some kind of terrible injury. Here he was, not yet married, thinking ahead not just about his own safety, but about what it would mean for the others he cared for and loved.

We don't know much about Richard's career as a scientist, we do know he was an impressive and important researcher and teacher. But the memory we have is of a generous and loyal son, husband, and father, whose pride in his parents, wife, and children knew no bounds.

Richard and Sunny's Wedding at the Protestant Chapel of Brandeis University with the founding president of Brandeis University, Dr. Abram Leon Sachar and his wife Thelma Horwitz.

Some Anecdotal Recollection of Richard Chang — A Stream of Consciousness

Arthur C.M. Chen, Ph.D.
Senior Executive Director, GE Research (Ret.)
Boston, MA, USA

Richard was a classmate of mine in MIT Class of 1961, where we both ended up in the Cooperative Program in Electrical Engineering. My motivation was the same as Richard's: financial benefits. I knew Richard in part because my younger brothers were also at the Stony Brook School at the same time as Richard.

Richard and I both applied to the Bell Labs co-op program. And among many applicants, we (ethnic Chinese) were the only ones asked to solve a technical problem as a part of the interviewing process. I did not answer the question correctly but was hired by Bell Labs anyway. Perhaps Richard was lucky to go to IBM as my Bell Lab experience was not that memorable. One excuse could be that we were the first "guinea pigs" of MIT's effort in the new totally revamped EE curriculum to be more science oriented. And thus, we were not exposed to many practical problems in EE.

Socially at MIT there were about 100 ethnic Chinese students from all over the world: Hong Kong, Singapore, Vietnam, Burma, Malaysia, and of course the U.S. The ethnic Chinese students in Boston formed the Greater Boston Chinese Student Association. Thus, our social life generally involved these organizations. However, there was a more selective organization, to which we belonged: the FF Fraternity, the first Chinese American fraternity founded in 1910. Both Richard and I were among the many other Chinese classmates recruited into this fraternity in 1957. FF has grown and now even has a lodge in Singapore. Many FF brothers attended Richard and Sunny's wedding at the Brandeis Protestant Chapel followed by a reception at the MIT Faculty Club.

Some recollections:

One of my strongest memories was that Kevin, their son, collected keys as a child. I wonder what happened to all those keys.

And of course, admission process to Yale. Given the recent complaint about discrimination against Chinese American in Ivy League admissions, Richard shared a story during our class reunion from his experience on the admissions committee. Someone wanted to donate \$7M to Yale to ensure that his grandson would be admitted. To Yale's credit, Richard said that the offer was declined, and the applicant was not admitted. At that time, the most important question the admissions committee considered was whether or not Yale would be the best place for the applicant to develop his/her potentials.

Finally, our shared experiences with Professor David Epstein, who must have had unusual experiences with Chinese students. Professor Epstein played a different, but significant role in both of our graduate school studies and our respective Ph.D. degrees.

It is very sad that Richard became ill later in life. In our youth he was a vigorous young man. But we are happy that he had a full life with Sunny and a loving family.

Richard K. Chang — Teacher 萬世師表

Yu-Chi Larry Ho
T. Jefferson Coolidge Professor Emeritus of Applied Mathematics
Harvard University
Cambridge, MA 02138, USA
ho@seas.harvard.edu

I like to think that I played a role in Richard's becoming a memorable teacher/professor. I gave him his first teaching job as a teaching fellow in a course I taught at the Harvard School of Engineering and Applied Science in 1961, when he first entered Harvard as a graduate student. When Richard worried about not knowing the course content well enough, I assured him that the best way to learn new material in a course was to teach it.

For Richard's memorial volume I can think of no better saying than the four word Chinese idiom (萬世師表) "teacher model for ten thousand generations." This will be my humble contribution to his memorial volume.

Reception at the MIT Faculty Club, Larry Ho is the second from the right, with Arthur Chen, third from the right.

Memories of Richard Chang

Peter Pershan
Frank B. Baird, Jr. Professor of Science (Emeritus)
Harvard University
Cambridge, MA 02138 USA
pershan@seas.harvard.edu

I first knew Richard, when I was an Assistant Professor in the Division of Applied Sciences at Harvard when Richard was a graduate student. He eventually asked to join my group with a request that I supervise his PhD research; however, as a junior faculty with no promise of tenure I thought it wouldn't be a good choice for him. Fortunately, Professor Bloembergen did accept him as a student and history tells that, this led to his very successful career.

In spite of our student/faculty relation, we became good friends and my memories include the fact that he introduced me to Dim Sum and a performance of Chinese Opera, when it was performed at MIT. The most precious memory was when he brought his father to my house with a dinner that his father prepared for our family. My son, who was only about 7 or 8 years old remarked that this was the "first time that a Chinese waiter" ever sat down to eat with us. Of course, I was embarrassed but Richard and his dad thought that was precious.

My principal regret is that although Harvard and Yale aren't much more that 100 miles apart my family and I didn't get to spend much time with Richard and his family. Richard was a fine friend and his loss has saddened me enormously.

Encounters with the R. K. Chang Family

Nicolaas Bloembergen†
Professor of Physics,
University of Arizona,
Tucson, AZ 85721, USA

These recollections are the subject of after-dinner remarks by Nicolaas Bloembergen at the dinner on June 12, 2008, in honor of Richard Chang.

I was very fortunate to have Richard as a graduate student and member of our research group in the Gordon McKay laboratory at Harvard University from about 1962 to 1966. His Ph.D. thesis described basic experiments on the reflection and refraction of second harmonic light in nonlinear optics. He was a co-author on more than a half dozen publications.

My wife Deli and I had Richard and Sunny over at our home in Lexington, Massachusetts, often together with other foreign-born members of my research team, who did not have the opportunity to go home during Thanksgiving and Christmas recess. Our children complained later that we never had an intimate family get-together on those holidays, but they enjoyed sneaking out of their bedrooms to peek at the visitors.

Soon after Richard had left Cambridge, Massachusetts, to join the faculty at Yale University, we visited their house in Hamden near New Haven, where he and Sunny still worked on plans for the unfinished basement. It was fun to see their son and two daughters grow up.

I enjoyed visiting Richard's active research team at Yale on several occasions in later decades, when I was invited as a colloquium speaker or a member of a visiting committee to discuss the status of engineering at Yale.

Deli and I enjoyed a visit by the Chang family when we lived in the

Netherlands in the spring of 1973. We had rented a furnished house in Waalre, where I was a consultant at the Philips Research Laboratories. The six year old daughter of the Chang's, Nicole, bumped into a large vase. The shards were a great cause of concern to them, but we told them not worry as we had very friendly relations with our land lady. Fortunately the hurt leg of the daughter did not prevent the Chang's from continuing their travel by car to France.

Deli and I also remember well, our joint participation with Richard and Sunny at the French-Israel Physics Conference in Elath, Israel, although none of us were either French or Jewish. We enjoyed a day-long excursion to Petra in Jordan. The Chang's presented us with some pictures of that trip when they came to Harvard to celebrate my 80th birthday. It was held in May 2000 while my actual birthday was in March, but the weather and the academic calendar are more cooperative in May or June. Reunions of my Gordon McKay lab research groups had also been organized in 1980 and in 1990 to celebrate my 60th and 70th birthday respectively. Richard and Sunny attended all of them and I cherished a present in the form of a paper rubbing of a huge slate of Chinese calligraphy dating to the third century B.C. It has one hundred and two characters arranged in six columns of seventeen each, wishing health and prosperity.

I also remember two encounters with the Chang daughters. Richard's father, who lived in the San Francisco area, wanted to send me a token of appreciation for having been his son's teacher. He chose to give me a huge smoked salmon. Not knowing my address, he sent it to his granddaughter, who was then an undergraduate at Harvard. She called me one January afternoon and asked if she could deliver the fish early the next morning. Since she had no refrigerator in her dorm room, she kept it overnight on the window sill outside in the cold air. She was glad to get rid of it, when she came to my office.

The other encounter took place in Moscow, where one of the Chang daughters, Natasha, has been living for many years. She took half a day off to assist at a ceremony at Moscow State University, where I received an honorary doctorate more than ten years ago.

More recently, Richard and Sunny invited me to a dinner on a terrace overlooking the Mekong River in Bangkok, Thailand. I had been invited in 2004 to give a lecture at the University there on the role of lasers in peace and war.

These worldwide encounters illustrate that the globalization of science has long preceded the present globalization of economics. It is a privilege and a pleasure to be able to attend the present reunion of the family and the research group of Richard Chang. We wish him all the best on his retirement from the Yale faculty.

†Passed away

Personal Reflections on Richard K. Chang

James L. Merz
Frank M. Freimann Professor Emeritus of Engineering,
Concurrent Professor of Physics,
Notre Dame University
Notre Dame, Indiana 46556, USA
jmerz@nd.edu

I have known Richard Chang since 1962, when we were graduate students at Harvard. Richard was one of the most genuine, humble, and sincere friends, I have known in my lifetime. Although we had different Ph.D. advisors (Richard was a student of Nicolaas Bloembergen, Nobel Laureate in Physics; I was a student of Peter Pershan, himself a former student of Nico), I was luckily fully incorporated into the Bloembergen lab as Peter's first and (for most of my time there) only graduate student. Richard and I shared an office with several other students, so I came to know him quite personally; my wife Rose-Marie and I enjoyed many happy occasions with the Chang family. Much has been written about Richard's enormous body of scientific accomplishments, but I would like to share some of my insights into the unique personality, that was Richard.

Nico never held weekly or monthly group meetings of his students. When he wanted to discuss a particular issue, he suddenly appeared in the Gordon McKay Laboratory looking for the student, who was working on that problem. Most often it was Richard Chang and/or Pierre Lallemand, both of whom were working on non-linear optics. It was clear that Richard's dissertation would be well-received.

Richard and Sunny, his wife, introduced Rose-Marie and me to Chinese food. They had us to dinner several times to enjoy Sunny's cooking. They took us to Chinatown to introduce us to Dim Sum, for which I have thanked them for the rest of my life. Richard organized a group dinner at the local Chinese restaurant, Joyce Chen's, where Richard taught us how to use chopsticks. I thought I had become so proficient with them, that I challenged Richard to a dual that I could take away the piece of meat that Richard was holding with his chopsticks. After a struggle I did come away with it, and said "I won!" "No you didn't," replied Richard, "just look at your tie!"

Sunny and Richard had three children, and at one point Richard decided that they should choose American names for themselves to simplify their lives in the States. We found it amusing that they chose Kevin, Natasha, and Nicole as their "American names." When Kevin was young, the Changs struggled with the frequent problem of their child's poor appetite. Richard tried to solve it by bringing Kevin to our house at feeding time for our German Shepherd, who could devour her dinner in nearly one bite. Richard then announced that Kevin was President of the "Mootsie Merz Clean Plate Club!"

When Sunny was near-term pregnant with her daughter Natasha, Bloembergen wanted Richard to attend a research conference overseas. They worried that Sunny would come time to deliver, with no Richard in town. They finally resolved the problem by asking me to stand by to take Sunny to the hospital should she start labor. I was delighted, as Rose-Marie and I were in the process of adopting our first baby, and this might be my only chance to bring a pregnant mother to the Emergency Room! Sunny, Rose-Marie and I made a trial run to the Boston-Lying-In Hospital where Sunny was to give birth. As it turned out, Richard returned just in time, so my initiation into the frights and fears of fatherhood was to be delayed.

Rose-Marie and I did eventually adopt four children; the fourth, a girl, who came all the way from South Korea. We asked Richard to be her Godfather. After explaining the significance of Godparents in the life of a new-born child, Richard happily accepted, and asked if there was a traditional gift that was given to a Godchild. We said Yes, it would be a sterling silver child's cup, but there was no need for him to bring a gift. When he and Sunny came for the baptismal celebration, his gift was the most meaningful of all — a pair of sterling silver chopsticks!

When Richard finished his dissertation at Harvard, he received several excellent job offers, one of which was as an Assistant Professor at Yale. He confided with me that Bloembergen advised him not to take it, because he would not get tenure at a place like Yale. So Richard accepted the offer to prove Nico wrong. He ended up spending his entire career at Yale, much of which as a Chaired Full Professor. His record of scientific contributions in optical phenomena is outstanding, for which he will be justly remembered. I knew him as a cherished man throughout his life, caring for his family, creative in his endeavors, fully joyful and human in his interaction with colleagues, relatives and friends. He will be missed.

In 1978, after working 12 years as a research scientist at Bell Labs, I decided to accept an offer for a faculty position at University of California Santa Barbara (UCSB). My most ominous concern was the search for research funding, a problem I never had to worry about at Bell. I was soon to learn that this would occupy much of my time as a professor, and I had no idea where to start! I immediately thought of Richard, my friend, the successful professor at Yale. He must know how to follow the money! Then I followed a two-hour phone tutorial as Richard told me the names and phone numbers of the program directors at the relevant agencies like the National Science Foundation and the Army, Navy, and Air Force, with tips on how to present my ideas. One unusual suggestion was that I call his contact at Hanscom Air Force Base outside of Boston, as they had funded a project of his for several years. That led to support for some of my projects for much of the time I was at UCSB. I certainly owe Richard (and Hanscom Air Force Base) for that support! Whoops, I should not be giving away our secret!

Letter to Richard Chang

Joan Bovarnick, Ph.D.
Royal Botanic Gardens Kew
London, TW9 3AE, England, UK

We all began at Brandeis, where Sunny and I were graduate students together in the Biology department — from September 1964. Sunny and I began by studying biochemistry together, often at your home where Kevin played with the typewriter, as a two (?) year old. Then came Pang-Mei, and we shifted study to the hospital for a while. Richard, you were finishing up at Harvard and Arthur was working in New York and coming up to Cambridge on weekends.

I recall (but can't find the important relevant photo) our excursion to Cape Cod with Kevin and Barbara Filner, to the herring run. Then there was the time, I had a party in Waltham and you and Sunny were there. Sunny was shocked that Arthur was still smoking and promptly sent him the Reader's Digest article about patients on a cancer ward, who were still smoking with a tracheotomy — that did it — he did quit.

We've shared many good times — Richard, I recall your mother at my flat in Waltham bringing a special rice cake for my birthday, with the lace handkerchief and a couple of small ivory elephants. I was recently able to return them to you as your heirloom.

From our apartment in New York, where one time you came with the three children, who sat quietly reading the whole time, we were all together — to Kevin starting his love affair with London at our home in London.

You both helped us celebrate our wedding, Andrew's bar mitzvah, Aaron and Silvie's wedding in Berlin, and we have shared with you many special occasions — certainly Nicole and Denis' wedding in Paris and Normandy was wonderfully memorable.

Bike riding in Hamden, Connecticut and Brattleboro, Vermont, as well as lots of time together in London.

Train ride in Borneo: Richard, Nicole and Sunny.

Dinner in the Long House — which included fern fronds — young ones.

Another of the train journey — Richard and Joan.

The famous Long House which Richard will not let me forget. Especially its amazing facilities!!! Awakened in the a.m. by the roosters roaming under the see-through flooring! Richard and Sunny in their 5 star accommodations.

And let's of course, remember Borneo, Malaysia in August 1995, where we travelled together with our son Andrew and Nicole as well. I know your favourite spot was The Longhouse overnight with that sophisticated 'out house' — guess I'll never live that one down.

Then there were the roosters crowing at dawn underneath our sleeping areas.

Do you remember us all escaping from monkey island — except for brave Nicole and the guide. And there you and Sunny were on your adventure battling your way around Mount Kinabalu on a 'well-marked' trail through underbrush and rain, while we slogged up Kinabalu a way only to slide down the mud in the rain. What good sports!!

All our times together over these years — although often geographically distant — we've kept close. We value our long friendship with you both and look forward to its long continuation. Richard, we have watched you over the years as a wonderful husband and father and now grandfather, as well as, a devoted son and brother.

Your boundless energies are an inspiration to all.

We wish you a long and relaxing retirement (although I can't think of relaxing and you going together) and continuing enjoyment with your wonderful family. You are blessed with a loving and devoted wife. Sunny, you too are amazing and shoulder all your responsibilities so capably and elegantly.

Arthur and I wish you all our best in the years to come and look forward to more time together with you both.

April 2008

In the market place — Richard, Sunny, and Arthur — great pineapples and bananas.

My Memories of Richard Chang

James J. Wynne, Ph.D.
Senior Research Scientist
IBM Research Headquarters
Yorktown Heights, NY 10598, USA
jjwynne@us.ibm.com

I first met Richard in the Fall of 1963, when I was a senior at Harvard College. I was a physics major, and the previous summer I had my first job working in a physics environment, where I was a lab assistant to a team working on lasers. Because of that great summer experience, I decided to pursue a career in experimental physics, rather than medicine, the career my parents wanted me to pursue.

When I got back to Harvard at the start of my senior year, I sought out professors who were working with lasers, and I found Nico Bloembergen. After meeting with Nico, he agreed to let me spend time working with his graduate students in the Gordon McKay Laboratory. With great fortune, I was assigned to work with Richard, who mentored me during that entire Fall term. I learned all about semiconductors, the current state of laser physics, and, of course, nonlinear optics. Richard assigned me to search through the AIP Handbook of Physics and find all the crystals that lack a center of inversion symmetry, since these were the materials that could respond to laser light and produce second harmonic generation. I knew about crystal quartz and gallium arsenide (GaAs), since these were two of the first materials to produce second harmonic generation, but there were countless other materials that also lacked a center of inversion symmetry, and I produced a very lengthy list of these materials. Richard oversaw what I was doing, coached me, and was always available to answer my questions and give me advice. Little did I know at that time that I would stay at Harvard for graduate school and have Nico Bloembergen as my PhD thesis advisor.

After 2 years of graduate school, taking courses, I was ready to embark on my PhD thesis topic. Nico directed me to build a carbon dioxide (CO_2) laser and use it to study the nonlinear optical properties of semiconductors, including GaAs. How fortunate I was to have learned about the panoply of semiconductor crystals under Richard's guidance three years earlier. After I built the CO_2 laser, I was off and running, measuring nonlinear optical susceptibilities in GaAs, many other III-V crystals, several II-VI crystals, plus silicon and germanium.

Richard had departed for Yale just before I started on my thesis project, but he was always available to advise me and point me in the right direction. After I completed my PhD project in 1969, I joined IBM Research, where I am still employed. I remember meeting

Richard's sister in Taormina, Sicily, in 1972, when I was attending a physics conference. I also remember being invited by Richard to give a seminar at Yale, in ~1980, on my work on high resolution laser spectroscopy of alkaline atoms. And I remember meeting him and his wife, Sunny, in Atlanta, GA, in 1999, at the celebration of the 100th anniversary of the founding of the American Physical Society. And finally, I remember meeting him and Sunny at Harvard, in 2010, where we were celebrating Nico's 90th birthday. On each of these occasions, we shared fond memories of the "good old days" in 1963 when I blossomed as an experimental physicist, under Richard's guidance. I couldn't have done it without Richard to steer me in the right direction.

Remembering Richard Kounai Chang

Y. R. Shen
Professor of Physics
University of California
Berkeley, CA 94720 USA
yrshen@berkeley.edu

Richard K. Chang was a pioneer, an innovator, and an esteemed educator in science and technology. He was among my most respected scientists, colleagues, and friends. I first met Richard 62 years ago at Harvard as a fellow student. It was an early day of June in 1962 when Professor Nico Bloembergen brought Richard to the third floor of the Gordon McKay Laboratory and introduced him as a new graduate student into his research group, which I joined one year earlier. My first impression of Richard was that he was very polite and friendly and must be from a well-educated family. Later, I learned his Chinese name was Chang Kounai. Kou (国) means nation and Nai (鼐) an exceptionally large tripod vessel with three legs and two ear loops on the sides for handling. Kou-Nai (国鼐) literally refers to a large, inscribed bronze tripod kept by a kingdom in ancient China to symbolize the establishment of the kingdom and could be interpreted as the corner stone of a nation. Richard's parents obviously had high expectation of him when he was born.

It was an exciting time around 1962 for modern optics. Maiman invented the first (ruby) laser and published his work in May, 1960.[1] Franken and coworkers using a borrowed ruby laser from industry first observed optical second harmonic generation in quartz and reported the result in August, 1961,[2] and subsequently, sum and difference frequency generation from quartz in January, 1962.[3] The classical paper of Bloembergen with Armstrong, Ducuing, and Pershan that laid the theoretical foundation of nonlinear optics and led to Bloembergen's Nobel Prize was submitted to *Physical Review* in April and published in September, 1962.[4] The first ruby laser purchased by Bloembergen's group arrived in the lab in early 1962. As Professor Bloembergen commented at the time, there were tons of nonlinear optics problems waiting to be explored experimentally; one only needed to go to some sections of Born and Wolf's "The Principles of Optics" and ask how the results would be different if the medium response to the optical field were nonlinear. With the new ruby laser, John Armstrong[5] and Jacque Ducuing,[6] a postdoc and a senior student from France, respectively, in the group were able to begin experimental studies on nonlinear optics. The rest of the group later learned with envy that Richard was assigned to join the project.

Richard's PhD thesis work was on the basics of second harmonic generation (SHG) in materials. It created huge, long-lasting impact in the field of nonlinear optics. He and Ducuing experimentally demonstrated that nonlinear susceptibilities are complex and dispersive.[7,8] The scheme they developed to measure the phase of the SH output for deduction of complex nonlinear susceptibilities is still being generally used nowadays to measure phases of outputs from nonlinear optical processes.[7] Richard did experiments to verify the predicted laws of nonlinear optical transmission and reflection at an interface.[9] He and C.H. Lee carried out the first electric-field-induced SH (EFISH) generation measurement.[10] They also followed Bloembergen's suggestion and showed that SHG from media with inversion symmetry could occur because of the presence of symmetry-allowed electric quadrupole contribution to the second-order nonlinearity in the bulk[11] and electric-dipole-allowed nonlinearity in the surface layer.[12] The latter is the basis of surface nonlinear spectroscopy currently being widely adopted for surface and interfacial studies.

After leaving Harvard in the summer of 1966, Richard went directly to Yale as an assistant professor, where he eventually spent his whole academic career. William R. Bennet, the inventor of the argon ion laser, was already there, and Richard's arrival greatly enhanced the strength of optical science research at Yale. He continued his pioneering research in nonlinear optics, reporting first observation of third-harmonic generation in dye solution[13] and stimulated Raman scattering in silicon.[14] He then branched into spontaneous Raman scattering (RS) studies of materials on many fronts, namely, electronic RS in rare-earth garnets,[15] dispersive RS,[16] resonant RS,[17–19] two-phonon RS,[20,21] and others, and began to explore the use of RS, Mie scattering, and fluorescence to monitor aerosols and other microparticles.[22] The latter prepared him for a future endeavor on aerosol detection in environmental science after the turn of the century.

The discovery of surface enhanced Raman scattering (SERS)[23] in 1973 and subsequent sorting out of the underlying mechanism[24,25] in 1977 created a great deal of excitement. Richard was among the first group of scientists working on SERS. He and Furtak edited the first book on SERS in 1982.[26] Around the same time, Richard started on optical spectroscopic studies of structural resonances (morphology-dependent resonances or whispering gallery modes) of microparticles[27] and microfibers.[28] It paved the way for his subsequent pioneering research on liquid droplets. He demonstrated that with resonance enhancement of whispering gallery modes, lasing[29] and nonlinear optical effects such as stimulated Raman and Brillouin scattering,[30,31] coherent anti-Stokes Raman scattering,[32] third harmonic generation and four-wave mixing,[32,33] self-phase modulation,[32,34] laser-induced plasma and breakdown,[35,36] etc. in individual droplets could be readily observed. The resonance spectrum of a droplet was found to be very sensitive to changes of the droplet size and morphology as well as the environment it was in, and accordingly, could be used to characterize droplets and sense environment. Richard's works created a lot of excitement. Many researchers in science and technology soon joined in and since then, microparticles and microfibers as optical microcavities have bloomed into a very active field. It was recognized that high-quality fused silica microspheres could provide whispering gallery modes with much higher Q-factors.[37] This prompted several groups led by eminent scientist to employ them for studies of cavity quantum electrodynamics that deals with field interactions with a small number of photons in a microcavity.[37–39] Richard edited a book on "Optical Effects Associated with Small Particles" with P.W. Barber in 1988,[40] and another one on "Optical Processes in Microcavities" with A. J. Campillo in 1996.[41] In ensuing years, Richard extended his investigation on microcavities to lasing and nonlinear optical effects in 2D microdisks.[42,43]

In his later career, as many accomplished scientists would do, Richard became interested in transforming his expertise to useful real applications. He focused on detection and characterization of airborne aerosols,[44] especially bio-related aerosols.[45] Using single-shot fluorescence spectroscopy with pulsed laser excitation, he demonstrated real-time detection and characterization of individual airborne particles.[44,46] The results were truly impressive; they would not be possible if it were not for his clever experimental implementation.

Richard had a research style many of us admire. He emphasized elegance, novelty, and originality. He planned experiments carefully. He always tried hard to combine theory and experiment in his works. He was an outstanding teacher who could convey hard concepts without difficulty. He was a beloved supervisor for members of his research group. He was a great helper and dear friend to many of his colleagues. He served generously in the academic and research communities.

Richard was probably the most favorite student of Professor Bloembergen. He and his wife, Sunny, had established strong bonds with the Bloembergens. When I had opportunities to visit Professor Bloembergen during his later years, "how Richard and Sunny are doing" was always a subject jumping out in our conversation. Richard cared about his pedigree. He visited China numerous times, lecturing, consulting, and advising at universities and research institutions. He was very close to the Chinese academic community. He hosted and helped train many students and young scientists from China.

I met with Richard only occasionally after we left Harvard, but I knew Richard was doing well from his publications. The bad news came in 1996 that Richard was diagnosed to have Parkinson disease, but he was very courageous in fighting the disease, struggling mightily to get back to a normal life while continuously working hard on teaching and research. The last time I saw him was at a 2004 laser science workshop in Li Jiang, China. Despite suffering from Parkinson at the time, he was attentive and interactive throughout the workshop. He even joined the group sightseeing tour and walked a fair distance. Richard was hit by another unfortunate incident in 2008, nearly drowned in a swimming pool, but fortunately, both Richard and Sunny were optimists and knew how to face hardship. They moved to Singapore in 2014 to escape the severe winter in New Haven and be close to their son, living happily ever after.

I have a vivid image of Richard in my mind, a handsome, well-educated, friendly gentleman, always smiling, ready to engage, and willing to help. This, I believe, is how Richard will be remembered on top of being a highly accomplished and well-respected scientist.

References

1. T.H. Maiman, "Stimulated Optical Radiation in Ruby", *Nature* **187** 493 (1960).
2. P.A. Franken, A.E. Hill, C.W. Peters, and G. Weinreich, "Generation of Optical Harmonics". *Phys. Rev. Lett. 7*, 118 (1961).
3. M. Bass, A.E. Hill, P.A. Franken, C.W. Peters, G. Weinreich, "Optical Mixing", *Phys. Rev. Lett.* **8**, 18–19 (1962).
4. J. A. Armstrong, N. Bloembergen, J. Ducuing, and P. S. Pershan, "Interactions between Light Waves in a Nonlinear Dielectric", *Phys. Rev.* **127**, 1918 (1962).
5. Former vice president of IBM Research Labs, Yorktown Heights.
6. Former director of Centre National de la Recherche Scientifique (CNRS), France.
7. R.K. Chang, J. Ducuing, and N. Bloembergen, "Relative Phase Measurement between Fundamental and Second-Harmonic Light," *Phys. Rev. Lett.* **15**, 6 (1965).
8. R.K. Chang, J. Ducuing, and N. Bloembergen, "Dispersion of the Optical Nonlinearity in Semiconductors," *Phys. Rev. Lett.* **15**, 415 (1965).
9. R.K. Chang and N. Bloembergen, "Experimental Verification of the Laws for the Reflected Intensity of Second-Harmonic Light," *Phys. Rev.* **144**, 775 (1966).
10. C.H. Lee, R.K. Chang, and N. Bloembergen, "Nonlinear Electroreflectance in Silicon and Silver," *Phys. Rev. Lett.* **18**, 167 (1967).
11. N. Bloembergen, R.K. Chang, and C.H. Lee, "Second-Harmonic Generation of Light in Reflection from Media with Inversion Symmetry," *Phys. Rev. Lett.* **16**, 986 (1966).
12. N. Bloembergen, R.K. Chang, S.S. Jha, and C.H. Lee, "Optical Second-Harmonic Generation in Reflection from Media with Inversion Symmetry," *Phys. Rev.* **174**, 813 (1968).
13. R.K. Chang and L.K. Galbraith, "Optical Third-Harmonic Generation in Dye Solutions," *Phys. Rev.* **171**, 993 (1968).
14. J.M. Ralston and R.K. Chang, "Spontaneous Raman Scattering Efficiency and Stimulated Scattering in Silicon," *Phys. Rev.* **B 2**, 1858 (1970).
15. B.E. Argyle, J.L. Lewis, R.L. Wadsack, and R.K. Chang, "Electronic Raman Studies of Antiferromagnetic Dysprosium Aluminum Garnet below T_N and with and without an Applied Magnetic Field," *Phys. Rev.* **B 4**, 3035 (1971).
16. R.H. Callender, S.S. Sussman, M. Selders, and R.K. Chang, "Dispersion of Raman Cross Section in CdS and ZnO over a Wide Energy Range," *Phys. Rev.* **B 7**, 3788 (1973).
17. J.M. Ralston, R.L. Wadsack, and R.K. Chang, "Resonant Cancellation of Raman Scattering from CdS and Si," *Phys. Rev. Lett.* **25**, 814 (1970).

18. R.L. Wadsack and R.K. Chang, "Resonance Electronic Raman Effect in DyA1G," *Solid State Commun.* **10**, 45 (1972).

19. D.G. Fouche and R.K. Chang, "Observation of Resonance Raman Scattering below the Dissociation Limit in I_2 Vapor," *Phys. Rev. Lett.* **29**, 536 (1972).

20. P.B. Klein, H. Masui, J.-J. Song, and R.K. Chang, "Selective Resonant Enhancement in the Two-Phonon Raman Spectrum of Si and GaP," *Solid State Commun.* **14**, 1163 (1974).

21. P.B. Klein and R.K. Chang, "Comparison of Second-Order Raman Scattering Measurements with a Phonon Density-of-States Calculation in GaSb," *Phys. Rev.* **B 14**, 2498 (1976).

22. E.-H. Lee, R.E. Benner, J.B. Fenn, and R.K. Chang, "Angular Distribution of Fluorescence from Monodispersed Particles," *Appl. Opt.* **17**, 1980 (1978).

23. M. Fleischmann, P.J Hendra and A.J. McQuillan, "Raman Spectra of Pyridine Adsorbed at a Silver Electrode". *Chem. Phys. Lett.* **26**, 163 (1973).

24. D.L. Jeanmaire, R.P. van Duyne, "Surface Raman Electrochemistry Part I. Heterocyclic, Aromatic and Aliphatic Amines Adsorbed on the Anodized Silver Electrode", *J. Electroanal. Chem.* **84**, 1 (1977).

25. M.G. Albrecht and J.A. Creighton, "Anomalously Intense Raman Spectra of Pyridine at a Silver Electrode", *J. Am. Chem. Soc.* **99**, 5215 (1977).

26. R.K. Chang and T.E. Furtak, eds., "Surface Enhanced Raman Scattering" (Plenum Press, New York, 1982).

27. R.E. Benner, P.W. Barber, J.F. Owen, and R.K. Chang, "Observation of Structure Resonances in the Fluorescence Spectra from Microspheres," *Phys. Rev. Lett.* **44,** 475 (1980).

28. J.F. Owen, P.W. Barber, P.B. Dorain, and R.K. Chang, "Enhancement of Fluorescence Induced by Microstructure Resonances of a Dielectric Fiber," *Phys. Rev. Lett.* **47**, 1075 (1981).

29. H.-M. Tzeng, K.F. Wall, M.B. Long, and R.K. Chang, "Laser Emission from Individual Droplets at Wavelengths Corresponding to Morphology-Dependent Resonances," *Opt. Lett.* **9**, 499 (1984).

30. J.B. Snow, S.-X. Qian, and R.K. Chang, "Stimulated Raman Scattering from Individual Water and Ethanol Droplets at Morphology-Dependent Resonances," *Opt. Lett.* **10**, 37 (1985).

31. J.-Z. Zhang and R.K. Chang, "Generation and Suppression of Stimulated Brillouin Scattering in Single Liquid Droplets," *J. Opt. Soc. Am.* **B 6,** 151 (1989).

32. R.K. Chang, S.-X. Qian, and J. Eickmans, "Stimulated Raman Scattering, Phase Modulation, and Coherent Anti-Stokes Raman Scattering from Single Micrometer-

Size Liquid Droplets," in *Methods of Laser Spectroscopy*, Y. Prior, A. Ben-Reuven, and M. Rosenbluh, eds. (Plenum Press, New York, 1986), p. 249.

33. W.P. Acker, D.H. Leach, and R.K. Chang, "Third-Order Optical Sum Frequency Generation in Micrometer-Sized Liquid Droplets," *Opt. Lett.* **14**, 402 (1989).

34. S.-X. Qian and R.K. Chang, "Phase-Modulation-Broadened Line Shapes from Micrometer-Size CS_2 Droplets," *Opt. Lett.* **11**, 371 (1986).

35. J.-B. Zheng, W.-F. Hsieh, S.-C. Chen, and R.K. Chang, "Temporally and Spatially Resolved Spectroscopy of Laser-Induced Plasma from a Droplet," *Opt. Lett.* **13**, 559 (1988).

36. J.H. Eickmans, W.-F. Hsieh, and R.K. Chang, "Laser-Induced Explosion of H2O Droplets: Spatially Resolved Spectra," *Opt. Lett.* **12**, 22 (1987).

37. L. Collot, V. Lefèvre-Seguin, M. Brune, J. M. Raimond, and S. Haroche, "Very High-*Q* Whispering-Gallery Mode Resonances Observed on Fused Silica Microspheres", *Euro. Phys. Lett.* **23**, 327 (1993).

38. V. Lefèvre-Seguin and S. Haroche, "Towards cavity-QED experiments with silica microspheres", *Mat. Sci. & Engg.* **B48**, 53 (1997).

39. D. W. Vernooy, A. Furusawa, N. Ph. Georgiades, V. S. Ilchenko, and H. J. Kimble, "Cavity QED with high-Q whispering gallery modes", *Phys. Rev.* **A57**, R2293 (1998).

40. R.K. Chang and A.J. Campillo, eds., Optical Processes in Microcavities (World Scientific Publishers, Singapore, 1996).

41. P.W. Barber and R.K. Chang, eds., Optical Effects Associated with Small Particles (World Scientific, Singapore, 1988).

42. Jens U. Nöckel and Richard K. Chang, "2D Microcavities: Theory and Experiments," *Experimental Methods in the Physical Sciences*, edited by R. D. van Zee and P. Looney (Elsevier Science B.V. 2002), p 185.

43. G. D. Chern, A. W. Poon, R.K. Chang, T. Ben-Messaoud, O. Alloschery, E. Toussaere, J. Zyss, and S.-Y. Kou, "Direct Evidence of Open Ray Orbits in a Square 2-d Resonator of Dye-Doped Polymers," *Opt. Lett.* **29**, 1674 (2004).

44. R. G. Pinnick, S. C. Hill, P. Nachman, G. Videen, G. Chen and R.K. Chang "Aerosol Fluorescence Spectrum Analyzer for Measurement of Single Micron-sized Airborne Biological Particles," *Aerosol Science and Technology* **28**, 95 (1998).

45. S. Hill, R. Pinnick, Y. Pan, S. Holler, R. K. Chang, J. Bottigter, B. Chen, C. Orr, and G. Feather, "Real-time Measurement of Fluorescence Spectra from Single Airborne Biological Particles," *Field Analytical Chemistry and Technology*, **3**, 221 (1999).

46. Y. Pan, S. Holler, R. K. Chang, S. Hill, R. Pinnick, S. Niles, and J. Bottiger, "Single-Shot Fluorescence Spectra of Individual Micron-Sized Bio-Aerosols Illuminated by a 351 nm or 266 nm UV-Laser, *Opt. Lett.* **24**, 116 (1999).

My Uncle Richard

Elizabeth Chang
Professor of Violin
University of Massachusetts
Amherst, MA 01002, USA
echang@music.umass.edu

In my childhood memory, Uncle Richard occupied a very particular place in our family constellation — he was the serious uncle whom you obeyed without question. When I think back, he was such a young man during those years when his authority loomed over us so imposingly! But in the years that followed, we gradually grew to know the whole, warm, sensitive person, endlessly dedicated to his family. I remember vividly playing the violin for him in the Brattleboro hospital after his near-death incident in a pool in Vermont. He was barely able to move at the time but was able to communicate through a tear rolling down his face. How miraculous it was that he had many years with his family ahead of him, sustained by Auntie Sung-Wen's indomitable spirit — surely one of the great love stories for the ages.

The Academic Family from 1966 to 2008

Yale University, New Haven, CT 06520, USA
Yale University
Becton Center
15 Prospect Street
New Haven, CT 06520 USA

Yale *Department of Applied Physics*

A. DOUGLAS STONE
Carl A. Morse Professor of Applied Physics,
Professor of Physics

June 1, 2008

Richard K. Chang
Henry Ford II Professor of Applied Physics, Professor of Electrical Engineering and Physics

Richard Chang, B.S. Massachusetts Institute of Technology, M.S., Ph.D. Harvard University, faculty member at Yale since 1966: as an optical scientist, you have spent your career revealing the underlying ***Veritas* of *Lux***. Your earliest work with Nobel Laureate Nicolaas Bloembergen laid a cornerstone of the discipline of nonlinear optics. Since that auspicious beginning you have made major contributions to our understanding of a broad range of optical phenomena, including the Resonant Raman Effect, Surface Enhanced Raman Scattering and, for the past twenty years, the interaction of light with microparticles and microresonators. Following your pioneering discoveries on laser emission from microdroplets, the field has grown enormously and it is clear that laser diagnostics for the characterization of small particles has become an important optical tool with many practical applications. Recently you have devoted yourself to one such timely and important application: the identification of airborne pathogens for the purpose of protecting the citizenry. At the same time your work on microparticle optics has had impact on fundamental studies of cavity quantum electrodynamics and on the invention of novel microlasers.

Your work has been distinguished by an elegant interplay between theory and skillful development of new experimental techniques. The more than 300 papers you have authored demonstrate your vigor and productivity, and the long list of co-authors reflects the influence you have had on your many students and colleagues.

Your teaching in the classroom, and training of students in the lab have always stressed the very highest standards and the importance of detailed understanding. Generations of undergraduate and graduate students have benefited from your wisdom and guidance. You have been a model of generosity, integrity and professionalism in all your work. Hence you are an admired and beloved mentor, who has engendered great loyalty in the students and collaborators whose careers have been launched or enhanced by their association with you.

In addition to all of this, you have been a superb community member. Many is the time when the University has needed to call on you for help, and it has always found you a deeply concerned, committed, and loyal member of this faculty, an articulate advocate for Engineering, a partner, a friend, and an ally of any Yale cause.

Your many ties with China have benefited Yale in recent times. You have been an official consulting professor at Beijing and Tsing Hua Universities and were on a panel of three American scientists to evaluate optics programs of major universities throughout China. We are proud to have you connected with us internationally because you embody those qualities we hope to transmit to our students: in you they can see the value of intellectual accomplishment, hard work, warmth and humanity.

For your lasting scientific contributions, your kindness as a colleague and mentor, and your long and distinguished service to the university, Yale salutes you and wishes you the very best in your retirement.

Sincerely Yours,

Alfred Douglas Stone

Alfred Douglas Stone,
Carl A. Morse Professor of Applied Physics

Thanksgivings at the Chang's in Hamden, CT

Chang Family Album

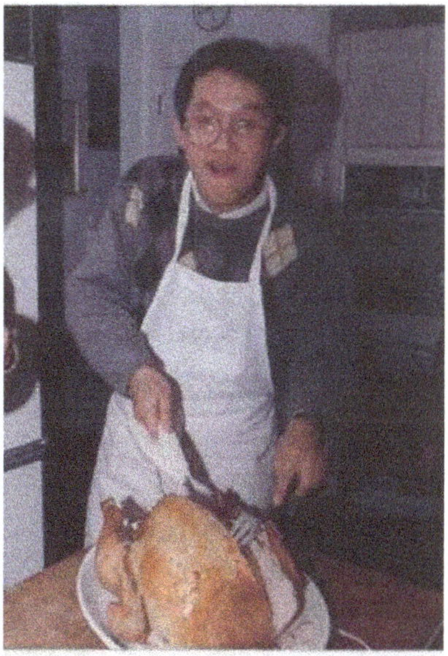

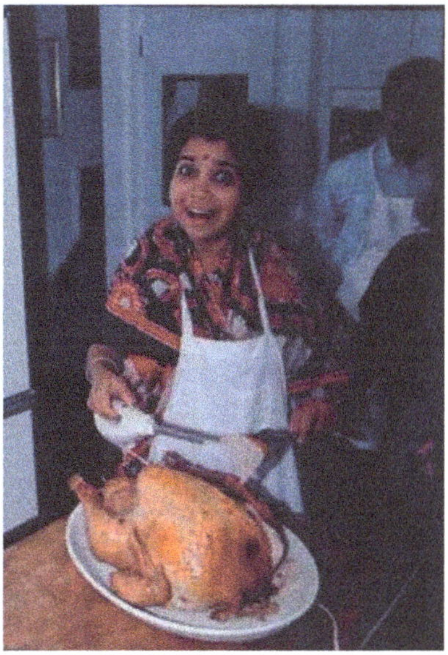

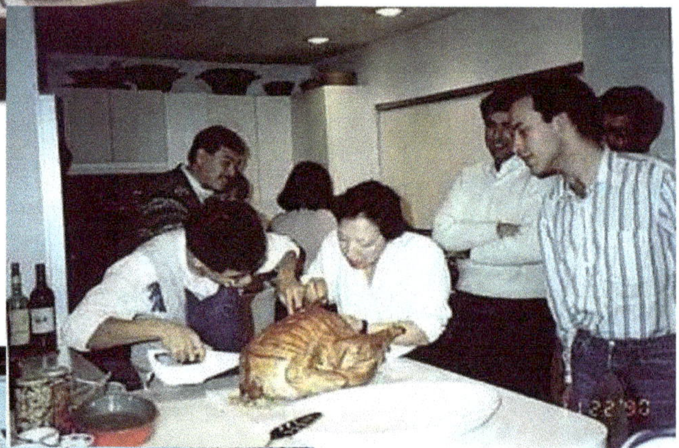

Yale University Retirement Dinner Speech on June 12, 2008

Richard K. Chang[†]

I am overjoyed to see all these familiar faces this evening, and would first like to thank everyone for being here tonight. I appreciate the time, energy, money, and trouble all have gone through to be in this room this very magical evening.

I would first like to take a moment to thank two individuals who made an enormous difference to making this gathering run smoothly: Giselle DeVito, and Maria Gubitosi. Their energy, dedication, and above all their positive attitude, and human touch have made this a truly enjoyable event.

Looking around the room, I am filled with pride and joy when I think about the many talented students, post-docs, and visiting professors I have worked with or the key people in my life who provided great personal support these 42 years at Yale. More important than the number of papers, or grants, or research dollars, are the **people** and the **relationships** that we have built together over the years. This is reflected in the printout at each table — this is our shared legacy. I have enjoyed your talks on "Life after Yale," and our private chats, and I am proud of your success in research at universities, the government, private sectors, or your own companies.

I also would like to thank Marshall Long for organizing this event, for his initiative inviting people, in personally reaching out and encouraging you to attend, and creating such a rich program of events. The program looked great on the website, and it's even more wonderful living it for real with you today.

We are all honored to have Prof. Nico Bloembergen among us today. We thank him and his wife Deli for coming. We are grateful for the wonderful lecture of this afternoon, but even more for the inspiration he gave me at Harvard from 1961 to 1966. He led by example, and in a way, we are all disciples of this great scholar.

Finally, I am also deeply indebted to my dear wife Sunny, who has accompanied me on this long and rich journey since we arrived in New Haven in 1966. Throughout my career, especially in recent years, she has been my steadfast companion, giving me strength and support, even in trying and challenging times. I have been the fortunate recipient of much affection people feel towards her. When I was once asked to be housemaster for one of the Yale residential colleges, I am sure the person they really wanted was Sunny — and I was just part of the package. I pray that my retirement does not drive her nuts.

Someone once said that "old professors never die — they just settle down somewhere between the nightmares and dreams of past graduate students." This could be interpreted in many ways, but the Yale Ph.D. experience in experimental physics/engineering is one of the most intense multiyear experiences I have ever known. No experience asks more of you, but then again, maybe no experience gives you back so much reward. One parent once said to me: "Our son came into your lab as a boy, and left as a man."

Just seeing you here today, I am reliving those moments — those late nights in the labs, the sessions at Clark's Dairy, Clark's Pizza, Mory's — where we discussed the details and celebrated our triumphs large and small. These moments were among the highlights of my life, and here in June 2008, they are coming back to me as if they were yesterday.

The Chinese speak of five fundamental relationships that govern society — Tian1, Di4, Jun1, Qing1, and Shi1… these correspond to Heaven, Earth, King/General, Parents/Family, and Teacher. But in my experience as a professor, I have learned from every single one of my students, post-docs, and visiting professors/scholars. So, in the end, you are also my teachers — and I respect you for all you have done.

Indeed, one of the great joys of this profession and my 42 years at Yale, has been that while my title is that of teacher, my true role has also been that of student. When I received the David Sinclair award, I wish it had been presented in the form of a jigsaw puzzle with many pieces — each piece would belong to one of my students, post-docs, and visiting professors.

As for the actual graduate students I have advised, I feel amazingly lucky. While our field is physics, I think our success also depended on chemistry, the human chemistry for each of you to have enough trust. You worked long hours, with the faith that dedication would bring intellectual reward. This field requires a kind of discipline and training — sometimes you must sprint, sometimes you run a marathon.

I remember someone once said that when you start your Ph.D., you think you're working for the professor, but as you get more deeply involved, you take ownership of your research, you're in the driver's seat, you've thought more about your subject than anyone else in the world, and all of a sudden, it's as if the professor and the lab are working for you.

My students nearly always exceeded my expectations. One piece of advice I frequently offered was never to be afraid to go beyond your limits — both in research, and in life. As you surpassed your expectations, so did I. I remember a colleague once asked why my group didn't do more "mainstream research" — well, we stuck to our new ideas, and as others followed in our footsteps, gradually what we did became "mainstream."

In this field, we love what we do. But more important, we are doing what we love. And that has been one of my greatest privileges of being at Yale, with the freedom to do what I love to do.

In research, we can travel across the world and find companionship even in a foreign land, as we speak a common language, the language of science. Travel has certainly broadened my horizons — and has been the catalyst for many lifelong friendships and professional collaborations (or usually both) spanning both the continents and the decades. Some of my colleagues have even said that I travel too much. Nonetheless, I believe that travel can bring us all closer to each other, both literally and metaphorically.

On this note, Sunny and I are establishing an endowment fund for Yale graduate student travel. I recently observed a graduate student who applied for a travel grant from Yale, having prepared 6 pages of rationale, and who was awarded $100. This suggests to me that it is easy to under-estimate both the cost and the benefit of international travel and scholarly exchange. We are calling the fund "The Sunny and Richard Chang Graduate Student Travel Fund" — and for those who share these views, any donations will be welcomed. As we who are in optics might phrase it ... "Others have lit the way for me, so I now hope that I may shine some light on the path of others."

So, as I end this leg of my journey, a wonderful trip accompanied by many of you, I wish all of you, my fellow travelers, a safe trip and stimulating adventure in your future. I am comforted and happy to know that Hui Cao is taking over my labs in Becton. In this profession, sometimes it's a marathon; sometimes it's a sprint. And in this case, it's also a relay. Today, I pass on the torch to all of you ... and I assure you that whatever you are doing in the future, I will be there with you, following its glow.

Thank you again for your presence today, and I look forward to seeing many of you tomorrow in Vermont.

†Passed away

60th Birthday Celebration in 2000

Richard K. Chang†

Index cards (numbers 1–7) Richard used in his talk to acknowledge the students that came to celebrate his birthday.

INTRO: Your presence here makes me accept the fact that I am 60 years old. I don't quite believe it but it must be true. One of my few regrets tonight is that I don't have the oratory skills of Allan Bromley. 1

As a young boy born in China, burned into my read-only memory is the exalted example of the life of Confucius, who at the age 70 wrote the following:

By age 15, He had already decided to take learning as his goal in life

By age 30, He had more or less established himself.

By age 40, He had not much doubt of his ability to think.

By age 50, He knew his destiny as a great scholar

By age 60, He became more receptive toward different opinions

By age 70, He did just as he pleased

2

With your patience, I'd like to compare my life with his.

By 15, Conf had already decided to take learning as his goal in life

Juvenile delinquent in Brazil. Sent to Christian prep school.
The Stony Brook School. 100 Bible verses
No goals, just hanging loose

By age 30, He had more or less established himself.

- In 1970 I had been at Yale for 4 years, my youngest child Nicole was then 3 yrs

- No clues whether I am "established'-could not read Bob Wheeler's facial expressions being that he is a such a good poker player. Werner Wolf is too British.

- EAS was downsizing, more or less knew no chance for tenure

- Just graduated my very first student, James Ralston

- I worked in the lab then. Bernell Argyle can attest we worked 'till 6 am in our underwear as the equip. gave off so much heat.

- Had a great group of graduate student:
  (Ralston, Wadsack, Argyle, Lewis, Parsons)

3

By age 40, He had not much doubt of his ability to think.

Had real doubts of my ability to out-think my students—

Had to deal with Fouche, Song, Klein, Black, Lee, handsome-Dan Murphy, forever young-looking Farrow, and Long.

Roger Farrow : you totally made me cast doubt of my ability to think----How can a 12 year old think so quickly and deeply? You still look like a 12 year old, but with some white hair. What a child prodigy.

- **Dan**: I could not find your thesis. You did submit one, didn't you?

Paul Klein: I have something to clear up with you. Mugging/crystals

Prof. Peter Barber: when you spent your sabbatical year in 1980 at Yale, you taught me all I know about μ-cavity resonances.

- **Jin-Joo Song** with your brain and organizational skills, I knew you were destined to be a Chaired Professor and Head of a large Laser Center.

4

By age 50, Confucius knew his destiny to be a great scholar

Me, I was still doing "drip-drip-drip" droplet physics and surface enhanced Raman scattering from colloids and atomic-scale roughness on the metal surface with von Raben, Wall, Hsieh, Acker, Leach, and J-Z Zhang.

- Hsieh, you spoke more tonight than all the years you were in the Becton Basement.

Acker you were always the leader and tonight a MC par excellence.

HueyMing Tzeng, your discovery that droplets can got me into combustion and nonlinear optics of droplets.

- David Leach – it's good to see you walking upright. I recall you typing your thesis lying on your stomach on an optical isolation table, because of a accident when you came down a hill at some 30 mph, much too fast for your bicycle.
- J-Z had to go to Konstanz<Germany> so that he could find the true love of his life, a wonderful Chinese lady from his home town in China.

5

- By age 60, the golden age I am tonight. I am not sure exactly what Confucious meant when he said He became more receptive toward different opinions

- I think he means, mellow out, chill out. No doubt I have mellowed and slowed down. The lesson I am learning from Confucius is that I should be more receptive when my group's paper submitted to Nature, Science, or PRL gets rejected with different nasty reviewers comments.

- In the past decade, I have had another fantastic group of students

Serpenguzal—you gave me an added dimension of what it means to be a professor, When broke you leg and stayed in our downstairs for three days because you needed a place without stairs. I never imagined that I would be bringing meals to my students! I am very glad that you now found a wonderful woman to take care of him.

Kwok - who thinks and speaks so fast that he finishes my sentences. I know I speak slowly (and long winded by now), but not that slowly.

Paul Chen – I realized just how ingenious he was when I walked into his shared office one day in June 1989 and saw that he had managed to hook up 10 fax machines to help in a brigade to spread news about Tiananmen.

Justin Hartings being Capt. in the Army, your posture has greatly improved since his basement days leaning over the optical table.

Fields—How is the German you learned in the dark lab room while aligning your equipment with Jergen Popp?

Seongsik Chang - you should count as one of your greatest achievement at Yale that you managed to HALF impressed Doug Stone. Most of us are on the QUARTER level.

6

By age 70, Confucius did just as he pleased.

Here finally is where I can beat Confucius's progress by a number of years. By retiring before age 70, I can "do just as I pleased" without worrying about grants and contracts, progress reports, faculty meetings.

If it pleases me I might write more papers with Steve Hill and Jean-Pierre Wolf.

I look at all the talent and brain power assembled here in this one room and think: what a lot we could get done together now, just one more time – all of us together.

THANK YOU

I am deeply grateful and moved at the energies and times it took to organize this event, everyone coming from different corners of the earth.

Marshall -- thank you for conceiving of this event, organizing it, and encouraging everyone to come through your webpage. If you look over the pictures==learning on the dinning room table and the degrees are awarded in the backyard.

Pat, thank you for all the extra work that you have had to do. And thank you in particular for your unending thoughtfulness.

I also want to thank each and every one of my former graduate students for coming so far. I am proud of every one of you. I am even proud of Yale students who I wish had been mine such as Armand Tanguay, El Hang Lee and Rufus Cone. Your presence has made this a real celebration.

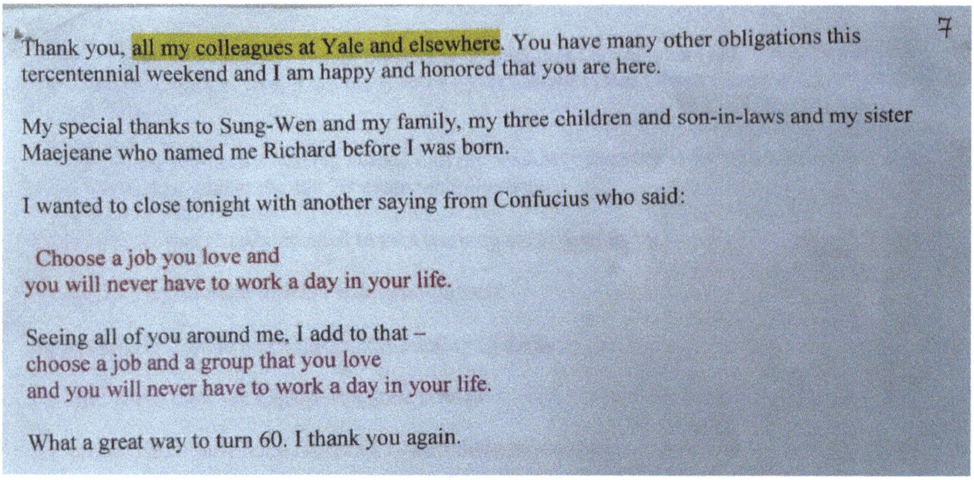

7

Thank you, all my colleagues at Yale and elsewhere. You have many other obligations this tercentennial weekend and I am happy and honored that you are here.

My special thanks to Sung-Wen and my family, my three children and son-in-laws and my sister Maejeane who named me Richard before I was born.

I wanted to close tonight with another saying from Confucius who said:

Choose a job you love and
you will never have to work a day in your life.

Seeing all of you around me, I add to that –
choose a job and a group that you love
and you will never have to work a day in your life.

What a great way to turn 60. I thank you again.

Students who attended the party at Yale Graduate Club: 1st row left to right: Paul Chen, Jian-Zhi Zhang, Richard Chang, Seongsik Chang, Alfred Kwok. 2nd row left to right: William Acker, David Leach, Ali Serpenguzel, Mitch Fields, Chris Swindal, Justin Hartings. Missing from photo: Bernie Argyle, Philip Black, Roger Farrow, Thomas Heldt, Wen-Feng Hsieh, Paul Klein, Marshall Long, James Moore, Armand Tanguay, El-Hang Lee, Daniel V Murphy, Andrew Poon, Jim Ralston, Nathan Rex, Jin-Joo Song, Huey-Ming Tseng, Bruce Webber.

Personal Reflections on Richard K. Chang

Paul Fleury
Frederic William Beinecke Professor Emeritus
Engineering and Applied Physics
Yale University, New Haven, CT 06520 USA
paul.fleury@yale.edu

Richard and I were physics graduate students in Cambridge, Massachusetts from 1962–1965 — he at Harvard and I at MIT just 'down the road'. Although we were working in 'rival groups' on lasers, spectroscopy and nonlinear optics — he under Nico Bloembergen and I under Charlie Townes — we did not meet until near the end of our grad school terms. The occasion was the first international Conference on the Physics of Quantum Electronics in Puerto Rico in the spring of 1965, where we both gave papers. Attending that meeting were several scientists from Bell Laboratories, who were also — as was the custom then — acting as recruiters for potential new BTL scientific staff. There I met Sol Buchsbaum, who became my first boss at the labs later that year. Sol also took careful note of Richard Chang, who was a young star of the Bloembergen team. Most likely because Richard had already decided to remain at Harvard for another year as a post-doc, he did not visit the labs until 1966, when he was vigorously recruited and quickly offered a position. I recall being involved in his visit and with others did my best to entice him to join the labs. Unfortunately for Bell Labs, but very fortunately for Yale University, he accepted a faculty position there in the Applied Physics department. He spent his entire career at Yale, where he was among their most outstanding and prolific research scientists, teachers, and mentors.

The dinner in June 2007, marking the end of my Deanship. Seated are Mary Tully and me, my wife Carol, John Tully, and Richard. Standing are my daughter Jennifer and my granddaughter Katie with husband David Metz, my daughter Laura Fleury, and Sunny.

Our fields of research were close, but not identical. We were, for many years, colleagues and friends. But the only time I visited Yale during the 20th century was for a Yale football game, together with Chuck Shank, also from BTL, at Richard's invitation. He was, as always, a most gracious host and we became closer friends over the next few years. While he excelled at Yale for the next four decades, I remained at Bell Labs for over 30 years, with two years at Sandia as 'time off for good behavior'.

From left to right: Wen-Feng Hsieh, Nicolaas Bloembergen, Mohiuddin Mazumder, Werner Wolf, Seongsik Chang, and Richard Chang in Brattleboro, Vermont, June 2008 after Richard's retirement party in New Haven, Connecticut.

Following the government's break-up of the Bell System and the subsequent fragmentation of Bell Labs, I moved to an academic position as Dean of Engineering at University of New Mexico. Several BTL scientists had also moved to academia — including John Tully and Pierre Hohenberg, who went to Yale. In 2000, when Alan Bromley retired from his deanship at Yale, I was given a chance to join the university as his successor.

Richard was most helpful in helping me find housing in nearby Hamden, Connecticut and integrating me into the university and into the Applied Physics department. He generously allowed me to set up my lab equipment in his own lab space, the commodity most jealously guarded by faculty, and offered me the help of his outstanding assistant Yongle Pan.

Richard and Sunny frequently invited me to dinner at their Hamden home and quickly made me feel a part of the Yale community and family. On many occasions over the next several years they also invited me to their beautiful summer home in Brattleboro, Vermont. Overlooking the Connecticut River and among scenic foothills, it was a fitting retreat for this wonderful family. There Richard gave me several lessons in trail bike riding in the nearby hills. I never succeeded in winning any of our friendly races — but he was as always a most gracious winner.

From left to right: Seongsik Chang, Alfred Kwok, Wen Feng Hsieh, Richard Chang, and Mohiuddin Mazumder in Brattleboro, Vermont, June 2008.

As Dean, it was my responsibility to expand and improve Yale Engineering. Critical to this endeavor was the recruiting and hiring of excellent new faculty. Here again Richard's help and advice was indispensable. A particularly illustrative example was his most positive role in convincing Hui Cao to join Yale from her position at Northwestern University — especially in view of other very attractive competitive offers. In several other such instances, I believe that, Richard's advice and example 'made the difference' in our being able to hire outstanding new faculty. I did not know until well after joining Yale in 2000 that Richard had been diagnosed with the devastating Parkinson's disease. His strength, courage and determination and the miraculous help and support of Sunny and their family kept him going when most others would have failed. He never complained, felt sorry for himself, or allowed his condition to defeat him.

Following the end of my term as Dean in early 2008, I was, as with all Yale faculty, required to teach two courses. I had not taught a class since 1962 — forty six years earlier. I was 'assigned' to teach Physics 448-Solid State Physics, which had been taught for many years by two of Yale's best teachers — Vic Henrich and Richard Chang. Well out of my depth, I fortunately received advice — and their course notes — from both Richard and Vic. Again, without Richard, I would not had a chance.

Richard Chang was the very best example of what the human race has to offer. He was kind, wicked smart, unselfish, sharing, strong, tough, relentless, good humored, generous, and loving. The world was a better place with him in it and is lesser for his absence. We need more like him and will remember him always.

Letter to Richard K. Chang on the Occasion of His Retirement in June 2008

Werner P. Wolf
Professor Emeritus, Yale University, New Haven, CT 06520 USA
werner.wolf@yale.edu

Yale University

Department of Applied Physics
Box 208284 Yale Station
New Haven, CT 06520-8284

Campus address:
419 Becton Center
15 Prospect Street

Werner P. Wolf
Raymond J. Wean Professor Emeritus

Phone: (203) 432-4282
Fax: (203) 432-2207
E-mail: werner.wolf@yale.edu

April 18, 2008

Dear Richard,

Trying to compose a suitable epistle on the occasion of your retirement is a formidable undertaking. Having known you for more than 42 years, and interacted with you in so many different ways over the years, there are a lot of things one might say. But let me try, starting at the beginning.

It must have been in the fall of 1965 when I first heard from my friends at Harvard about this very bright young guy who had recently finished his PhD with Nico Bloembergen, and who might be just the kind of live wire we needed here at Yale to get the new Department of Engineering and Applied Science going. I invited you to come down and immediately recognized that they were right. The only problem was how to persuade you that Yale would be the right place for you. I knew that you were interviewing at other places too, and Yale wasn't very strong in E&AS in those days. But we did have people like Bill Bennett, and Bob Wheeler, and you could see that there really was a great opportunity here to build up a flourishing research group. So you accepted, and started to build.

But there was one big obstacle that you only learned about indirectly. After you had been here five years it came time to promote you to tenure, but that was just the time when the University was going through a really traumatic budget crunch, and all the great plans to expand were suddenly cancelled. Instead of being able to consider you together with the five others who were also up for promotion, the University told the department that we could only promote *one* of the six! Bob Wheeler was chairman then and he called a meeting of the senior faculty on a Saturday morning and announced that we weren't going to adjourn until we had picked *the one*.

I think you can imagine how that meeting went. Everyone had their favorite candidate, and most of them were very good. But they were all in different fields, and so were the faculty members trying to make the decision. How to choose? And I am happy to say that after a while it became clear that there really was one candidate whose research stood out from the rest, and we voted for you.

And now, looking back over 37 years, I think it has become even clearer that we were right. Your career here has been really splendid, and we could not have wished for anyone more outstanding. You have produced an impressive amount of truly significant research, trained many excellent students, and you have contributed in many other ways to both the department and the University.

I cannot enumerate all of these contributions, but I do want to mention some of the ways you have made my own time at Yale better. You were always the voice of reason, the person one could approach with a problem and who would offer constructive suggestions, and generous help whenever it was needed. Twice you served as Acting Chairman so I could take a triennial leave, and I knew that I could just leave everything to you. Looking through my files the other day I came across four pages of notes I took during a meeting you and I had in September 1990. I had just been asked to take on the chairmanship of the recently recognized Department of Applied Physics, following its ambiguous eight-year "Section" status, and I needed advice on how we should proceed. According to my notes we met for only 45 minutes, but during that time you offered a lot of good advice.

One interesting suggestion was that we should all go for an off-campus retreat, and you organized such a day for us at the Trumbull Marriott. I don't recall what happened at that retreat, but I know that we all agreed that we should ask for a departmental review committee, and such a committee indeed came in the spring of the following year. And what a good thing that turned out to be! The committee wrote a glowing report that provided us convincing support the next year during the infamous "Restructuring" turmoil. Without that support we might well have been swallowed by the Physics Department, and Engineering would have been seriously set back.

And thinking of the Restructuring times brings back a memory I am sure you have not forgotten either. At your suggestion Liz and I were spending the Spring break at Guavaberry Spring Bay in Virgin Gorda, trying to recover from the trauma of the battle against the powers of darkness. When we arrived there, we were greeted by a fax from you telling me that the Provost, who had initiated the whole Restructuring process, had resigned and now we could really enjoy our holiday. It was a great moment.

Of course your recommendations for great places to visit were not restricted to Virgin Gorda. Over the years you sent us to St Johns, Tobago, Brunei, Kota Kinabalu and Sipadan, to mention the most memorable. Each time you told us exactly where to stay and what we should do, and every time it worked out perfectly. Our stay in Sipadan in the "end hut that had been used by the BBC," as you specified in precise detail, was one of the highlights of all our travels. We still haven't been to Palau, but we'll get there.

For our part we haven't been as creative in recommending places for you and Sunny to visit, but we like to think that your stay in Oxford in 1969 was also memorable, as were your visits to the Ricketsons and to my parents' house in Zell. We found some photographs of

View towards the Baths from Guavaberry Spring Bay.

Accommodations with Borneo Divers in Sipadan.

the visit to Zell, and also my mother's description of that day, which may bring back some more memories. Here is an excerpt from my mother's diary of July 28th 1971:

Wednesday 28th July.

W.&.El.meet the CHANGS at the Railway,sightseeing in Salzburg,Lunch Peterskeller
in the open courtyeard with Salzburger Nöckerln,get room for them at the "AIDA"
for S 150.-than drive to St.Gilgen and go up to the Swölferhorn,buy cake in Mondsee
Omi at the hairdresser,dinner on the terrace,take photos,very animated atmosphere
lighnings,cool air,they drive back to Salzburg through rain,than see the view from
Winkler(restaurant closed)

But as soon as one starts to think about meals with you and Sunny there are so many that it is impossible to remember them all. What we do remember is the warmth of those occasions, and the good food. Dinners in Paradise Avenue, with your friends at Shangri-Lee and in your lovely house in Vermont were always such happy events. It was

With the Wolfs in Zell am Moos.

Salzburger Nockerln.

After Salzburger Nockerln.

so good to hear about your great trips all around the world, and we always admired your energy and your organization. It was always most impressive to see how you managed to combine pleasure with getting so much work done.

It is hard to think of you not managing half a dozen (or more) grants and contracts simultaneously, but I am sure that you will find a lot of interesting things to do. Maybe you will follow your father and write a memoir of your life. That would be quite a book! But whatever you decide to do next, both Liz and I wish you many contended and happy years of retirement and we look forward to continuing to share both memories of the past and plans for the future with you and with Sunny.

With great admiration and affection from us both,

Werner and Liz

Letter to Richard

William Bennett[†]
Professor of Applied Physics and Physics (Emeritus)
Yale University, New Haven, CT 06520 USA

Yale University *New Haven, Connecticut 06520* DUNHAM LABORATORY
10 Hillhouse Avenue

April 22, 2008

Dear Richard,

It seems impossible that time has passed by so rapidly. I recall a faculty meeting that must have taken place at least 30 years ago as if it occurred only yesterday in which Peter Wegener was summarizing all the retirements that would occur by the year 2000. Now it has caught up and passed us both!

I remember with great satisfaction the outcome of the faculty meeting in which you were promoted to tenure. As I recall, there was only one slot available with about seven candidates to occupy it. Among many comments, there was one special one by George Schultz in which he extolled your excellent style in setting up and running a laboratory. I'm afraid the other candidates didn't have a chance! We are all indebted to you for the skill with which you enhanced our department.

I do hope that we will get to see you and Sunny more frequently in the future — perhaps in Brattleboro or Colrain this summer. I hope that we will be able to get to your celebration on June 12th, although I am not sure how my chemotherapy will be scheduled at that time.

With all best wishes,

Bill Bennett

[†]Passed away

Memories of Richard Chang

Frances C. Bennett
Teacher & Facilitator (Ret.)
Independent Study and Seminar Program (ISSP)
Hamden High School
Hamden, CT 06514 USA

It is often said that marriages are made in heaven. Well, that certainly was the case for Sunny and Richard. I cannot separate the two and probably that is because my view of Richard is firmly anchored in his life with his family.

Richard was my husband Bill's colleague, but I certainly knew little or nothing of his scientific work. What I did know is that he was a devoted family man. I picture him arriving to pick up Kevin from our house on his way home from the lab, and Kevin's eager lookout for his dad's arrival. I remember his pleasure in the birth of Nicole. (Natasha was already part of the entourage.) I think of his warm greetings and perfect hosting at their parties. His consummate good manners were always evident, even in the waning years of his life when he withstood great physical difficulties with grace and humor. In all of this he was complemented by Sunny who shared his enthusiasm for science, teaching, and good cheer. It is a pleasure to have known him.

Left to right: Jean Blanning, Sunny, and Fran Bennett, ISSP-TAG, Hamden and New Haven Public Schools, 1979.

Sunny and Richard with Fran Bennett at the memorial for William Bennett, principal oboist for the San Francisco Symphony Orchestra, May 2013.

Letter to Richard K. Chang on the Occasion of His Retirement in June 2008

Robert Wheeler
Professor Emeritus
Yale University, New Haven, CT 06520, USA
robert.wheeler@yale.edu

April 15, 2008

Professor Richard K. Chang
Henry Ford II Professor of Applied Physics
Yale University
New Haven, Connecticut

Dear Richard,

I am including some photographs which will remind you of a few events which occurred only because of your sincere interest in people. I remember the time that you and your dear wife came to my office in Dunham laboratory suggesting the Saturday Morning Series for High School juniors. This program since that time has been a great success. Recently your concern for Cliff is giving all your old colleagues an excuse to see one another. These two moments separated by nearly forty years illustrate your humanity all the while your science continued to be magnificent.

After all these forty-two years at Yale I do expect our friendship to continue unabated.

Congratulations on taking this next step in your life.

My very best,

Bob

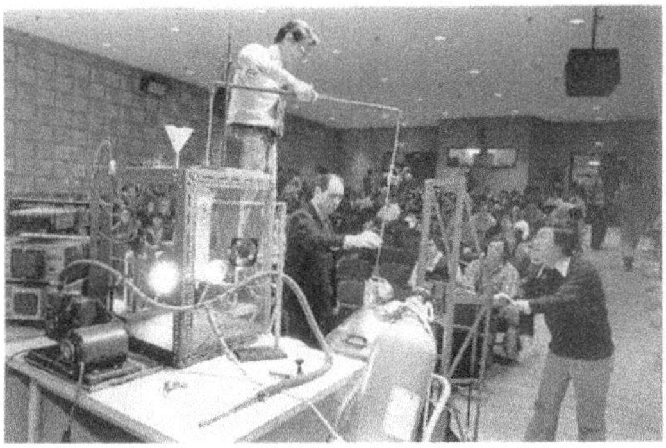

One of the Saturday Morning lectures, R.G. Wheeler, W.P. Wolf , and C.C. Hu.

Frank Firk, Robert Wheeler, Richard Chang, Jane Miller, Clifford Snyder, Stanley Mroczkowski, Richard Barker, Werner Wolf, at Westbrook, CT, 2007.

Frank Firk, Robert Wheeler, Stanley Mroczkowski, Clifford Snyder, Richard Chang, Richard Barker, Werner Wolf at Westbrook, CT, 2006.

Richard Chang, Richard Barker, Clifford Snyder, Robert Wheeler, Stanley Mroczkowski, Werner Wolf at Westbrook, CT, 2005.

Richard Chang, Richard Barker, Clifford Snyder at Westbrook, CT, 2005.

Richard Downing and Gill Vogel.

Clifford Snyder at 99, April 7, 2008.

Letter to Richard K. Chang on the Occasion of His Retirement in June 2008

Peter Kindlmann[†]
Professor Emeritus, Yale University, New Haven, CT 06520 USA
peter.kindlmann@yale.edu

Yale University

Department of Electrical Engineering
Box 208284
New Haven, CT 06520-8284

Campus address:
503 Becton Center
15 Prospect Street

Peter J. Kindlmann
Professor (Adjunct)

Phone: (203) 432-4294

April 23rd, 2008

Dear Richard,

You arrived at Yale when I had just finished my Ph.D., and you are retiring at the same time as I'm phasing into retirement. We both began at beginnings, you at the beginnings of laser technology, I at the beginnings of the semiconductor technology that made possible so much instrumentation and other electronics. We have occupied different quarters on the ship "Yale" on our 42-year journey on it, but even when research collaboration didn't join our purposes directly, we were well aware of each other's contributions and viewed them in terms of greater goods, the values underlying research and education. I fully intend to have us continue our contact, our friendship and our shared values, though now on a more personal scale.

In deciding how to make a statement of tribute to you on your retirement, many thoughts are flooding toward me, many concerns about the attrition of fundamental values, the ever greater segmentation and specialization within research, the effects of which have now reached down to fray even the educational beginnings and the mentoring of future researchers and teachers. You have heard me voice those concerns on previous occasions, with whatever degree of eloquence I was able to muster. This is not the time to repeat myself. So at the risk of embarrassing you, let me reflect on what it has meant to me to have you at Yale.

The words that come to mind are closely linked: values, integrity and quality. We, of the same generation, are still likely to attribute similar meanings to those words. In the present day, the prisms of technology and competition have decomposed them into many individual spectral lines, unlikely to ever again fuse into a single source of inspiration.

In the laboratory you set an example by working with the graduate students and post-docs, navigating them through the many overt and subtle complexities of experimental research, at the same time ready to derive a physical explanation of complex processes from first principles. There was an exemplary directness in your approach to experimental method and to the theory to which it coupled. You insisted on data "honestly owned" through thoroughness of procedure and thoroughness in understanding of the underlying physics.

Over the years, on all past occasions when prospective graduate students consulted me about research opportunities in Engineering and Applied Science, I always mentioned your group first: as the best run, best organized, and as the most effective environment within which to learn and grow.

Your teaching in all senses, in your individual mentoring of graduate students and junior faculty, and in the classroom, was similarly suffused with an attention to fundamental values, how to ask the best questions and what it means to understand something, and how those are the fundamentally most important aspects of learning.

I also speak from a viewpoint unique to my long dual personality as an adjunct. With my one foot in industry and another in academia, giving me something of an "observer's distance" from each (and questions whether I had 'dual citizenship,' or none), I actually became more sensitive to the important distinctions between academia and industry, more demanding of the best qualities in each domain, rather than blurring them together as many have assumed. It has been a source of dismay for me to see some at Yale seek to become more entrepreneurial, more industrial, often rather naively so, and in the process neglecting to understand, retain and evolve the values important to academia.

That does not mean shying away from having academia address problems of concern to industry. That kind of reflexive "academic purity" has also been part of Yale history, but it is a sterile approach, naïve in its own way, incapable of evolution. When you addressed "real-world" problems, the academic standards of integrity and quality of the research, and what it can teach, were never compromised by opportunism and entrepreneurial rush. This is not easy, but I have always been proud of how well you have steered that difficult course. Some of your colleagues were in the past hostile or just puzzled about, for example,

your early combustion diagnostics work. Their viewpoint was too narrow to ever really understand what you were doing. I very much appreciate what you have done.

Since you are among the rapidly vanishing minority that still values modesty, I have surely embarrassed you by now. For this I am sorry. But I could find no other words to pay tribute to what I always appreciated as your affirmation of academic values, both in their traditional form, but also in an extended form where academia can evolve to address "real-world" problems without compromising its core values.

In retirement I wish you and Sunny all the best, in the full knowledge that you will have no lack of interesting and enjoyable things to do.

With warmest regards,

Peter

†Passed away

Letter to Richard

Arvid Herzenberg[†]
Professor of Applied Physics (Emeritus)
Yale University
New Haven, CT 06520 USA

New Haven,
May 2008

Dear Richard,

Thank you for your efforts to tie your colleagues into a community.

I particulaly remember an occasion when I was feeling miserable at a memorial meeting for George Schulz, whom I had come to work with. You took the trouble to congratulate me for keeping my name out of it.

And thank you for the bicycle. It still gets regular use, week in and week out, thirteen years later.

Arvid Herzenberg

[†]Passed away

Letter to Richard K. Chang on the Occasion of His Retirement in June 2008

Daniel Rosner
L.W. Jones Professor of Chemical Engineering
Yale University, New Haven, CT 06520 USA
daniel.rosner@yale.edu

Yale University
HIGH TEMPERATURE CHEMICAL REACTION ENGINEERING LABORATORY
e-Mail: *Daniel.Rosner@yale.edu*

CHEMICAL ENGINEERING DEPARTMENT
Mason Laboratory
P.O. Box 208286 Y.S.
New Haven, CT 06520 -8286 USA

To: **Richard K. Chang** **April 17, 2008**
From: **Dan Rosner** (L. W. Jones Prof. Chemical Engineering, Yale U.)

Dear Richard:

Your formal retirement provides a wonderful opportunity to look back at your fruitful and enjoyable years at Yale, and our mutual interests in aerosol transport and characterization. It is true that upon my arrival at Yale/EAS in 1969 I first heard your name mentioned in connection with the failure of one of my ChE colleagues to gain tenure — the slot going instead to some "hot shot" in Applied Physics! But the evolution of EAS and ChE at Yale not only "survived" this episode — it has led to some remarkable research, much of it collaborative.

My career was not only strongly influenced by "role models" and friends within Yale (like John Fenn, who recruited me and went on to win the '02 Nobel Prize in Chemistry for work done in our Department), but people like Sheldon Friedlander, who put aerosol science "on the US map" and helped found the American Association of Aerosol Research (*AAAR*). As an earlier winner (supported by Prof. Friedlander *et al.*, primarily for my work on aerosol transport THEORY) it was therefore a special pleasure to take the lead in nominating you for the 2005 D. Sinclair Award of the *AAAR*. I was not really surprised by the strength of the supporting letters I received from two additional *AAAR* members: my former Yale PhD student: Juan Fernandez de la Mora (who did his Post-doctoral research with Friedlander at UCLA) and Dr. R.G. Pinnick (Army Research Laboratory), each of whom had the pleasure and good fortune to collaborate with you in recent years.

[Indeed, Pinnick remarked that you were technically one of the strongest scientists with whom he had worked in his (then) 30-year career, significantly adding that "it's been both stimulating and enjoyable".]

While I am personally unable to grasp many of the details of your experimental accomplishments I could not help observing how often the work of my respected colleagues at other laboratories have exploited optical methods derived from your ground-breaking research. As one relatively early example, I immediately think of the fruitful applications that my ChE friends: Jim Davis and A.K. Ray made of morphology-dependent resonances, not only for remote but precise droplet sizing, but also the interference of otherwise inaccessible physico-chemical properties. At AFOSR meetings I learned that many of the techniques currently being used to characterize evaporating multi-component fuel sprays — e.g., in air-breathing combustor applications, also derive from your earlier work — some in collaboration with our own Marshall Long (who continues to be a key player/collaborator in our Center for Combustion Studies). Another creative aerosol researcher who I admire (both personally and technically) is our mutual friend Chris Sorensen (KSU/Physics). Together you have demonstrated the enormous power of optical techniques to characterize the size and morphology of individual particles and particle populations.

I was pleased to find among my personal collection of Yale photos, one showing you, me and Marshall Long in the same smiling "line-up" — taken on the occasion of Randy VanderWal's visit/seminar, probably ca. 1 decade ago. [For the record, the other two people shown (far left) are Dr. Andrey Filippov (who collaborated with me in the late 90s and who is now at Corning Glass Research Labs) and, to his right, grad student Manoj Kumar (who dropped out and has not been heard from since.] At the time we were all actively working/publishing on the subject of "laser-induced incandescence" (LII) for inferring aerosol size distributions — another now widely used laser-based technique. But

Manoj Kumar, Andrey Filippov, Daniel Rosner, Randy VanderWal, Marshall Long, and Richard Chang at Yale University.

I always insisted on complementary "immersion" measurements (to confirm morphology) using "thermophoretic sampling" — another now popular technique which we extended and exploited in the decade of the 90s.

In the more distant past I can recall occasionally teaching a graduate course entitled: *Aerosol Science and Technology*, to which you contributed a valuable lecture on the use of optical techniques for aerosol characterization. Actually, this elective course has not been taught for too many years, and we should probably recommend that our younger colleagues (Juan, Marshall, Sandro, ...) re-offer (at least every 3–4 years) an up-dated version. Perhaps we could serve as "consultants"!

I consider myself fortunate in many ways — not the least of which is that our presence at Yale/EAS was "strongly correlated in time". These have been "interesting" and productive times — and I think we each came "on-stream" at just the right moment. As usual, the future is not readily predicted, but your "legacy" is intact and there is no doubt that your impact will persist for *many* decades to come.

Sincerely,

Daniel E. Rosner,
L.W. Jones Prof. Chemical Engineering
Yale University

RK Chang Reminiscences

James Ralston, Ph.D. Yale '70
Fairfax, VA22031, USA

Yale 1966–1970

I have often reflected on my great good fortune as a graduate student having Richard Chang as my thesis adviser. His encouragement and frequent attention to my work no less than his personal example were enormous assets to me in my career at Yale. It was clear that Richard was ambitious and his graduate students were fortunate that much of his ambition and drive to accomplishment "filtered down" to us! In subsequent years, I have come to appreciate Richard's achievement in building up a team of mutually supporting graduate students. My progress was helped immeasurably by interaction with my colleagues in the atmosphere of hard work and mutual respect that he fostered.

I will never forget how I met Richard. I had arrived at Yale as a graduate student in 1966 with some vague ideas that "lasers" were a hot research area. At that time, the Engineering and Applied Science department included the preeminent laser scientist, the esteemed Prof. Bill Bennett. Consequently, I made an appointment with Prof. Bennett in the hope of impressing him with my prospects for doctoral research. At the appointed time, I presented myself at the great professor's office in Dunham Lab. I was impressed to see that the professor had not only a private secretary but moreover, a personal Xerox copier! Overcoming some diffidence I introduced myself and went into some detail about my undergraduate performance, my interest in optics and my laboratory experience. To my delight, Bennett seemed very impressed. "Let's take a walk," he said. I accompanied him down the hall, anticipating a tour of his labs. I kept talking about my experience, and he kept nodding in the most encouraging fashion. "Wow," I thought, "this is really going well!" At the end of the corridor, he paused at an office with door ajar. I noticed the name on the office nameplate:

R.K. Chang

Without ceremony Bennett opened the door, greeted the occupant, and announced "Richard, this fellow wants to talk to you." With no further ceremony, he pushed me through the door! Whatever momentary disappointment I might have felt quickly disappeared in the successes of the following years. No doubt, I owe Bill Bennett a debt of gratitude after all!

I was only briefly Richard's sole research assistant; Ron Wadsack soon joined the team. At that time in the 1960s, my fellow grad students and I were obsessed with the need to pass the dread Qualifying Examination for doctoral studies at the end of our first year's work. The exam was a daylong ordeal and by evening, I was anxious to find out whether I had succeeded. Rather than let me stew Richard kindly called me at home to tell me I had passed. He mentioned that, Ron Wadsack and I had earned the two top scores. He seemed as pleased as Ron and I were by that. Having passed the "Qualifier" I was able, with Richard's guidance, to map out an experimental research program that would lead to the doctorate.

By 1969, I had obtained some interesting experimental results on optical limiting in several different semiconductors. With characteristic generosity, not to mention trust in my presentation ability, Richard used some of his own Alfred P. Sloan fellowship funds to send me, solo, to the Conference on Non-Linear Optics held at Queen's University, Belfast. The following year I was able to complete my dissertation and was graduated.

Physicists of my era probably remember that 1970 was not a time when newly minted Ph.D.s were in high demand. With the assistance of Richard's contacts and prestige, however, I was able to land a couple of attractive offers and started working that fall at Bell Telephone Laboratories. I spent five years there getting my technical career off to a good start before moving to Washington, where I have spent the rest of my career working on national defense issues and technologies.

It is a pleasure to have this opportunity to reminisce on the start of my scientific career and acknowledge the influence that Richard Chang has had on it.

RKC Lab 1968–1970

Prof. Chang's research at Yale was initially conducted at Dunham Lab on Hillhouse Ave. His first two graduate students were Ronald Wadsack and me (James Ralston). Ron worked primarily with a multiple-wavelength Argon laser that he developed himself. I had joined the team a bit earlier and spent most of my time developing experiments with solid-state Nd:YAG lasers. Here are a few photos from the Dunham laboratories in the 1968–70 timeframe.

This is a view of your correspondent (Jim Ralston) with his pulsed YAG laser experiment in Dunham Lab circa 1969. We used this laser to measure induced absorption in several semiconductors and, later, for stimulated Raman scattering in Silicon. Note data collection via Polaroid camera. In this era before fast analog to digital data collection, the traces recorded on film were 'digitized' with a ruler.

In later work we built a CW Nd:YAG laser that was used to measure Raman scattering efficiency in Silicon. By swapping cavity reflectors, we were able to induce the CW YAG laser to oscillate at several different wavelengths. This enabled us to demonstrate Raman resonance effects in Silicon.

Another view of the pulsed Nd:YAG laser collection setup. The vertical cylinder in the center is a cryogenic dewar containing the semiconductor sample. This was evacuated using the vacuum system on the right. Operating and maintaining that vacuum system occupied a good fraction of my laboratory time. The smaller horizontal cylinder in the foreground contained the photomultiplier used to detect the light transmitted through the double monochromator on the left (not shown).

Some of the collection optics used with the pulsed laser setup.

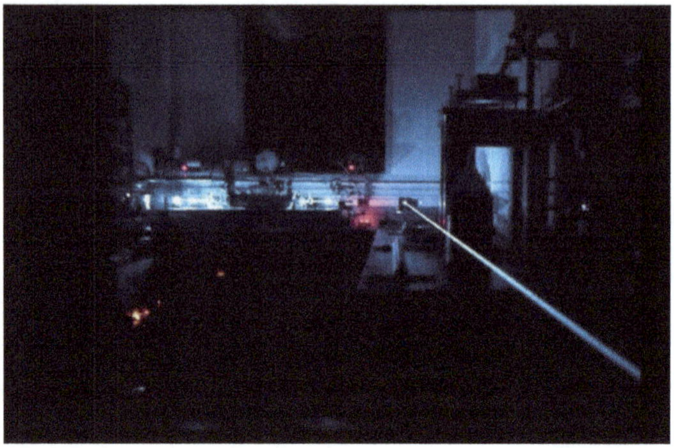

Here is a shot of the continuous wave (CW) Argon laser in the laboratory. This CW Argon laser was a built by fellow student and esteemed colleague Ron Wadsack with some help from skilled technician Dave Keating (to whom we both owe much thanks!). It was capable of operation at several different wavelengths and was used by Ron and me for, among other things, demonstrating interference effects in resonance Raman scattering in Cadmium Sulfide.

The same CW Argon laser showing its multicolor capability. These were separate exposures with the camera moved slightly as the laser was tuned to different wavelengths, thus leading to a double-exposure effect on the background.

Ron Wadsack adjusting laser current. Note the sophistication of the current controls! The bank of rheostats had been salvaged from an even earlier era. The laser tube and optical cavity are in the background.

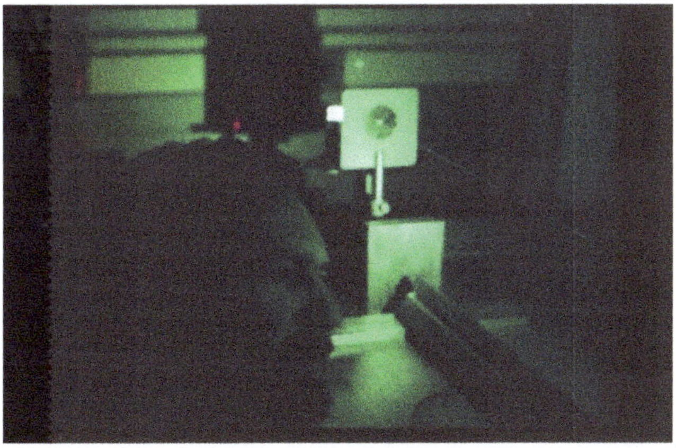

Ron adjusting his apparatus.

The diffraction pattern of the 5145 Angstrom line. (If you want to sound old — just keep using the term "Angstrom"!)

Letter to Richard on the Occasion of His Retirement from Yale in June 2008

Bernell Argyle, Ph.D. Yale '71†
IBM Corporation,
Yorktown Heights, NY 10598 USA

Dear Richard,

Barbara and I are delighted to have the opportunity to help you celebrate your retirement. We have many fond memories of our relationship with you and Sunny while I was an 'aged' grad under your tutelage. Now that you are retired we hope we will be able to persuade you to visit our humble home, renew our relationship, and perhaps even meet our children in their adulthood.

I personally am very grateful that you took me on as an advisee. You showed me a level of leadership and experimental prowess that I had never seen before even though I had already completed eight years of experience working with colleagues at the esteemed IBM Research Division in Yorktown Heights, NY. That influence on me made the biggest difference when I returned to IBM; you had raised my self confidence and my own prowess to a level that made me more competitive and successful in the areas of specialty I took on: micro magnetism, optics and magneto-optics. I am confident the results of my work in these areas aided IBM to develop and maintain its leadership in magnetic recording technology.

As for more personal stories, you and Sunny were especially kind to invite us into your home, introduce us to your very gifted children, and create opportunities to meet other professors at and above your own level at that time. I will never forget Kevin telling me expertly how a typewriter works. Barbara and I will never forget the outstanding hospitality and food, and we'll always remember Shama. We were especially honored when you invited us to your home for dinner following the graduation ceremonies. You even included both our parents and other family members.

Barbara and I wish you a very happy and healthy time in this new stage in your life.

Bernell Argyle

†Passed away

Letter to Richard K. Chang on the Occasion of His Retirement in June 2008

Charles Long
Professor of English, Yale University
New Haven, CT 06520, USA
charles.long@yale.edu

A Memory of Richard Chang

Richard and his wonderful family have been sporadically in and out of my life and my family's life for many years, and all our contacts have been warm and rewarding.

Richard and I joined the Yale faculty together in 1966, he to engineering and I to the English Department. But we did not meet until we shared a presidential assignment in the fall of 1971, when President Kingman Brewster called us separately into his office and asked us to take some time away from our faculty activities to serve part time for a year on the Standing Committee on Undergraduate Admissions. He wanted us to get deeply involved in the process in order to ensure a consistent faculty perspective to the admission of undergraduates. Privately, he also wanted us to tell him whether we believed that the Admissions Office was paying appropriate weight to academic quality in selecting students. It appears that a young member of the English Department, A. Bartlett Giamatti, later Yale's President and at the time of his early death many years later Commissioner of Baseball, had served on an admissions committee and had told Kingman that faculty had little or no say in admissions and that admissions officers exerted their own views about who should be offered admission.

This was no light assignment. It was to be half time: one third time in the fall, getting to know the process, and two thirds time in the spring, reading hundreds of folders and voting on thousands of students. It was a lot of work, but we both enjoyed it and both came away with great respect for the process and especially the wisdom and institutional commitment of the Dean of Admissions, Worth David, and his key staff members, at least one of whom still serves Yale in that office. We came away with the conviction that the careful reading of folders was the only way to avoid the bias of sheer scores on the SAT's scores that could not separate the rote-learning grind from the creative, imaginative student, scores that could not separate the students who announced an interest in "science" from those whose talents and interests would make them good science majors at Yale.

In the process, which was a little like Boot Camp in its focus and intensity, we both learned a lot about Yale, and about ourselves, and we formed several Yale friendships that lasted many years. Richard has become emeritus and I am not far behind, and it is a pleasure to be able to remind him of our first interaction nearly four decades ago.

Chip Long, Deputy Provost of the University.

Memories for Richard on the Occasion of His Retirement in June 2008

Rose Schulz
Office of Undergraduate Admissions
Yale University, New Haven, CT 06520 USA
rose.schulz@yale.edu

Sunny and Richard Chang and T.P. Ma on a Cruise with the EAS Dept on the CT River, 1971.

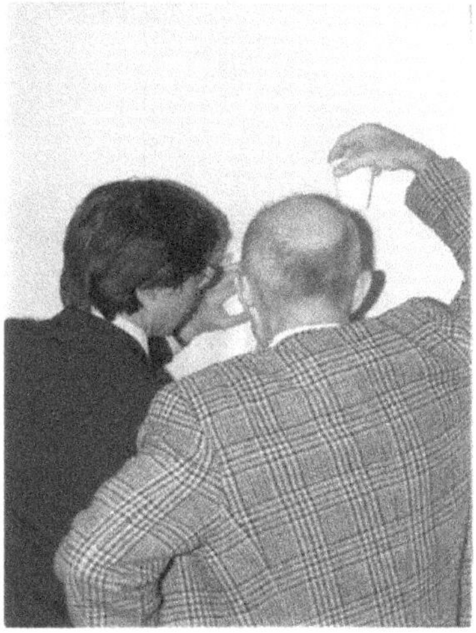

Richard Chang and John Fenn at the George J. Schulz Symposium, 1978.

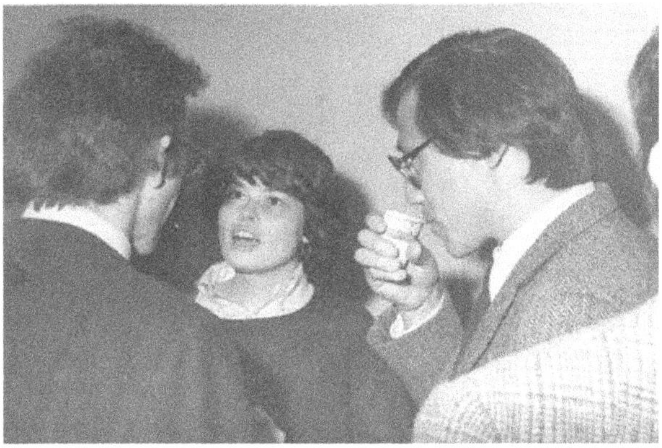

Richard Chang on the right at the George J. Schulz Symposium, 1978.

Letter to Richard on the Occasion of His Retirement in June 2008

Joseph F. Dillon†
Adjunct Professor of Applied Physics, Emeritus
Yale University, New Haven, CT 06520, USA

19 April 2008

Dear Richard & Sunny,

It is a great pleasure to join in the shower of congratulations as you step gently into retirement. You've had a rich and wonderful time at Yale, and I've been privileged to touch your career every once in a while.

One big thing was our shared student/coworker, Eddie Chen. After his student days with you he worked with me for 6 or 8 exciting years. Big things in those years were magnetooptical studies of the metamagnets DAG and $FeCl_2$. We saw fascinating, insightful, photogenic things that no one had ever even glimpsed before. Building on the computer expertise Eddie had amassed working with you, we were amongst the leaders in my part of the Labs in computerizing experiments. There was collaboration and overlap of interest with your colleagues at Yale, and we were up to New Haven quite often. We always made a point of seeing you, and delighted in learning of your work and your exotic travels.

After I left Bell Labs, and came to Yale, I was privileged to give a few lecture series in some of your courses. I enjoyed it immensely.

We were all amused when my son Drew realized that he was teaching Chinese to your daughter Natasha.

I am confident that the two of you will slide gently into a fulfilling retirement. Doris will join me in the celebrations, and we both wish you a long healthful time together.

Joe

Joe Dillon

Five Blue Fish Lane
Brewster, Cape Cod MA 02631
508 896 9073
bluefish05@verizon.net

†Passed away

Letter to Richard K. Chang on the Occasion of His Retirement in June 2008

Philip J. Wyatt, Ph.D.
CEO, Wyatt Technology Corporation
Santa Barbara, CA 93117 USA
wyattpj@mac.com

Some Memories of Richard Chang

Given my "advanced" years, well beyond Richard's, it is not surprising that memory of days and events may have eluded me. However, thinking about Richard invariably brings a smile to my face! The best estimate of when we first met seems to go back to the early '70s, perhaps a year or so after the infamous "Oil Embargo" that began in late 1972. It would be another 7 years before my oldest son Geof began his undergraduate stay at Yale which would mean many more opportunities to visit with Richard and another new friend at the time, Ed Adelberg. By the mid '70s another close friendship would have been established with David Rich, but he was the "Patent attorney" from Boston who would soon play an interesting role in Richard's career.

The most vivid memory of Richard was a very cold one. Apparently we had become good enough friends at some time in the late '70s or early '80s that Richard and Sunny kindly invited me to stay with them overnight at their home. At the time Richard was struggling to support his family on a very minimal Yale salary. One of the most costly items associated with surviving at Yale during those years was heating oil and Richard told me that at night they turned the temperature down significantly. Nevertheless, there were adequate blankets! That may have been a memorable night, but when that event was discussed later with David Rich, a new adventure was born.

It didn't take David Rich (brother of MIT's Alex Rich) very long to determine that Richard was indeed an exceptional physicist. After meeting Richard, perhaps during a visit to New Haven with Ed Adelberg and others that were within his ken, David became personally incensed at how grossly underpaid Richard was. This was no time to be polite and David really let Richard know his opinion about how underpaid he was! So, I was told eventually, David convinced Richard that he had to request (demand) a significant raise. Apparently, it worked and the Chang household was never cold again! Again, these are just anecdotal accounts. Only Richard would know what really happened.

There were other connections with Richard. I lived in Santa Barbara and for many years Richard's wonderful parents ("...the apple doesn't fall far from the tree!) used to spend part of their winters in the City. On occasion, Richard would drop by to spend a few days or hours with them. During those brief visits, I too got to talk briefly with Richard (and sometimes Sunny) on a few occasions, as well.

Because of our similar interests and quite a few common friends outside of Yale, we did meet from time-to-time at various meetings. Sometimes Richard would gather some friends and scientific colleagues at Yale and was kind enough to invite me to join them were I in the neighborhood! Many of Richard's scientific achievements were legend and I, trying to run an under funded company based on the concept of laser based instrumentation as it might have application for the classical physics "Inverse Scattering Problem" applied to bacteria and spores, followed them with considerable interest. (My little company didn't last much beyond a decade despite the heroic attempts to save it by David Rich and his friends.)

Of all of Richard's scientific achievements so far (after his retirement, I intend to keep badgering him in new areas!), one has always been for me the exemplar of his innovative brilliance. But first, a little background, which began quite innocently when a physicist at Purdue University, Petr Chýlek, had at his disposal a large computer (in the mid 1970s) and the funds required to use it. It was well known to those of us actively using Lorenz-Mie theory, that the total scattering cross section of a homogeneous sphere of radius a normalized by its geometrical cross section πa^2 displayed slight ripples when displayed as a function of the size parameter $x = 2\pi a n_0/\lambda_0$, where n_0 is the refractive index of the medium surrounding the sphere upon which was incident light of wavelength λ_0. Van de Hulst had suggested that this ripple effect was related to interference between the forward diffracted wave and surface waves and Nussenzveig [*J. Math. Phys.* **10**, 124 5 (1969)] had reasoned that the ripple could be attributed to partial wave resonances only when x was less that or equal to 4. Chýlek began his work in 1975 by recalculating the normalized scattering cross section by smaller and smaller intervals of Δx. His first results [*JOSA* **66**, 285–287 (1976)] showed some remarkably sharp and very narrow partial wave resonances in the form of double peaks extending indefinitely for even very large values of x. These had not been noticed earlier because such numerical calculations had been made with relatively coarse intervals of Δx. Chýlek repeated his calculations [*Applied Optics* **16**, 3019–3020 (1978)] for larger values of x and showed that at increments of $\Delta x = 10^{-7}$ new "super narrow" resonances were to be found. There were no experimental results at the time of Chýlek's *JOSA* paper that confirmed their presence either. By 1977, Askin and Driedzic [*Phys. Rev. Lett.* **38**, 1351 (1977)] confirmed experimentally using optical levitation that some resonances did occur.

Other experiments were performed further confirming the existence of these "ripples," but my favorite was Richard's 1980 paper.

In a paper published in *Physical Review Letters* [*Phys. Rev. Lett.* **44**, 475–478 (1980)] entitled "Observation of structural resonances in the fluorescence spectra of microspheres," Richard and his colleagues confirmed unequivocally that the theoretical predictions were correct. I have always referred to this paper whenever I discuss Richard's remarkable career with friends. But perhaps of greater importance than the mere existence of these resonances was the realization that they corresponded to regions of extremely intense surface waves. Thus if one could place specific molecules on the surface of the proper objects, one might be able to see some interesting interactions with the light waves incident upon them. Raman Scattering? Certainly. But that meant the beginning of many adventures and scientific discoveries for Richard and his seminal experiments and developments with metallic particles and surface enhanced Raman scattering or SERS. In my mind, the tremendous flow of papers from Richard's lab concerned with the interaction of small particles with laser light took off from that early fluorescence study. I realize that I may well have slighted his even greater experiments and analyses with this "tiny" window into his work that included so many people and their associated talents. Nevertheless, according to the great Chinese philosopher, Confucius, "Every journey must begin with a single step," and certainly all of Richard's other colleagues will have their own recollection of where that step began.

Philip J. Wyatt

Message to Richard Chang on the Occasion of His Retirement in June 2008

Edward Yi Chen, M.S. Yale '72
Bell Laboratories
Murray Hill, NJ 07974 USA

Chen's Family

Left to Right: Yi-Lan, Ed, Han-Tee and Yi-Bin

Yi-Lan and George Lazarou

Julie and Yi-Bin Chen

Letter to Richard K. Chang on the Occasion of His Retirement in June 2008

Daniel Fouche, Ph.D. Yale '72
MIT Lincoln Laboratories
Lexington, MA 02421 USA
fouche@ll.mit.edu

Lexington, Massachusetts
April 2008

Dear Professor Chang,

Thank you for

- Agreeing to be my thesis advisor
- Allowing me to do research that was in line with my interests
- Having creative and fruitful ideas for me to investigate
- Providing me the support that I needed — facilities, equipment, technicians, stipends, and guidance
- Helping me get loans and a National Science Foundation Fellowship
- Writing papers that described our research and that were published in prestigious journals, and making me first author
- Not upbraiding me for making an error in a paper published with you as co-author
- Allowing me to describe our research at conferences, and assisting in preparing the talks
- Helping me get through my orals and thesis defense
- Giving me job leads, and writing recommendations that helped result in job offers
- Ushering me through my thesis work quickly in concern for my family and financial situations
- Always being enthusiastic, energetic, helpful, and friendly
- Taking an interest in my family and personal life
- Inviting me and my family into your home, and letting us get to know your family.

In other words, thank you for being a great and caring thesis advisor.
Best wishes to you in your retirement.

Sincerely,
Daniel Fouche, Ph.D. '72

P.S. Thanks also to your wonderful wife, Sunny, for her warmth, friendliness, and hospitality.

Letter to Richard K. Chang on the Occasion of His Retirement in June 2008

James G. Moore, M.S. Yale '75
Senior Research Scientist, Ret.
Presco Engineering
Woodbridge, CT 06525 USA

June 2008

Dear Richard,

Congratulations on the occasion of your retirement. Now the real work begins, to enjoy yourself as much as possible. Your enthusiasm for your work has always impressed me, along with the ability to transmit that pleasure to your students. In combination with Armand Tanguay, you taught by far my favorite course as an undergraduate.

I especially enjoyed your tradition of having students to your house for dinner. It was Sunny's cooking that introduced me to star anise and other spices that I continue to use to this day. I was also able to invite you, Sunny, and Armand to dinner at my apartment and have every dish come out the best I had ever made them, what luck!

I greatly appreciate being invited to attend your festivities and look forward to seeing you.

My best wishes,
Jim

Sylvia Bianco, Jayne Miller, and Jim Moore at a dinner at the Chang's.

Richard at the Department picnic at Rocky Neck, circa perhaps late 1980's.

Letter to Richard K. Chang on the Occasion of His Retirement in June 2008

Jayne V. Miller
Senior Administrative Assistant (Ret.)
Yale University, Department of Applied Physics
New Haven, CT 06520, USA

June 2008

Dear Richard,

I send my warmest congratulations and best wishes to you as you embark upon this new chapter in your life.

We first met at a party that you and Sunny hosted at your home for the students in your lab.

Jim was pursuing his graduate studies in your group at the time and he brought me along as his guest so I actually met you before I came to work in Becton.

After Werner hired me, it was Susie Engelman, your trusted assistant, who showed me the right way to get things done working for professors running busy labs. For that I will always be grateful to you and Susie.

I always looked forward to your return from exotic trips to hear, in detail, all of your experiences. Even though your schedule was unbelievably full because of teaching and research commitments, you were always most gracious to stop in to see me and tell me all about your trip and to share photos — and there was almost always a very funny story to share.

Jim Moore and Jayne Miller by the Atlantic Ocean.

It has been a great joy to me to follow the paths your children took as they grew up. Natasha, Nicole, and Kevin are all extremely bright and very talented individuals and you and Sunny must be very proud of them. Hopefully once you are retired you'll have even more time to spend with them and your grandchildren.

My best wishes to you and your whole family. Jim and I hope there will still be occasions to see you and Sunny. Please keep in touch.

Fondly,
Jayne

Memories of Richard K. Chang

Dorreth Cole
Assistant Business Manager (Ret.)
Engineering & Applied Science
Yale University, New Haven, CT 06520 USA

I first met Professor Chang in 1973 when I was Assistant Business Manager in the Engineering and Applied Science Department at Yale University. I worked with Prof. Chang in acquiring Grants and Contracts for his research. He was always very thankful for my service. I have a collection of various gifts he gave me from the different countries he visited. I miss attending all the parties at his house. He would be delighted to hear that I visited Hong Kong and loved it! Rest in Peace dear Richard!

On the occasion of his retirement in June 2008, I wrote the following:

Hello Prof. Chang,

We have been friends for 35 years.

The memories I have are loving, beautiful and kind.

I will always remember your smile and calm demeanor and jokes when you visited me in the Engineering Business Office.

However, there was no joking when it came to balancing your budgets for your Grants and Contracts; it was absolutely clear there had to be zero balance.

I always had a wonderful time at the many luncheons, dinners and wedding I shared with you, your family and friends.

The beautiful gifts you gave me and your sincere encouragement to me and my family, these will never be forgotten.

Always,
Dorreth

Letter in Support of Richard K. Chang for AAAR Award

Ron G. Pinnick, Ph.D.
Senior Research Scientist (Ret.)
Army Research Laboratory, Computing and Information Sciences Directorate
Adelphi, MD 20783 USA

June 6, 2005

RE: Letter in support of Richard K. Chang for AAAR award

Daniel E. Rosner
Professor of Chemical Engineering Yale University
New Haven, CT

Dear Professor Rosner,

Richard Chang's contributions to the optics, laser diagnostics, and chemistry of aerosols, droplets, and microparticles have been both extensive and continuous for the last 30 years. And they have had a large impact. Experts in light scattering by small particles and in optical diagnostic techniques for small particles have had to keep abreast of Chang's advancements to stay current in the field. In optics and optical diagnostics of microparticles, as revealed by Chang's numerous papers and citations (16th world-wide in *Optics Letters*), he is one of the world's leaders.

Chang's group was the first or one of the first to explore a variety of linear and nonlinear effects in microparticles, including fluorescence emission, lasing, Raman scattering, stimulated Raman scattering, coherent Raman mixing, coherent anti-stokes Raman scattering, multi-order stokes emission, laser-induced plasma breakdown, explosive vaporization, stimulated Brillouin scattering, third-order sum frequency generation, and two-photon pumped lasing.

Chang's group was the first to observe: (1) morphology dependent resonances (MDRs, sometimes called Mie resonances or whispering gallery modes) in the fluorescence and Raman spectra of microparticles, (2) the time delay and angular fine structure of stimulated Raman scattering in microparticles, and (3) a number of nonlinear effects in microparticles mentioned above.

My search using the ISI SciSearch engine reveals 3989 citations to Chang's scientific journal papers (not including books or book chapters; Pinnick's search on his own papers reveals the ISI SciSearch database misses about half his [mostly older paper] citations, and suspects the same under-estimate of citations for Chang). Most cited papers according to this search are: his 1985 *Optics Letters* paper on Stimulated Raman Scattering in water and ethanol droplets with 200 citations; his 1986 Science paper on lasing in droplets with 136 cites; and his 1986 PRL paper on multiorder Stokes emissions from droplets with 101 cites.

In the last 5 years Chang and his group have worked intensely on developing techniques to rapidly interrogate and sort, particle by particle, potential biowarfare agent threats from innocuous background aerosol. The techniques exploit the intrinsic fluorescence of biological particles and recent advances in diode-pumped solid state (DPSS) lasers and microcavity laser diodes, multi-anode photomultiplier tube detectors, and aerodynamic valves.

This effort has been quite successful and demonstrates the strength of Richard Chang: he is able to attract, motivate, and lead a group of students and postdocs to press the limits of technology to advance our knowledge of the optics of small particles and aerosol diagnostic methods.

Chang is technically one of the strongest scientists with whom I've worked in my 30 year career, and it's been both stimulating and enjoyable. He is deserving of an AAAR award.

Ron G. Pinnick
Army Research Laboratory
Computing and Information Sciences Directorate
2800 Powder Mill Road
Adelphi, MD 20783
Currently on assignment at New Mexico State University, Department of Physics, Las Cruces;
Tel 505 646 1248

Summary of search on "Chang RK" with address "Yale Univ or Harvard Univ" February 13, 2008 by Ron Pinnick:

Results found: 208
Sum of the Times Cited [?]: 5,455
View Citing Articles
View without self-citations
Average Citations per Item [?]: 26.23
h-index [?]: 39

Published Items in Each Year

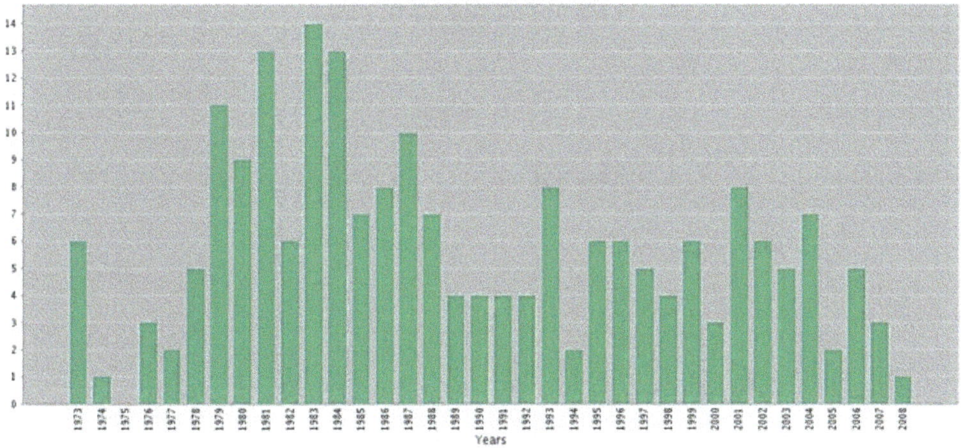

Citations in Each Year

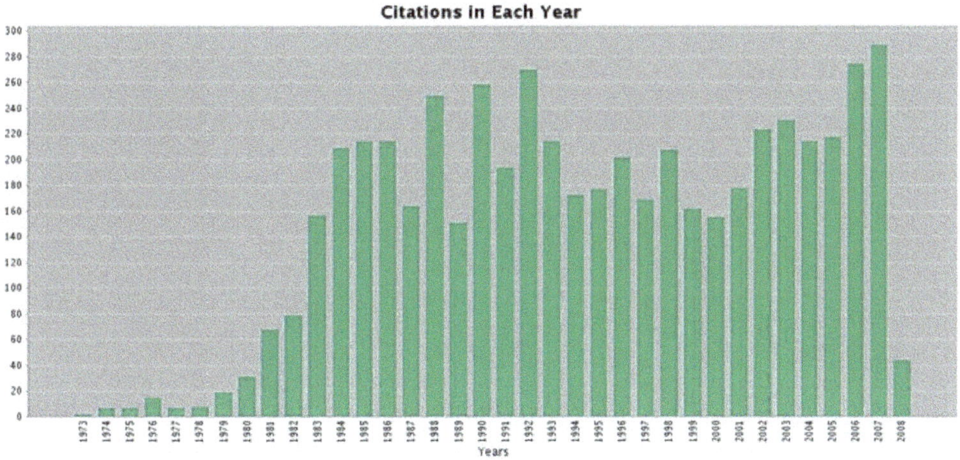

Author	Record Count	% of 208
CHANG, RK		100.0000%
PAN, YL	30	14.4231%
HILL, SC	26	12.5000%
OWEN, JF	24	11.5385%
BARBER, PW	23	11.0577%
LAUBE, BL	18	8.6538%
BENNER, RE	17	8.1731%
PINNICK, RG	17	8.1731%
CHEN, G	15	7.2115%
VONRABEN, KU	14	6.7308%

(169 Author value(s) outside display options.)

Memories of Richard K. Chang

Jin-Joo Song, Ph.D. Yale '74
Professor, University of California
San Diego, CA 92093 USA

Richard toasting everyone in Chang's Hamden home.

Left to right: Ali Serpengüzel, Sevin Serpengüzel, Seongsik Chang, Sunny Chang, and Richard Chang at Chang's Hamden home in 1992.

Left to right: El-Hang Lee, Richard Chang, Jin Joo Song, and Seongsik Chang in the foreground, Roger Farrow can be seen in between El-Hang and Richard at Richard's 60th Birthday celebrations at Yale Graduate Club in 2000.

*And several contributions to the Chang family photos

My First Meeting with Professor Richard Chang

Professor: Can you work 14 hours a day?
Student: Yes

It was one sunny afternoon in the fall of 1970. A timid Korean girl student knocked on the door of Professor Richard Chang's office on the 4th floor of Becton Center. She was greeted by a young professor. For this meeting, she put on a 3-piece suit (pleated skirt, jacket, and vest), custom-made in Korea. After all, she had to show her respect toward a professor… any professor. So, she put on her best suit for the meeting. It seemed that the professor had done his homework and had already checked up on her with other professors who had taught her in the first year of her graduate studies. There was not much discussion of science but he repeatedly said how demanding he could be and asked if she could work 14 hours a day. She without hesitation said "yes" with a timid voice.

Thus, this young Korean girl became his first Asian student, his first student from abroad, and the first female student in his laboratory! Neither of the two knew that this encounter would lead to many "first"s in her life.

Fifty-four years later, the meeting is still vivid in my mind. The sunlight from Prospect Street was exceptionally bright and pretty, I recall. But where did I have the nerve to become an experimental physicist? The old Korean saying "A day-old puppy is not afraid of tigers" applied to my case.

The fact that I had a career as an experimental physicist in the US may be the 10th wonder of the world, if not the 9th. And, when I received a Ph.D from Yale, another Ph.D should have been given to Professor Chang for his patience, courage, insights, and so on. To make a point that this is not an exaggeration, I need to talk about myself and the Korean situation back then.

I am the youngest child from a family of 5 children. In my family, there was no scientist, and no one was in academia. My sisters majored in music or fine arts. My mother was a full-time housewife, but an accomplished painter. Many male members of my family and relatives majored in economics or business administration. In the late 1960s, Korea was still a very poor country with very low GDP per capita ($243 in 1969). My family did not know what was going on in the scientific community either in Korea or abroad. The affirmative action or women's movement was never heard of. There were less than 5 Korean women who had passed the Korean Bar Exams throughout the entire country. Some of my academically outstanding friends, could not go to Seoul National University (the top university in the country) since their fathers did not allow them to study with boys. I wanted to study physics since I thought it was very pure and honest. But I had no role model or advisor around me. I vaguely knew that to study physics I had to go to America. When I came to Yale in 1969, Yale College just turned into a co-ed school from the full boys' school. New Haven was still a very much male-oriented town. In 1969, the year I entered Yale, there was no student from Korea at Yale undergraduate colleges, and in science areas, just one married Korean male graduate student entered the Chemistry Department. There was no one I could turn to in our department, to seek any advice or get tips on the professors, exams, or how to navigate the Ph.D. program.

Starting from "Could you draw a picture of that?"

On my first day in the lab located in the basement of the Becton Center, I showed up in a silk dress made in Korea which I wore often to my classes. My first assignment given to me by Professor Chang, after seeing me in a dress, was to clean the Precision Scientific mechanical pump and change the oil. I cannot think of a better assignment for me as the first task since from the next day forward, my attire changed from a dress to blue jeans and a pair of sneakers — I even got a white lab coat. I probably spent 2–3 days cleaning the pump thoroughly not knowing that the oil pump does not have to be so clean.

Professor Chang made it clear to the group members that I should not be helped except when I had to lift or move heavy objects. This reflects Professor Chang's determination to make me an "experimentalist". However, I was hopelessly unprepared to work in the lab. One day, Professor Chang was working in the lab with us. He asked me to bring a tester (multimeter) from next door. Not knowing what it was, I asked him to draw a picture of that. I do not know what was going through his mind at that moment. Possibly, "God, please help me. What do I do with this girl?". He probably could not laugh or get mad at me since I looked so sincere and serious. I really wanted to bring what he needed, but I did not know I was making a fool of myself.

Gradually, I learned how to turn on the continuous wave (cw) Nd: YAG laser which was home-built. When the laser output power went down, I had to change the pumping lamp and clean the laser cavity. Changing the lamp made me nervous since it was pencil-thin, easy to break, and quite expensive (I recall it was comparable to our monthly stipend of $233.33). Since the YAG rod and the lamp were water-cooled, I also had to do some plumbing and change O-rings to make sure there was no water leak inside the laser cavity. This laser was used in the Raman spectroscopy of rare-earth iron garnets cooled to 2K in the glass dewar. The narrow dewar tip was in between the magnet tips so that the magnetic field could be applied to the sample. The liquid helium (LHe) needed to cool the sample was brought up from New York and was quite expensive. Therefore, we could not do the experiments as often as we wanted. Since the 1.06 micron YAG laser beam is invisible, it was quite a challenge to focus the laser beam on the sample in between the magnetic tips inside the LHe dewar. Another difficulty was that there was no photodetector with low noise and high sensitivity in the 1.1 micron infrared region. This experimental setup was a great training ground for me. For LHe experiments, we needed lots of liquid nitrogen (LN_2). One evening, when there was no one around to help me, I had to bring the LN_2 tank from Mason lab to our Becton lab through the basement tunnel. It took a long time but I did it. I still do not know how I could manage that as a 110 lb girl. (Of course, I weigh more now.)

It is ok to be at the bottom

Most Ph.D. students learn the theory and experimental skills related to the research topic from their thesis advisors. They probably also learn that the quality, not the quantity of the published papers is important, among others. These pieces of knowledge and lessons they learn have a life-long influence on them. In my case, however, I learned something others did not have to learn. I learned that it is ok to be at the bottom. My experimental skills or laboratory experience were at the bottom in our group. I do not know why or how I got to think it was ok to be at the bottom though. I give credit to my professor who created the environment such that I did not feel embarrassed all the time. He let me climb up the research ladder little by little. I was crawling for a long time and then gradually started walking in the lab.

During one Thanksgiving holiday, I decided to align the SPEX double monochromator which was way out of alignment. Calling the SPEX engineer for the alignment costed a lot. I spent 2–3 days sitting on the top of the optical table. Finally, I could observe the beautiful (well-resolved) mercury UV doublet, an indication that my alignment was a big success. I felt that finally I had done something for the group and was no longer at the bottom.

During my career, I competed in several different research areas, each time starting from the bottom. One example is the nonlinear optical four-wave mixing experiments I did at the University of Southern California. As a newlywed, I was looking for a job in Southern California, which was my husband's job location. I was hired as a postdoc by a young professor who had done his Ph.D. work at Stanford and postdoctoral work in nonlinear optics at Harvard. I had no background in nonlinear optics, either in theory or in experiments. The main experimental setup was a four-wave-mixing apparatus based on the Molectron nanosecond nitrogen laser-pumped tunable Hansch-type dye lasers. We had 3 such dye lasers and the two or three beams were overlapped in the sample with phase matching. There were no commercial dye lasers at that time and all three Hansch-type dye lasers we had were home-built. Since I worked with a home-built YAG laser and did a lot of tricky alignment at Yale, I could quickly adapt to the new laboratory environment. In the evenings and on weekends, I worked on the theory of nonlinear optics. Just like I climbed up from the bottom at Yale, I climbed up again little by little. Soon, our lab was at the frontier of four-wave-mixing and the professor and I coauthored a chapter in the book "*Coherent Nonlinear Optics*" (Ed. by M. S. Feld and V. S. Letokov).

Another example is my entry to semiconductor multiquantum well (MQW) studies. In the early 1980s, I was quite intrigued by that semiconductor quantum wells can be tailored to have the desired material properties. I felt that the applications of these materials would be limitless and many new physical, particularly optical phenomena could be observed in this new category of materials. Again, I had to start from the bottom. I purchased a high-quality telescopic camera lens, and at conferences, I took pictures of the slides with MQW data, whenever possible. Fortunately, at the beginning of my picture taking, there were not many papers and presentations on MQWs. Soon, MQW areas were exploding. My MQW and superlattice studies were done initially with GaAs/GaAlAs MBE (molecular beam epitaxy) samples and then later with wide-gap GaN/Ga(In, Al)N MOCVD (metal-organic chemical vapor deposition) structures. Eventually, I was growing nitride samples with a high-cost Thomas Swan MOCVD machine purchased with Department of Defense (DoD) funding. Whoever would have thought that a Korean girl who was at the bottom in her group at Yale would have such a large operation involving a variety of lasers (ranging from cw to femtosecond) and a thin film growth MOCVD system as well as other sophisticated nanotechnology tools.

What do we learn from our thesis advisors other than science and problem-solving?

As a professor, I tried to "feed" my students very well. I also tried to send my students and postdocs to many conferences and workshops. I included my students in my meetings with other scientists as much as possible. I learned all these at Yale from Professor Chang.

One evening when we were working in the Becton lab with the professor, he took us to his Hamden house in his blue Van. We had good fusion-style Chinese food and returned to the lab and continued to work. This was just one of many good meals we had at the "Chang Restaurant" prepared by Sunny and/or Shima (Professor Chang's family maid since his baby days in China). So it may not be surprising that "feeding" students very well was high at the top of my priority list as a professor. Sometimes, I held my group meetings at a Teppan Yaki place (we faxed the order the day before, to save time) or at the country club.

Every year at the APS (American Physical Society) March meeting, I treated my group to a place of their choice. So, the students started picking the place they wanted to try (usually a very fancy place), one year before the APS meeting.

At Yale or at conferences, when Professor Chang had a discussion or just social meetings with other scientists and professors, he included his students. Later I found out that this is not easy for a professor to do since some meetings could be very political and confidential. I do not think we were included at every meeting, but enough to get to see the value of those meetings. My horizon was definitely broadened after those meetings.

When I was still young in my late 20s and early 30s, I visited many laboratories, both in the US and abroad. When I went to a conference in Germany, I made an appointment to visit Professor Cardona's lab at Max Planck Institute, Stuttgart. Professor Chang was very happy to hear that. I continued to visit other professors and scientists abroad. In the mid-1980s, I visited several representative Laser laboratories in Japan and wrote a report to the Office of Naval Research.

Research is not done in isolation. Particularly in semiconductor MQW research, the MQW samples should be obtained from the MBE and MOCVD labs. I worked with many MBE or MOCVD professors and industrial labs to get the samples. I also got to collaborate with semiconductor theorists both in academia and in industrial research labs. With some of my collaborators, we could obtain some large DoD funding. I tried to be a good team player and could maintain good relationships with other scientists for a long time. The foundation of the long-term collaborative relationship was probably built when I was at Yale by watching how Professor Chang interacted with others.

Touching the hearts of others

The professors are evaluated by 3 categories: teaching, research, and service. If you do all three very well, you get tenured. If you are judged to be outstanding in these areas, you can get titles such as endowed chair professor, University Professor, Distinguished Professor, and Institute Professor. The titles vary with universities. Professor Chang excelled in all three categories, but what truly separates him from others is in the fourth category which is not in the Faculty Book of any universities and is not easily described. I will call that category "touching the hearts of others", particularly, his graduate students in the Ph.D. program. I cannot speak for all the students, but the Professor Chang, I saw was interested in the personal well-being of each student and tried to accommodate the difficulties in the students' personal lives far more than other professors. As I look back on my professor days, I think I attained a reasonably good reputation as a scientist and received enough recognition to have the titles such as Regent Professor of Physics and Noble (endowed Chair) Professor of Photonics, in addition to Director of Center for Laser and Photonics Research at Oklahoma State University. But when I ask myself whether I have touched many people's hearts, I find myself feeling so small next to Professor Chang.

My "First"s and Professor Chang's Legacy

As mentioned before, I was Professor Chang's first Asian student, the first student from abroad, and the first female student in his lab. In fact, I do not recall seeing other female students working in the Becton Center labs when I was there. There were some at the Mason Lab. Being Professor Chang's first female student in the lab led to many other "first"s in my career.

I am hesitant to talk about my "first"s since they are not great achievements in an absolute sense and, I may come across as a bragging type. Susie Engelman (Professor Chang's secretary) once said "Jin-Joo, you are the least aggressive person I have ever met". She probably had met lots and lots of people when she said that since she was probably in her 50s at that time. I do not think I have changed that much since then. I might as well gather up the courage, however, and talk about some of my "first"s since they were "achievements" in a relative sense and, in my opinion, they are a part of Professor Chang's legacy.

I may be the first Korean woman Ph.D. in experimental physics, but I am not completely sure. I was the first Korean woman to become a fellow of both APS (American Physical Society) and OSA (Optical Society of America). Even now, I do not know of any other Korean woman who has dual fellowships in these organizations.

In the leadership roles, I was the first Korean ever elected a Councilor of APS and the first woman elected the President of the Association of Korean Physicists in America (AKPA). This probably is no big deal now but in the 1990s, Koreans were hardly on the APS ballot, let alone being elected, and AKPA was still a male-dominated organization.

Oklahoma Professor Jin-Joo Song's day ceremony picture. On the right side is Oklahoma Governor Frank Keating.

On numerous occasions, I served on various committees of APS, OSA, and international conferences as a member or chair. Often times, I was the first Korean or the first woman serving in that capacity.

When I received the Most Outstanding Asian in Oklahoma award, many young Asian girls wanted to have pictures taken with me. At that time, I felt that I might be influencing other girls' lives. Professor Jin-Joo Song's Day was declared in Oklahoma in February 1999. I received some honors in Korea, too.

As mentioned above, it is not easy for me to talk about all these recognitions. But I am saying this to emphasize that these could not have been possible if Professor Chang had not had the courage to take me as his first female student in his lab. Professor Chang probably passed away without knowing how much impact he had made, and how many lives he had changed by training his first female student who knew absolutely nothing in the lab, and went on to become an experimental physicist.

Jin-Joo Song
August, 2024

Memories of Richard

Daniel Prober
Professor of Applied Physics, Physics, and Electrical Engineering
Yale University, New Haven, CT 06520 USA
daniel.prober@yale.edu

My wife and I met Richard and Sunny, when I joined the Yale faculty in the summer of 1975. Sunny and Richard were wonderful mentors, guides and hosts for many years, and we felt part of their larger family when we joined them at home for many happy times. We admired all that you did for others, and we wish your family our fondest best regards. Below is the letter I wrote to Richard on the occasion of his retirement from Yale in June 2008.

Yale University

Department of Applied Physics
P.O. Box 208284
New Haven, CT 06520-8284

Campus address:
417 Becton Center
15 Prospect Street

Daniel E. Prober,
Chairman Professor of Applied Physics and Physics
Director, Graduate Admissions
Faculty of Engineering

email: daniel.prober@yale.edu
Phone: (203) 432-4280
Fax: (203) 432-2207

April 14, 2008

Dear Richard

Your retirement from active teaching is exciting for the time and opportunities you will have with your family, and we all wish you well in this new stage of life. You and Sunny have been good friends for over three decades, from the first time I came to Yale with Sharon, to the present, when our 'kids' are adults, older than Sharon and I were when we first arrived. The warm and continuing hospitality and friendship that Sunny and you shared with us has been most meaningful, and in the early days for us — pre-children, we did not appreciate how much time you were taking from your busy schedule. We enjoyed all the visits at your home, and also seeing your own family grow into the wonderful adults they are.

You have also been a wonderful role model for me as a science teacher and mentor. It is clear from your long years of dedication to training great students that their loyalty and devotion is very much deserved. And, as your group members have gone out to the wider world of science and government, we all have had great admiration for your creative and thoughtful leadership. I wish I had arranged my science goals to have more photons in them; maybe I will be able to do this in the coming years with Hui and Paul, and the new members of the department.

Sharon and I wish you and Sunny well in all your endeavors, travel, sharing with family, and continuing work at Yale. Thank you for 33 wonderful years of knowing you all. Our very best wishes.

Fondly,

Dan and Sharon

Dan and Sharon Prober

My Relationship with the Chang Family

Peggy Huang
Diabetes Prevention Awareness Educator (Ret.)
University of California San Francisco School of Nursing
San Francisco, CA 94143 USA

How do I describe my relationship with Richard Chang and his family? I consider myself a friend of the Chang family. I came to know three generations of the Chang family well over the past years (too many years to count).

I first met Richard's sister Maejeane through their cousin, my classmate from Hong Kong era. Maejeane was on her way to Paris to take up her position at UNESCO. I happened to be visiting her cousin at the cousin's mother's apartment in NYC. Maejeane was also there to say goodbye to her aunt and cousin. Richard and Sunny were also there to see their sister off to Paris. That was the very first time that I met the 2nd generation Chang family: Richard, Sunny and Maejeane.

In 1969, I moved to San Francisco, Richard's parents had already settled in San Francisco. Over the following many years, I became a part of Chang family. Whenever Richard and Sunny, or Maejeane, with or without the grandchildren visited the senior Changs, I also was present. In fact, there were visits every two months or so by members of the family. I remember one summer vividly, when Richard's two daughters were still pre-teenagers, the daughters were to stay with their grandparents for the summer. Richard's parents were very proper traditional Chinese, almost imperialistic. The Chang girls were American pre-teenagers. This led to certain anxieties! Fortunately for the girls "Aunty Peggy" came to the rescue! After many years, the younger daughter mentioned to me this incident, still vivid in her memory. I was pleased to have been able to help!

Rotation visits by the younger Changs to Richard's parents became more frequent as the elder Changs advanced in age and their health deteriorated until they passed away. I was privileged to be part of the entourage.

Two more memories I treasure: I also count it a privilege to be a witness to Richard's retirement party from Yale in 2008. After Richard and Sunny moved to Singapore, I also was able to visit them on two occasions. I did not know Richard professionally. However, having attended his Yale University retirement party, I was privileged to have heard all the accolades showered on Richard by his former students from all over the world. I came to know that he was not only an accomplished academic and scientist, but an excellent mentor to his many mentees. Through the years of my contacts with Richard and his family, I would categorize him as a filial son to his parents, a caring brother to his sister, a wonderful loving husband to Sunny, together with Sunny, loving and skillfully raised three accomplished children. The three Chang children all have children of their own now.

One final recall: I was the last person to bid good-bye to Elder Mr. Chang at a hospital in San Francisco, as well as to Richard's sister Maejeane in Paris, and among one of the last to say farewell to Elder Mrs. Chang at home. It's an unbelievable friendship!

Contribution for the Richard Chang Memorial Book

Paul B. Klein, Ph.D. Yale '75
Research Physicist (Ret.)
Naval Research Laboratory
Washington, DC 20375 USA

I first met Richard in 1968, when I was enrolled in two courses that he was teaching with a few other professors: EAS207b, "Special Topics in Applied Quantum Physics", which took place in Mason Lab; and EAS215b, "Introduction to Solid State Physics", which was held in Dunham Lab. I also remember having a discussion with Richard that Spring in his office in Dunham — I was probably advisor shopping at the time. He took me on a tour of his lab, which lived in space borrowed from Bill Bennett. It was the first time that I had ever seen a laser lab — or a laser for that matter. We talked about nonlinear optics and the use of optical probes to investigate magnetic materials, and his enthusiasm really inspired me. That Spring I was classified IA and had to leave Yale. Two years later I reapplied and was accepted, although I had to start from scratch, repeating a few of the courses. This time advisor shopping was not an issue — I knew that I wanted to work with Richard — if he would have me. What attracted me most to him was his enthusiasm and obvious love of physics, as well as his way of interacting with people. I felt very comfortable around Richard, even though I didn't know him that well at the time, and I sensed that this was someone who would be straightforward, and even caring. My years in the basement of Becton Center only confirmed these feelings.

When I arrived in the summer of 1970, I found that Richard now had his own space in the basement of the brand-new Becton Center. Basement space was prime real estate because of the reduced vibrations: I was never quite sure how Richard managed to get it. James Bond movies were still the rage then, so it was extra special to have one's desk in Room 007. Richard's first group of students had all left, except for Fred Parsons, who was doing his own thing. The "second wave" included myself, Jin-Joo Song, Dan Fouche and Eddie Chen. I am not sure what the others' prior experience was, but at one point early on, understanding that we were raw recruits, Richard brought us into his office and began a detailed lecture in laboratory practice. He went on to describe the construction of a vacuum system, the role of all of the components, and how to operate the system and how to avoid catastrophes. In another session we learned about the structure of a photomultiplier, how and why it worked, how to optimize its use, and — of course — what not to do. I thought it useful then. What I didn't realize was how unusual it was for a young assistant professor, concerned with funding, academic assignments and the prospect of tenure, to take several hours of his time to educate graduate students on how to survive in the lab. I have told this story a number of times, and always gotten an amazed response from colleagues who have had to pretty much fend for themselves. Yes, that is another way to learn, and often a painful one. This episode is just an example of the real concern that Richard showed for his students and their wellbeing.

Richard was always very much the gentleman. However, he could be aggressive when he thought it necessary. What brings this to mind is Richard's secretary (I don't remember her name) who was there when I arrived. She was young, and I assume competent — at least I don't remember hearing anything specifically negative. But she must not have been exactly what he wanted — maybe worked too slow, or made mistakes — I'm not sure what his dissatisfaction was. But at one point he got wind of someone who was available who was older, experienced, and incredibly competent. He acted quickly. Our secretary came up to me one day to tell me that, out of the blue, she had been transferred. Understandably, she was upset. In the end, he hired Susie Engelman and she was with Richard for many years, and undoubtedly made his academic life so much easier. When I was writing my thesis, and with the deadline the next day, the secretary that I was paying to type it had to go home — her husband didn't like her working after hours. Susie took over and had it done in no time. I can understand Richard's need to snatch her up before someone else did. And Susie became more than a secretary — she was family.

Richard had his own way of interacting with his graduate students — it was an interesting mix of warmth, familiarity and friendliness with serious conversations when those were needed. And there were definitely times when they were necessary. But we all had tremendous respect for him — I never saw anyone attempt to take advantage of Richard's gentle nature and his real concern for his students. This approach was not the norm for academics in his situation — at least not from what I have heard from colleagues over the years. How many thesis advisors take their students on field trips? The first physics meeting that I ever went to was in Rochester, NY. Richard piled us into his little Fiat and we drove to the meeting together. In the middle of the meeting the speaker was interrupted to announce that vice-president Spiro Agnew had resigned. There was loud applause. Richard even let me drive his beloved Fiat part of the way home. Another time we were all in the Fiat on a trip to Bell Labs in Murray Hill, N.J. — Richard knew a number of people there. We visited John Worlock — in our eyes a god of Raman spectroscopy — and I remember going into his lab. I was shocked to find that it was much smaller than Richard's space. I think there was a lesson there. On the way back he took us to a Chinese restaurant that he liked in Manhattan. It was there that I first learned that Chinese food was not all Cantonese — and could be fiery hot!

In 1973, Richard went on a 6-month sabbatical, covering numerous labs in all parts of Europe. He was concerned that we (Jin-Joo Song and I) not get slowed down by his absence, and there was a continuous train of letters back and forth (alas, no email) about what was going on in the lab. While we worked on rare-earth iron garnets, we were not making much headway. Meanwhile, he was seeing all this great research in places like Max Plank. When Richard returned, he must have decided that something had to change if two theses were to come out of this work. He switched us to working on semiconductors! At the time this was very disappointing for me, as I really liked the idea of magneto-optics. Semiconductors were like — transistors and diodes — what could be new there? (Hah!) But we wound up focusing on resonant effects, which turned out to be interesting in their own right. Jin-Joo and I were both able to write up acceptable theses. Ironically, I spent my entire career in semiconductors and optics. Of course, Richard knew the value of semiconductor research and the current interest in them. That was something I would understand later. So, this was one of those times when Richard had to be tough, as he did meet some initial resistance. In the end, of course, it was the right decision.

I spent my entire career at one institution (just like Richard) — in this case the Naval Research Labs. And who was responsible for this? Richard, of course. My thesis work involved intervalley scattering and Kia Ling Ngai at NRL had just published an interesting paper on it. So, Richard arranged that we would go down to DC and talk with him. At one point Richard asked Kia "are there any job openings at NRL?" The answer was something like "no, we generally hire through the NRC post-doctoral program, but you pretty much have to connect with someone here that works in your field and is willing to sponsor you". He pointed us to their local Raman spectroscopist, Larry Verble, and it was decided that I would send in a proposal. The NRC post-doctoral fellowships were very competitive, and Richard was very helpful in kicking around ideas over which direction to go. In the end I went to NRL as a post-doc in July 1975, and Richard supplied the HgCdTe samples that were required for the NRC proposal. Had Richard not spoken up for me 46 years ago (I was too intimidated at the time) I might have wound up — well, who knows? But this was so typical of Richard and his relationship to his students.

Then there was dinner at the Changs. I don't recall how often these occurred, but they were certainly memorable. With all the students there, along with Richard and Sunny and the rest of the family — it was like we were all one big family. And in many ways, there was this sense of togetherness in the group. Of course, the food was wonderful, and again I learned that Chinese pepper steak is exactly what it sounds like! Once I got there a little early — probably the first one there — only to be entertained in conversation by an incredibly adult five-year-old Nicole, while her parents finished getting everything ready. I have seen accounts of similar Chang get-togethers that occurred many years after I left, and that sense of warmth and family had not diminished.

When I left New Haven for DC, we had plans to do some GaSb multi-phonon Raman scattering measurements, and of course Richard supplied me with the semiconductor crystals. The measurements were done at NRL, and I did the analysis keeping Richard informed via mail, and we had several exchanges over that. When things were worked out, so that we could think about writing it up for publication, I sent the results to Richard. When he wrote back, it wasn't with "that will work", "nice job" or "this is good, but —". What he wrote was: "It gave me a sense of pride and joy to read your recent exposition on the phonon spectrum of GaSb." I think that quote epitomizes the emotional attachment (in my view) that Richard had for his students. It is consistent with everything that I know about him. It is why so many people were happy to take the time to come to his 60th birthday celebration, and to celebrate his retirement eight years later. As I mentioned at Richard's retirement, I thanked him in my thesis acknowledgements for his guidance ".... not only in the role of thesis advisor, but also as a teacher and a friend." Richard taught me more than science. Watching him over my four-plus years at Yale, I learned what it was to do science, what it meant to do it right, how to think like a physicist, and maybe more important — how to be a "mensch".

As for myself, I remained at NRL, officially retiring in 2012, after a career investigating defects in various semiconductor systems using primarily optical probes. From the time of my retirement until the beginning of the current pandemic, I continued to work at the lab as a contractor. Now that things have improved to some extent, I am beginning to return to the lab. My wife Barbara retired five years ago from the Civil Rights Division of the Justice Department, and is a mosaic artist, also enjoying gardening and two grandchildren that we can now see "in person". My son Matt is now a cantor, and his wife Elyssa a rabbi. They live in Akron, Ohio with their two children Georgie and Beatrice. My daughter Sarah is an art therapist and her husband Isaac an acupuncturist. They live in Fort Collins, Colorado, where we plan to relocate someday.

February 2018. Left: daughter Sarah and husband Isaac. Center: Paul and Barbara. Right: Son Matt, wife Elyssa and son Georgie.

Letter to Richard Chang on the Occasion of His Retirement from Yale, June 2008

Dear Richard,

On the occasion of your retirement as a full-time faculty member (which I guess means you will still be around "part-time"?), I feel it appropriate to reflect on my years as an RKC grad-student, and how that has affected my life.

As one of the 2nd generation of RKC students, my experience was of a relatively early stage in your career. In fact, our focus at that time was entirely on linear optics and solid state samples that you could hold in your hand or mount in a cryostat — not the ephemeral aerosols of a later time. But while the direction of your research changed with time, I know that your concern for your students did not. What I remember most of that time was that you were an advisor who cared. I guess I sensed that when I was "advisor shopping". That — and the bright pretty laser lights — were instrumental in helping me decide.

As a young assistant professor, much of your concern then was on obtaining funding to support your students and your lab. Yet, you took the time to give several lectures in your office to two green students (myself & Jin-Joo Song — possibly also Dan Fouche?) on some basic laboratory essentials — how a vacuum system works and how to use it, the science behind photomultipliers and how to use them in an experiment, ...etc. It is true that with your 1st generation of experienced students now graduated, vacuum systems and PMT's were probably among the most vulnerable items in a lab now run by newbies, so this could be looked upon as self-preservation. But leafing through the very first yellowed pages of my first lab notebook (you know I am a pack-rat), where I had carefully recorded as much of these sessions as I could, it is apparent that the level of care and detail reflected more than just the fear of damaged equipment.

You may also remember that I did some of my thesis work at CCNY in Bob Callender's lab, as he had the dye laser we needed. I am sure that you remember that on my way from the lab late one night I was mugged — not hurt, just relieved of cash. While there was little you could do, you did what you could, by reimbursing me for what I had lost. I think it may have been charged to a government contract — at least we used to make jokes about that! As a poor grad student I was very appreciative, but not surprised.

Oh yes — and how many thesis advisors would pack their students into their (tiny) car and take them on a field trip to Bell Labs? We thought then that John Worlock was like a God,... but I remember thinking ... "Richard's lab is bigger than this!" Then, on the way back we stopped for a Chinese dinner in Manhattan. Who knew then that Chinese food could be hot??? And an innocent-looking whole fish at that!

Although that period of my personal life was very difficult, I still have very fond memories of life in your group. Dinner at the Chang's was always a treat, as I think it has been for all of your students over the years; the food, the company, and some remarkably intelligent conversations with a 5-year-old Nicole. Group meetings were always fun — unless you were the one presenting something. I do remember a meeting that the grad students had with Bob Wheeler when he was Chairman. He was telling us that these new-fangled solid-state watches were the wave of the future, and that eventually we could buy one for $25! I don't think we believed him.

Your 6-month sabbatical (I called it the "Gypsy sabbatical" because of all the different places that you traveled to in Europe) was a real experience for us — or at least for me. Looking back at my old lab books, it looked like we didn't get that much done — in spite of a constant flow of letters back and forth. As I recall, what measurements we did get done convinced you that theses for two students would not be forthcoming from rare-earth iron garnets. Before we knew it, we were working on (yuch) semiconductors! But during that period I learned to be more independent, and it made a real impression on me. That experience came in handy a few years later when my post-doc advisor at NRL left for a year in Brazil, leaving me to deal with the laboratory move to another building. And I had this experience in the back of my mind when I went on sabbatical myself (Ann Arbor, 1993–1994) — I was sure that my two young kids would survive their apprehension of being plucked out of the only existence they had ever known and dropped into something totally new. Of course they thrived in Michigan, and came back much more confident in themselves. Can I say that they owe it all to you? Well ....that would be a stretch. But I can say what I owe to you.

In the acknowledgement section of my thesis, written 33 years ago, I thanked you for your guidance "...not only in the role of thesis advisor but also as a teacher and a friend". You certainly could be tough when it was necessary, but I knew that you cared. You taught me not just science — but by example I learned how to do science ... and how to do it right. I learned how to think as a physicist. It's something that I use every day — much like the silver Cross pen that you gave to me at the celebration of my thesis defense.

Having begun to think a bit about my own future, I'm trying to imagine what retirement will mean for you. Going into the lab only four days a week? Well, maybe not quite that bad. But I am sure your interest and involvement in science will be as keen as ever. And I can easily imagine you spending more time in Vermont. However you choose to spend your time, I wish you and Sunny much joy and happiness.

As my current family life began after I left Yale (I think you met Barbara only a couple of times), I am including a recent photo of our family — and a short bio: Barbara is

William & Mary College, 2006: Matt and Sarah Klein, Barbara Elkin and Paul Klein.

currently working as an attorney involved in fair housing for the Dept. of Housing & Urban Development, Matt graduated from William & Mary in 2006 majoring in music, and is currently studying to be a cantor at the Jewish Theological Seminary in New York (who ever thought I would have clergy in the family!), and Sarah is an accomplished violinist and a sophomore at Northwestern, studying film production. And of course I am still at NRL in the same group as when I started my post-doc in 1975. Truly amazing.

Best regards always,
Paul

A sample of written communication between Richard, while he was away, and his students before the internet.

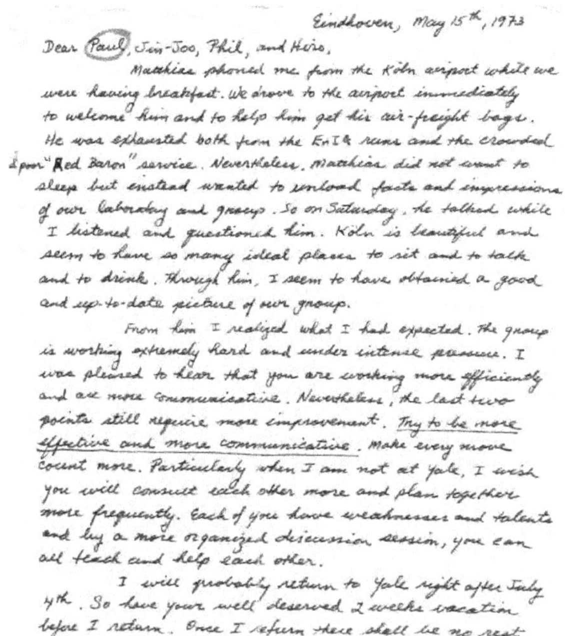

Eindhoven, May 15th, 1973

Dear Paul, Jin-Joo, Phil, and Hiro,

Matthias phoned me from the Köln airport while we were having breakfast. We drove to the airport immediately to welcome him and to help him get his air-freight bags. He was exhausted both from the EnI& runs and the crowded & poor "Red Baron" service. Nevertheless, Matthias did not want to sleep but instead wanted to unload facts and impressions of our laboratory and group. So on Saturday, he talked while I listened and questioned him. Köln is beautiful and seem to have so many ideal places to sit and to talk and to drink. Through him, I seem to have obtained a good and up-to-date picture of our group.

From him I realized what I had expected. The group is working extremely hard and under intense pressure. I was pleased to hear that you are working more efficiently and are more communicative. Nevertheless, the last two points still require more improvement. Try to be more effective and more communicative. Make every move count more. Particularly when I am not at Yale, I wish you will consult each other more and plan together more frequently. Each of you have weaknesses and talents and by a more organized discussion session, you can all teach and help each other.

I will probably return to Yale right after July 4th. So have your well deserved 2 weeks vacation before I return. Once I return there shall be no rest and letting up until Jin-Joo and Paul hand in their Ph.D. theses. So get a good rest. Before you go away, make sure the needed things are ordered.

-2-

From M.S. I learned that the best effort to see the electronic Raman lines in ErIG has only produced marginal results. You can not use more Nd:YAG power due to crystal–liquid He heating. The Biomation with the laser power monitor mode does not seem to be that much better. You need a ~~factor~~ 5 improvement somewhere, before meaningful rotational shift data can be obtained. I saw the raw data and hand plotted data on ErIG, which M.S. brought.

Before I go into the details of certain aspect of your individual experiments, let me give some structure to Jin-Joo and Paul's thesis. Perhaps, ~~which~~ with an overall picture, you can then decide for yourselves the relative merits and weights of individual experiments. The time limit to finish all these still should be September 1973.

Jin-Joo Song

I. Introduction
II. Phonons in Iron Garnets
III. Electronic Levels in ErIG
   a) Raman scattering
   b) Near IR absorption
IV. Quantum Counter
   a) Modulation of Sensitivity in iron garnets
   b) Image Conversion
V. Conclusion
Appendix
   Two-phonon Resonance in Si.

Paul B. Klein

I. Introduction
II. Phonons in Garnets (Lu and Iron)
III. Electronic Raman Effect in DyGG
   a) g-values
   b) relative intensity
IV. Magnetic Tunable Ho:YIG laser
   a) Fluorescence studies
   b) Laser characteristics
V. Conclusion
Appendix
   Wavelength Dispersion in CuCl & CuBr.

——— x ——— x ———

These are guidelines which I hope will bring your

-3-

present experiments in better perspective.

Let me comment in more depth as to the various chapters.

Jin-Joo

II. Included the iron garnet work and the considerations for scattering configurations. All that connected with the iron garnet work found in our JOSA paper should be included.

III. a) State ratio of electronic Raman intensity compared to phonons in ErIG. Try to justify why it is so small. Compare with Wadsack's case in DyAG and DyGaG and YbAG and YbGG. It would be much better if you can measure the ratio of electronic to phonons in an ErAG or ErGG. I believe we have a piece in one of the collections Argyle collected. Tell us if we have one!

Present your best Raman data on ErIG. That is the only thing you can do, unless InAs detector is really 5× better than your S-1, I think you are at a "stalemate". I had once hoped that you can use the Biomation (properly operated in a laser power monitor mode) to take repeated scans over the two peaks that you think are electronic lines. But the Biomation seems to waste more of your time than being useful.

b) To salvage all your hard work, near IR absorption data on ErIG at 1.5°K can be useful. Maybe you can still design an absorption geometry by which you can rotate the crystal in the $\vec{H}$ field. Otherwise, you may just have to pull your crystal out for each θ.

However, make sure you have a calibration wavelength source for each scan of the spex, as you will need good wavelength accuracy.

-4-

IV a) If you can just see some intensity up-conversion in an iron garnet and have the sensitivity change with magnetic orientation, I believe this will suffice for your thesis. Use any wavelength that suits your iron garnet.

b) Discuss image up-conversion. If you can demonstrate it in a LaF_3 or $LaCl_3$ crystal for the visible that will be excellent. If not, state your best effort result.

V. Conclusion

Appendix. The Si work does not fit in, but you might as well report on it as something exciting that came your way when you were doing your thesis.

Write me if you have questions. I hope you will at least see my view, but of course you must decide for yourself.

Paul B. Klein

II. Since you deal mainly with transparent garnets, why don't you include all that part in our JOSA paper on transparent garnets. The Lu^{3+} results and geometry.

III. a) g-values in DyGaG and their connection to Xtal field calculations.

Since you are all doing so well at 1.5°K, maybe you want to do ~~Bernie~~ ^more of^ Bernie Angyle's DyAlG in the ordered-disorder state for the $^6H_{13/2}$ and $^6H_{11/2}$ manifold. Bernie ~~just~~ ^only^ did the 0' and 1' lines. Lately Wolf, Dillon, Chen, and Stanley of M.I.T. are hot on the phase transitions in DyAG. Maybe ~~you~~ ^want to^ do the exchange field splittings for the

-5-

$^6H_{13/2}$ and $^6H_{11/2}$.

b) Relative intensity for the $^6H_{11/2}$ level in DyGG and maybe DyAG. Don't bother about any calculation for this part. Hopefully your g-value measurement and Joan's thesis will convince others why relative Raman intensity data are useful.

IV a) Fluorescence data and tuning range will be included.

b) Laser Characteristics pumped by Nd:YAG. M.S. told me that when he visited Van der Ziel at Bell, he was secretly working on a thin film of Ho:YIG.

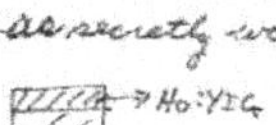

However, Van der Ziel was not willing to tell M.S. why and Matthias does not know. I can tell you why right away. The interest is in integrated optics. This is also why $NdO_{1x}P_5$ ~~that~~ ^which^ I wrote you about (Danielmayer) is so hot. Van der Ziel always work on hush hush hot projects. I am glad there is still excitement in this Ho:YIG, and thus your work can be very timely, ~~if~~ you ~~should~~ beat Van der Ziel before he does laser pumped thin-film Ho:YIG. Danielmayer and Weber at BTL is working on semiconductor laser pumped—$NdO_{1x}P_5$ (still 1.06 μ). Knowing Van der Ziel it must also be laser pumping.

V. Conclusion

Appendix

If you can do something exciting in CuBr and CuCl, this should be put into your thesis. Particularly using discreet laser lines before the first exciton, to see the wavelength dispersion. I will try to get us a CuBr. Can

-6-

you get some laser lines from our laser lower than 4545Å? (λ<4545Å).

This I hope will give you all an idea what I think you two can include and achieve for your thesis.

<u>Some minor points</u>

<u>Jin-Joo</u>. I believe IR-Raman selection rules to hold well in garnets. At K≃0, the phonons have the point system of the crystal.

For your IR absorp. work, can one use this geometry?

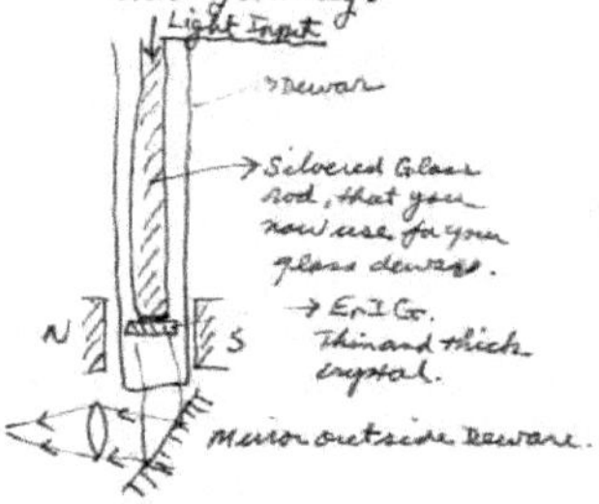

This way, you can rotate the Xtal in the magnetic field. Just a thought.

<u>Hsia</u> forget about the Iodine exp. Then don't worry about SHG for 1.06μ. Helping the students with their theses and some effort on the pulse dye laser are more than enough. Thanks for your delightful card.

<u>Paul</u> Thanks for helping on the AFOSR reports and for being effective on the lock-in.

-7-

<u>Phil</u> The RANN/EPA reads well. I will comment on it in my next letter. Thanks for your comments on Tobin. I sincerely appreciated them.

I hope you will all try to answer my letters more directly. I find many of my questions unanswered. Discuss my letters among yourself.

My schedule until May 28th is packed daily. Nevertheless, I hope to share with you my thoughts and my observations as I come across them.

Sincerely,
Richard

My Teacher and Mentor at Yale

Victor E. Henrich
Eugene Higgins Professor Emeritus of Applied Science
Department of Applied Physics
Yale University, New Haven, CT 06520 USA
victor.henrich@yale.edu

I was never a student at Yale. My degrees are from the University of Michigan, and my first job was as a staff member at Lincoln Laboratory, MIT. My only experience with students before coming to Yale was as a Teaching Fellow in graduate school. I came to Yale because I wanted to teach physics — just doing research was fine, but something was missing.

I arrived at Yale in August 1978 in what was then the Department of Engineering and Applied Science. Werner Wolf, who was the Chair, told me that I would be teaching Solid State Physics in four weeks! I panicked. But Werner had arranged that I would be teaching it jointly with Richard Chang, who had taught the course for many years. That was the beginning of a close and wonderful relationship with Richard.

Richard spent a great deal of time that first term teaching me how to teach. He gave me the easier parts of the course, and he spent whatever time was necessary answering my questions and making suggestions. He taught me pedagogy and how to connect with the students. We taught that course together for about twenty years, and I finally had learned enough from Richard to teach it by myself. I couldn't have had a better teaching teacher!

Richard also took me under his wing as my mentor in the department. I had never written a research proposal while at MIT, and Richard led me through the process. He also helped steer me to the proper sources of funding, showed me how to write progress reports, etc. His guidance was crucial to successfully building my research program. Whenever I had a question about almost anything, Richard was the first person I would go to for advice.

There were many administrative issues over the years in which Richard played an important role. Our physics section of the Department of Engineering and Applied Science was first split off as the Section of Applied Physics, and ultimately the Department of

Applied Physics. There were, of course, issues at each stage, and Richard was a reasoned and influential voice who helped those changes occur smoothly. His wisdom helped guide Applied Physics to what it is today — the strongest physical science departments at Yale.

Every bit as important as my professional relationship with Richard was a personal one. Richard and Sunny immediately became some of Janet and my dearest friends at Yale. They included us and our children as part of their family. We enjoyed visiting them at their home in Vermont, cycling with Richard through the hills of Hamden, seeing their grandchildren in musicals, and much more. All of the Changs are wonderful people, and we will always be close to them.

Below is the letter to Richard on the occasion of his retirement from Yale in June 2008.

Yale University
Department of Applied Physics

Victor E. Henrich
Eugene Higgins Professor of Applied Science
Phone: (203) 432-4399
Fax: (203) 432-4283

327 Becton Center
15 Prospect Street
P.O. Box 208284
New Haven, CT 06511
E-mail: victor.henrich@yale.edu

June 12, 2008

Dear Richard,

I guess we all must retire sooner or later, but your retirement is a major loss for your friends, our department, and Yale. You have been a very important part of both my and Jan's Yale experience over the past thirty years, and Yale simply won't be the same without you.

Even before I met you, I knew that you were someone special. On the day that I first visited Yale and gave my job hunting talk, you were out of town. The department seemed to like me, but they asked me if I could please come back again in a couple of weeks when you were around; they would not seriously consider my appointment until you had approved it. So I figured then that you were a crucially important member of the department (and I was right!). Fortunately, you seemed to like me also, and the rest is history.

As soon as I arrived at Yale, I was assigned to teach Solid State Physics I with you. I had not taught at all at MIT, and you taught me how to teach. That turned out to be a fantastic introduction both to teaching and to Yale, and, although we haven't taught together for several years, I still cherish the memories of our intimidating oral mid-term exams, etc. Since I have been teaching the course by myself, I continue to give oral mid-terms whenever possible, and I still use what you taught me in those early years.

Although our research has not overlapped very much, I have thoroughly enjoyed the interactions that we have had (although the Gas Research Institute wasn't very enjoyable to deal with!). There you also taught me important lessons in scientific integrity, how to interact with graduate students and postdocs, and much, much more. I am forever in your debt for all the wisdom that you have shared with me.

What Jan and I are most grateful for, however, is your and Sunny's friendship. It has meant a great deal to us over the years, as it will for many years to come. Although Yale is a very special place for many reasons, it is people like you that are really important. Both of us wish you and Sunny the very best of everything in your retirement.

With warmest wishes,

Vic

Which Do You Prefer? Salty or Sweet?

El-Hang Lee, Ph.D. Yale '77
Professor of Electrical and Electronics Engineering (Ret.)
Inha University, Inha 22212 Korea
ehlee@inha.ac.kr

On a summer day in the mid-1970s, in the middle of my graduate thesis research, I had a chance to visit Clarkson College (now Clarkson University) in Potsdam, New York, with Prof. Richard K. Chang. We were going to meet Prof. Milton Kerker, a world-renowned authority in theoretical studies on light scattering from aerosols and monodispersed particles. Prof. Chang had much interest in studying the same in experiment and I was at the center of that experiment. So, Prof. Chang proposed a trip to Clarkson. We started out early in the morning because the drive to Clarkson would take a whole day, mostly along the country roads, valleys, and rivers. I did all the driving with Prof. Chang on the passenger side as I had learned it to be polite for a younger person to drive for a senior.

As we set off, we talked about our research and about how to make the visit most fruitful. About halfway, we decided to have lunch. Then, Prof. Chang asked me which taste I preferred, salty or sweet. I generally prefer sweet taste, but I had also been very much accustomed to salty tastes, which I had acquired from heavily salted foods in Korea since my early age. Over lunch, we talked not only about food but also about many other things like history, cultures, traditions, customs of various countries and places. Though I may not recall all other details, this particular question on taste preference has remained in my memory, even after nearly 50 years have passed. It sounded quite fresh even then, because unlike the usual research and science talks in the laboratory, we started talking about personal things like this in a very friendly way. Back in my home country, I had been taught to respect and uphold teachers and professors even like a father or a king. This is the same even now at my late age. But, here, he was not like a strict professor that I had imagined him to be in all those previous years from back home. I still feel more comfortable addressing him as "Prof. Chang." But, this seemed to be changing. Was "Prof. Chang" becoming "Richard"? Or, was I beginning to recognize him as "Richard?"

Over the next near-50 years, we became good friends and colleagues. Prof. Chang and Mrs. Chang ("Richard" and "Sunny") used to visit South Korea for Prof. Chang's lectures at various places. And, naturally, we would have get-togethers, sharing various stories around our lives. Also, almost every year, we exchanged Christmas cards and letters, carrying the stories of our families and children. One of Prof. Chang's daughters also visited us in South Korea with her family. Our conversation in a traditional Korean restaurant naturally included the taste and the story of the famously hot-spicy-salty Korean food "kimchi" along with its history, varieties, culture, and even science around it. How did I get to have such an eminent scientist and his family as such a wonderful friend? It was simply a blessing, a mysterious blessing!

In the early 1970s, I very much wanted to do my graduate study at Yale, which was well known for its rich tradition of interdisciplinary and multidisciplinary teachings and research. With a college major in Electrical Engineering, I was looking into Applied Physics for my advanced study. Prof. Chang was a professor in Applied Physics and was leading an array of cutting-edge laser applied physics research. This is how I was attracted to Yale and to Prof. Richard K. Chang.

For my thesis, I studied elastic and inelastic light scattering from monodispersed micron/submicron-size spherical particles using plane waves and evanescent waves. The experimental results of mine matched remarkably well with the independently carried out theoretical results from Clarkson and so these two separate results were published together, but independently, side by side, later in Applied Optics. The trip to Clarkson that Prof. Chang had proposed for my thesis research had thus come out to be one of the most important and memorable milestones in my life at Yale.

After Yale, I continued my study in the science of light. And, I owe it all to the great teachings from Prof. Richard Chang. In 2004, Sigma-Xi, the Scientific Research Honor Society, contacted me about their fund-raising campaign to honor the mentors of Sigma-Xi members. I replied without a moment of hesitation and made a humble donation. Through the testimonial I said I honor Prof. Richard K. Chang for his great personal teachings, guidance, and inspirations that led me to a beautiful world of science and to many new visions where we can make our lives and the world a little better and brighter. Sigma-Xi has honored the mentors with "Paving Stones" and an "Honor Wall" in the building along with the "Honoree Plaza" on their website.

The world may be sometimes salty and sometimes sweet, frequently uneven, and perhaps never in balance. But as long as we are thankful for what we receive and do our best to serve all those around us, everything will turn out most blissful. I learned all this from my dear teacher, mentor, and advisor, Prof. Richard K. Chang. All his teachings, guidance, inspiration, and visions have been deeply honored all through my life. And, now, the sweet memory of him as a dear teacher and a friend will remain in all its sweetness for all the days to come.

Memories of Richard Chang

Pin-Fang Lin, Ph.D.
Research Scientist (Ret.)
Bristol Myers Squibb
Guilford, CT 06437 USA

In 1971, during my early graduate school days, I had heard that the Yale Electrical Engineering Department had a distinguished professor named Richard Chang, who was also related to a famed Chinese Poet, Xu Zhimo.

Richard and T.P. Ma in Harvard Crimson and Yale Blue robes getting ready to march on the faculty line during a Yale Graduation in the 1990's.

One day, to my pleasant surprise, my engineering friend Tso-Ping Ma (who would later become my husband), asked me whether I would like to accompany him to Professor Chang's home for a dinner party. In fact, this party was for students of the class that the professor taught. What a kind gesture for a professor to extend to his students! At this first meeting, aside from a delicious dinner, Richard and Sunny impressed me with their efforts to engage every guest in conversation and make everyone feel at home.

Since then, I also noticed that Richard and Sunny would always attend departmental activities, despite their very busy travel schedules. They were so loved by their Yale colleagues. Richard was always happy and smiling and able to strike meaningful conversations when bumping into others in the hallway or in an elevator, which was truly remarkable to me. It was amazing that this professor of engineering would remember me, a friend of one of his student's, enough to engage in such a warm conversation. I was not in frequent contact with Richard throughout the years, but in the year 2001, when I encountered a difficult situation and needed his help, he took time to give me proper advice and support, which is a testament to his kind and generous personality, for which I am forever grateful. Richard has shown me how to treat others kindly and how to make others feel comfortable.

Memories of Richard K. Chang from Fifty Years

Andreas Schlachetzki, Ph.D.
Research Scientist
Deutsche Post Foundation
Germany

From 1970 to 1971 I spent my postdoc year at Yale University's Department of Applied Physics. My office, shared with two other researchers, was on the third floor of Becton Center, an impressive building just completed in 1970 in the style of brutalism. A few doors down the hall was the office of then Assistant Professor Richard K. Chang. Each afternoon the residents of the hall met for a quick cup of coffee to discuss topics of interest and to hurry back to work. It was there that I got to know Richard. I remember a young man who concentrated on his work to achieve results most efficiently. At least that was my impression as seen from outside.

My training was in ferromagnetism and ultrasonics, at Yale I studied magnetization processes at very low temperatures in Werner P. Wolfe's group, and later I worked in semiconductor devices. There was no overlap with Richard's field of interest. Yet there was mutual sympathy which deepened over the decades, during which my esteem for him grew. That is how I came to know Richard personally.

Visit to Darmstadt

After my time at Yale I accepted a job with the Research Institute of the German Post Office in Darmstadt, a city near Frankfurt in the center of then West Germany. For about five centuries Darmstadt was the capital of the autonomous state of Hesse-Darmstadt which merged in 1918 with numerous other German states to form the Weimar Republic. Up to that year, the Grand Duke Ernest Ludwig of Hesse ruled the country. He was an open-minded person who believed that art was good for his country and in the long run would be profitable for its economy. Beginning in 1901 he initiated a series of exhibitions during which a hill near the city center was remodeled to become a unique example of Art Nouveau (or Jugendstil in German). Attached was the Artists' Colony where famous artists of the time could build their houses to their own taste. Now the ensemble, called "Mathildenhöhe" (Mathilde's Height), is a must-see for every visitor who comes close to Darmstadt and is interested in Art Nouveau.

It was in 1973 that Richard and his family visited Darmstadt. We started our sightseeing tour at another attraction of the city, the Hesse State Museum Darmstadt with its rich collection ranging from prehistoric to modern times, including of course, famous Art

Nouveau pieces. Naturally it was of no interest to the children, who instead found it much more fun to ride the two large bronze lions, guarding the stairs leading to the museum's entrance.

We continued our tour to Mathildenhöhe with its brilliant display of flowers. When approaching one first notices the Russian Chapel with its golden onion-shaped towers. Next to it is the large exhibition building. This church for orthodox rituals was built for the Tsar Nicholas II whose consort was Alexandra, the sister of Grand Duke Ernest Ludwig, to make them feel at home during their visits to Darmstadt. In 1918 the whole family was murdered by the communists.

Nothing reminded us of this tragedy when we strolled in the park. Everything from sculptures, benches, iron fences, pavilions, even the flower pots are in a homogeneous style. The children climbed one of these huge vases, decorated with Art Nouveau patterns.

My apartment in an old quarter was just a five minutes walk away from Mathildenhöhe, and also very near the Technical University of Darmstadt. When I moved in, I had to adapt myself to the bare floor until my furniture arrived. I slept on a piece of foam rubber cut to my size, my belongings in a suitcase with its lid leaning against the wall, and a simple camping chair. To understand the following better, a short description of this apartment, now furnished, might be useful. Its floor measured about 3 m × 10 m, plus a kitchenette and small room for WC and shower. A double bed with bedside table and a closet filled the bedroom, the door leading to the other room framed by bookshelves. This half of the apartment served both as dining and living room with a table plus four chairs and a simple sofa. As a kind of separation between these two parts, four slim wooden beams, extending from floor to ceiling, carried a number of square boards so that the whole structure appeared like an open tower. The board at the bottom was filled with bottles of liquor, collected from various regions and countries. Whoever wanted to taste one of the liquors was free to do so, but mainly the arrangement was just decoration. Many of the liquors survived the next twenty or thirty years.

The Chang family in the park of Mathildenhöhe with the Russian Chapel and part of the exhibition building in the background.

I invited the Chang family to stay overnight with me under the condition that the children should refrain from touching the liquors. Richard, pointing at the assembly of bottles, ordered in a commanding voice: "Listen children! This here is off limits." It worked perfectly.

I do not remember how we six people were packed in the small apartment. I just guess that Sunny and Richard slept in the bed together with one of their children. The other two tried

Richard's family in front of the Russian Chapel in Darmstadt.

to make themselves comfortable on the sofa cushions on the floor, while I unrolled my foam rubber as I had done when there was no furniture yet.

Next morning the "breakfast lady" came, as the children called her. It was Marianne bringing fresh rolls from the bakery. So everything was set for a good start of the day. I had asked the secretary of our department to take care of the children for the day. We handed the children over to Klaus and "Gili" Riedel (short for Gisela) who obviously did such a good job that Kevin remembered their names even after almost fifty years.

The four adults, Richard, Sunny, Marianne and I (we married in 1976), took off on a tour which we used to offer distinguished and discriminating guests, as the Changs surely were. We drove an hour east through the Spessart mountains to Würzburg on the Main River, the capital of Lower Franconia. The huge hospital Juliusspital, founded and financed almost five centuries ago by the then-ruling prince-bishop, was good for a country-style lunch. Subsequently the splendid Residence followed with its masterpieces of Baroque frescos and stucco works and the Court Garden. Napoleon, who stayed in the Residence several times on his campaigns, is said to have called it "the most magnificent parsonage in Europe". Rising above vineyards on the left bank of the Main River, the Marienberg Fortress was for almost 500 years the medieval home of the local prince-bishops.

These were the main sights we used to show to our visitors, finishing off with dinner in a good restaurant a few miles down the river, since, as the saying goes, in a wine-growing region with a Catholic population, good food is served. Whether this was precisely our program I do not remember, but it fits in the picture what Sunny wrote in a recent email. We had white asparagus (it was the season for "the queen of vegetables" as the Germans believe), then wild boar (most likely from the dark forests of the Spessart, we crossed while going to Würzburg), and all this topped with truffles. We drove back to Darmstadt to collect the children from the care of Gili and Klaus. All together, we spent another night in my cramped apartment.

Hamden in winter, San Francisco in summer

Around the turn of the year 1973/1974, Marianne and I travelled for several weeks in Bolivia and Peru. We chose that season because it was summer in the southern hemisphere.

Chinese breakfast in Chinatown, San Francisco: Richard and Andreas Schlachetzki, framed by Sunny and Sarah Schlachetzki.

The main means of transportation on our backpack tour were the local Indio busses, looking like American school busses. Although in the rugged Andes it was sometimes near freezing the coastal area near Lima was comfortably warm. From there we went directly into the New England winter to see Richard and Sunny.

We enjoyed their hospitality in their cozy Hamden home. Obviously, I had visited there before and had eaten English muffins in large quantities. This had led Richard to remark drily that I was eating muffins by the dozen. Now Sunny had ample provisions so that any shortage was avoided.

We remember with pleasure a young German shepherd obviously new in the Chang family. As is often the case this young dog had only one ear standing upright whereas the other one still hanging, a cute view to disappear when the dog grows up. Richard was so embarrassed: a German shepherd in the presence of German visitors with a hanging ear, supposedly appearing as evidence of failing education by Richard! We felt sorry for Richard taking a natural phenomenon so seriously.

Many years passed before our next meeting, and many changes had occurred during that period. For several years I was with the Technical University Braunschweig in Northern Germany. In 1992 Marianne and I rented an RV to tour the natural wonders of America's West, very much to the pleasure of our children Constantin and Sarah. We planned our travel so that we could meet Richard in San Francisco on July 4. We parked our RV at Candlestick Park near the Giants' Stadium. Richard and Sunny collected us from there to introduce us to a Chinese breakfast in Chinatown. We gathered around a table whose revolving central part was crowded with small dishes containing all kinds of Chinese delicacies. Thus we could taste comfortably this or that specialty which we all enjoyed very much, although not so much the duck feet.

Then Richard took us on a sightseeing tour, Sunny, Marianne and our children squeezed into the back seats while I sat in the passenger seat. After a while Richard stopped next to a taxi to ask for directions to Lombard Street. The driver, an Indian, eloquently explained the way in what he thought was English. Richard nodded understandingly and thanked the man. While continuing our ride Richard turned towards me: "Did you understand him?" Upon my no, Richard said: "I did not understand a word." It was comforting for me, another

Having a drink in the View Lounge of the San Francisco Marriott: Sunny and Richard, Marianne and Andreas Schlachetzki.

non-native speaker. Nevertheless, we made it somehow to Lombard Street.

Sunny and Richard stayed at the Marriott, which at the time offered a special price for Yale faculty. In the View Lounge at the top of the hotel, we had a vivid conversation while sipping our drinks. The impressive view of San Francisco's skyline did not distract us. We concluded a wonderful day with a dinner the Chinese way, together with Richard's mother and Jorge who took good care of her. Richard's warning of numerous bones in a special fish is still in our ears.

A nice byproduct of our visit to San Francisco was the first and only baseball game we ever saw: the Giants vs. the St. Louis Cardinals. I was fortunate that the lady next to me gave me some basic explanations of the complicated rules. We watched closely a gentleman swinging a club, another one throwing a ball towards him. Hitting the ball, he dropped the club and ran from cushion to cushion, as other gentlemen might also do. This procedure was followed several times. Then all of a sudden and to our surprise, everyone got up and left the stadium, so did we. In short, this is baseball seen through European eyes.

Epilogue

Fifty years of friendship with Richard is a long time. It was my privilege to have met him several times and to exchange our Christmas letters with him and Sunny year after year. We learnt how our families fared and how we developed in our profession. From 1983 to 1987 I headed a group of eventually 80 researchers working on integrated optics in Berlin. It was a period of extreme tension for me.

During these years Richard visited Berlin. He did not realize that I was in Berlin. He spent his nights in a boring hotel room, so did I in my lonely apartment not far from him. We both were away from our families. We missed a great opportunity to meet. Later Richard told me about his deep regret, I felt the same. We both were too much involved in our work.

In 2015 Sarah, our daughter, did research at New York University, funded by the Swiss

National Science Foundation, a good occasion for Marianne and me to visit her. It was no question that we three drove north to Hamden to see Richard under Sunny's care. When returning to New York at night along winding narrow streets and busy highways in pouring rain, the unfriendly side of the New England weather, we could not know that it was the last time we would see Richard. It was our farewell visit.

February 2021

The Influence of Richard K. Chang

Armand R. Tanguay, Jr., Ph.D. Yale '77
Departments of Electrical Engineering, Chemical Engineering, and Materials Science, Biomedical Engineering, Ophthalmology, and Physics and Astronomy, Neuroscience Graduate Program, University of Southern California, Los Angeles, California 90089 USA
armand.tanguay@usc.edu

Richard K. Chang was a remarkably able scientist, articulator of science, scholar, teacher, and mentor. Whenever interacting with Richard, one could not help but be struck by his unique combination of intellectual acumen, originality, focus on penetrating insights, clarity of thought, ability to ask just the right question, depth of knowledge, and perhaps above all, alacrity of response.

Often in a group discussion Richard would remain quiet for a time as an idea or concept was first being explored, and then cut to the quick with a rapid-fire question that pointed out the key element of the idea that needed to be fleshed out. He was equally quick at pointing out possible avenues of attack to crystallize the idea.

Those among us who were privileged to work with him can also attest to the warmth of his personality, his engaging laughter, his excitement at the revelation of a new idea, and his sheer delight at the confirmation of a counterintuitive hypothesis. None of us will ever forget the many lessons that we have learned from him, and the impact that he has had on our lives as well as on our careers.

Programmed Failure: One Key Step Along the Path to Success

No one, but no one, likes to fail at anything. Even more so for scientists and engineers, a failure in an experiment or project is often considered to be an indication of *failing*, rather than a key step along the path to success. Yet most if not all scientists and engineers have encountered far more failures than successes in their careers.[1] We propose that it is one's *response* to failure, even repeated failure, that eventually guarantees success.

To this end, we developed an experiment in a lab course at Yale University that incorporated *programmed* failure, in which the very design of the experiment guaranteed that the goal of the experiment could not possibly be achieved given the instructions and equipment initially provided.[2] Even the following set of suggestions and addition of key equipment were designed to guarantee that a successful outcome could not be obtained. Only in response to a deeper analysis of the experiment itself by the student participants following the first two failed measurements was a third stage of suggestions and further key equipment provided that finally allowed for accurate and meaningful measurements to be completed and analyzed.

This is but one of many lessons along the path to success that one learned working with Richard Chang. In this case, the lesson was focused on teaching through example a combination of persistence, analytical insight, fundamental understanding, and an attention to detail, all incorporated in a single experiment.

Description of the Initial Experimental Configuration

Richard taught the course EAS 50a, Physical Electronics in Fall, 1972 and Fall, 1973. I was privileged to be asked to be his Graduate Teaching Fellow, developing homework sets, exams, and laboratory experiments from scratch. The course met on Mondays, Wednesdays, and Fridays from 1:30 p.m. to 2:20 p.m. each week in 105 Becton Center, with laboratory experiments set up in one of the Becton Center basement rooms within the Chang Laboratory.

In the very first experiment, the goal was to measure the response times (in particular, the decay times) of three common types of photodetectors: a CdSe photocell, a Si *npn* phototransistor (photo-Darlington pair), and a Si *p-i-n* photodiode that consisted of a *p*-type layer, an intrinsic region, and an *n*-type layer. The decay time is a measure of the time that it takes for the photodetector to respond to the offset of a pulse of light.

The initial description of the experiment and its associated equipment involved the use of an incandescent light source, an optical chopper, the three types of photodetectors, and an oscilloscope (an electronic instrument capable of displaying the output of one or another of the photodetectors as a function of time). The optical chopper consisted of an electric motor that rotated a spoked wheel, with regions of the wheel designed to block the light beam from the light source, and interspersed gaps to allow light to pass through.

The optical chopper was set up by the students in between the light source and one of the photodetectors, so that the light striking the photodetector would be interrupted at a regular rate, thereby creating the opportunity to observe the response times (rise and fall times) of each type of photosensitive device. The photodetector was incorporated in an appropriate bias circuit to create a measurable output voltage that depended on the incident light intensity. In this experiment, the light was essentially either on or off, creating a binary output state that nonetheless had intervening transitions from one state to the other that could be visualized and measured using the oscilloscope connected to the output stage of the photodetector bias circuitry.

The experiment was set up in one of Prof. Chang's laboratories in the basement of Becton Center. The students were asked to read the description of the experiment provided, set up the associated equipment, complete the experimental measurements, and then come to my office on the 5th floor of Becton Center to tell me their results when they were finished.

The First Key Realization

After a couple of hours, the student team arrived in my office to exultantly report that all three photodetectors had decay times that were approximately the same, and were around a few milliseconds (ms) in each case. They described how much they enjoyed the experiment, as well as the opportunity to use state of the art optical and electronic equipment for the first time.

Of course they were a bit crestfallen to learn from me that although this was a reasonable value for the CdSe photocell decay time, it wasn't even close to being reasonable for either the phototransistor or the photodiode.

They all went back to the laboratory somewhat confused by this turn of events, but convinced that they were right, and that they would just remeasure everything and then come back to prove me wrong.

About an hour later, they came back upstairs to see me, and claimed that while remeasuring all of the decay times, they had realized that there was a flaw in the experimental design. When asked why, they went to the white board and diagrammed how the combination of a broad optical beam and wide gaps in the chopper wheel meant that the light pulses did not in fact turn on and off instantly, but instead increased and decreased somewhat linearly as each gap crossed the broad optical beam. When pressed, they were able to use the dimensions of the chopper wheel in combination with the rotation rate to calculate that the decay time of the light intensity itself was, sure enough, just about equal to a few ms. As a consequence, they now understood, they had actually been using the photodetectors to measure the rise and fall times of the optical intensity, and not at all the rise and fall times of the three photodetectors themselves.

The Second Key Realization

Given their new understanding, the students stated that what they now needed was a light source that turned on and off much faster than the rise and fall times of the various photodetectors that they were testing. At this point, we then gave them a high speed light emitting diode and a pulse generator that were together capable of producing rise times of the light intensity on the order of a few ns, with slightly longer fall times.

All three photodetectors were remeasured to determine their decay times using this new source configuration. The students then reported that the decay time of the CdSe photocell was still around a few ms, but that the decay times of both the phototransistor and the *p-i-n* photodiode were actually much shorter, around 150 μs for the phototransistor and about 10 ns for the photodiode.

They were then told that they were right on track for the case of the phototransistor, but were still a bit off for the *p-i-n* photodiode. This revelation caused considerable concern, as now they were convinced as a group that they had done everything "right" and thought of all possible sources of error. But back to the lab they went.

Another hour passed, and then a triumphant group tracked me down on the 5th floor of Becton Center to announce that they had (finally) consulted the manual for the oscilloscope that they were using, and that the rise and fall times of the oscilloscope itself (with the probe they were using) were only around 10 ns, and so now they were actually using the experiment to measure the decay time of the oscilloscope and its probe instead of the decay time of the photodiode.

At this point we wheeled into the lab a higher bandwidth oscilloscope with a high bandwidth probe that had a response time of 1 ns or so. With this new instrument, they determined that the decay time of the phototransistor was still approximately 150 μs, but that the decay time of the *p-i-n* photodiode was actually of order a few ns.

All in all, the combination of experimental design and available instrumentation yielded a sequence of measurements that ranged over six orders of magnitude!

The Law of Unintended Consequences

When they left the lab, there was great enthusiasm for the fact that they had finally conquered this experiment, and understood quite a bit about the limitations of experimental design as well as electronic and optical components and instrumentation.

By the time of the next experiment, which actually was not designed to incorporate the concept of programmed failure in any way, the Law of Unintended Consequences rose to the fore. The student group was now highly suspicious of any and all aspects of the experimental design, as well as of the operational limitations of the components and instrumentation. They were not at all satisfied with the lab setup and experimental description alone, but instead demanded to see all of the component specification sheets as well as the instrument manuals. In this second and each succeeding experiment throughout the semester, they were able in each case to convey one or more concerns about how each experiment was being conducted, with suggestions of how the experiments could be improved.

We had not anticipated this consequence at all, but were delighted to see that the lesson, once learned, carried over into everything else that followed. Richard was just as amused and delighted by this unintended outcome as I was.

Further Observations

Failure as Endemic in Science and Engineering

Although most scientists and engineers have been trained to focus on success, failure along the way to success is a nearly everyday, and in some cases essential, occurrence. Many types of failures happen in all three primary modes of research and engineering: theory and design, experiment and prototype development, and computational modeling and simulation. Failure can be evident in the choice of appropriate theoretical assumptions, in the initial design of a prototype against a stated set of constraints, in the design of an experiment to test a key hypothesis, in the implementation of the experiment itself, in the initial implementation of a prototype, in the choice of a computational model for a given phenomenon or design, in the implementation of the computational model or simulation, and perhaps most importantly in the interpretation of results, to name just a few. Careful analysis of each failure met along the way, however, can yield not only an improved theory, design, experiment, prototype, or computational simulation, but also often a new understanding of the problem itself as well as a well-formulated restatement of the problem. As such, a readiness to anticipate failure, immediately analyze its causes, and redirect the effort is a hallmark of every successful scientist and engineer.

It should be mentioned that the capability to recognize failures, appreciate their significance, and generate appropriate responses is also just as applicable and important in the humanities and social sciences, in medicine, in business, in law, and in many other disciplines as well.

Programmed Failure

In order to break even freshman students free from the fundamental misconception that most have leaving high school, namely, that education means learning what is already known, instead of experiencing the excitement of discovering what is yet unknown, unusual techniques can often be employed to great advantage. To this end, programmed failure has been one of the most interesting and successful concepts that we have developed over the years, as specifically implemented in courses with associated laboratory experiences. In this approach, based on the initial experience described above at Yale in EAS 50a, we design a series of laboratory exercises that require familiarization with key pieces of research equipment as well as an understanding of basic laboratory technique. In the programmed failure method, we design specific "traps" into such experiments that most students will unthinkingly fall into if they are not actively challenging the assumptions of each experimental step. As a result, they consistently report obviously erroneous results during the exercise, usually without comment.

When the magnitude of the (often quite large) error is pointed out, the students are asked to reconsider their experiment and try to figure out what went wrong. When they arrive either individually or as a group at the correct conclusion, they are then provided with additional apparatus or guidance to continue the experiment, this time much more warily. By the third such instance, they become remarkably crafty, and on their own read each and every component data sheet and equipment manual for the remainder of the semester, using back-of-the-envelope calculations to estimate all of the anticipated results. In my experience, as a result, they become highly skilled experimenters in only one semester, and remain justifiably suspicious of any and all "programmed" experiments. In fact, they usually end up gaining extra credit by pointing out shortfalls in the experiments that even the professor and the teaching fellows had not anticipated.

Programmed Failure in the Curriculum

The concept of programmed failure can also be extended to other aspects of the curriculum as well. For example, homework sets are typically designed to increase problem solving aptitudes and skills, to illustrate concepts discussed in lecture, and to investigate certain critical phenomena in depth.

In the undergraduate and graduate courses that I teach, several of the problems within each homework set will be straightforward, and are designed to familiarize students with a particular key concept. Often, these problems will couple together more than one concept in order to demonstrate the multifaceted nature of the subject matter, and to forge connections among seemingly unrelated topics that do in fact have a common underpinning or conceptual root.

In addition, two unusual features are incorporated in the various homework sets, each designed to implement a form of programmed failure. First, in consideration of the fact that real-life problems in science and engineering are almost never well-specified, some problems will either present data that is unnecessary for the solution of the problem, or will not present enough information for a satisfactory solution. In the former case, students are expected to identify the extraneous piece(s) of information, and explain why it is (or they are) not needed. In the latter case, students are expected to make a reasonable assumption, carefully justify their rationale for the assumption they've made, and then proceed to solve the problem. This feature develops an ability to ascertain the key elements of a problem and its solution, and then to discriminate among possible parameters that are either necessary or irrelevant.

The second unusual feature is that the homework sets will also include numerous design problems, in which students are asked to consider a particular design goal, make necessary assumptions, couple concepts together as necessary, develop a coherent design that satisfies the design goals, and in some cases analyze the system performance given their particular design. Examples of such design problems in our graduate optics course might include a wavelength dispersing prism design, an optical imaging system design, or the design of a polarization component such as a quarter wave plate. In each case, what is emphasized is the recognition that a given initial set of assumptions does not always yield an appropriate design that meets the stated design goals, such that either relaxed assumptions or additional assumptions must be made. This iterative process is a form of programmed failure without the pre-specification of a given failure mode, and hence develops enhanced skills for recognizing and adapting to failed initial attempts.

Teaching by "Infection" *vs.* "Injection"

To a certain degree, teaching is both a vocation and an avocation. My first teaching experience was gained as an Undergraduate Teaching Fellow for Prof. Carver Mead (a master teacher) at the California Institute of Technology in EE 3, Introduction to Solid State Electronics, a new semiconductor device physics course for Caltech freshmen. Throughout more than 52 years of ensuing experience to the present, I have had the privilege and opportunity of either teaching or tutoring students in every grade level from pre-school through sixth year graduate students, including teaching English as a foreign language (using science as a teaching tool) to foreign students at Albertus Magnus College while a graduate student at Yale University.

The fundamental goal in all of my teaching to date has been to teach by "infection" of an enthusiasm for learning (leavened by a heavy dose of self-organization and research technique), rather than by "injection" of factual information related to a particular subject. From my perspective, the specific subjects of classes act primarily as vehicles for personal intellectual growth and enfranchisement, and only secondarily as advancements in one's knowledge base. As such, I have endeavored to convince over five decades worth of students that they can quickly and efficiently learn almost any subject they desire *without* resorting to traditional classroom instruction, in hopes that they will carry away from the university a lifelong passion for learning as well as practical tools for its achievement.

Teaching is an exhilarating yet highly demanding and at times intensely frustrating experience. Inspiration and renewal in teaching, in my view, often derive from the use of novel points of view, novel techniques, new course development and revision, and, of course, from interactions with truly outstanding students. Programmed failure has proven over all these years, perhaps counterintuitively, to be a powerful technique for "infecting" students with enthusiasm and passion.

Teaching and Learning as a Nonlinear Dynamical System

It might be noted that the joint process of teaching and learning is highly nonlinear. For example, it is well known that multisensory perception is critical to the long term retention of information and understanding. In the classroom environment, seeing an illustration of a concept, hearing it described, and consciously deciding to take note of the key features and record them (a motor response with a tactile component) yields far longer term memories than any one of the three perceptual modes separately. As yet another example, reading a book chapter through first as one would read a novel, learning the vocabulary and recognizing how various concepts are linked, and then reading the same book chapter a second time more carefully is an effective method for achieving the "Aha!" reaction of sudden understanding.

Programmed failure is an exercise in "multisensory" perception insofar as it tends to engage a number of different senses (*e.g.*, constructing a diagram (involving a tactile response), visualizing the key elements of the diagram (involving a visual response), and perhaps discussing the problem or experiment with others (involving an aural response)). In addition, programmed failure tends to elicit multiple emotional as well as analytical responses, such that taken together the experiment or homework problem in question forms a memorable experience.

As with all nonlinear dynamical systems, which if pushed to the limit can often descend into chaos, careful choices of the initial conditions are necessary to attain (reasonably) stable behavior.[3] Nonetheless, the greatest opportunity for breakthrough concepts and ideas often occurs at the edge of chaos, where even small changes in the initial conditions (or assumptions) can lead to very different outcomes.[4]

References

1. Andrew Moseman, "The Transformative Power of Failure", Caltech Magazine, Summer 2021, 16–23, (2021).

2. Armand R. Tanguay, Jr., "Filling the Great Teaching Void: Unusual Teaching Concepts", Caltech Project for Effective Teaching (CPET), Feynman Lecture Hall, 201

E. Bridge, California Institute of Technology, Pasadena, California, 22 April, 2015, Invited Presentation, (2015).

3. James Gleick, *Chaos: Making a New Science*, Penguin Books, New York, (1987).

4. Roger Lewin, *Complexity: Life at the Edge of Chaos*, 2nd Edition, University of Chicago Press, Chicago, (2000).

Letter to Richard on the Occasion of His Retirement from Yale University in 2008

Richard K. Chang: Light Speed…

3 May, 2008

Once upon a time, in a Becton Center lab far beneath the bustling Yale University campus, Phil Black (Philip C. Black, Ph.D., 1977) and Armand Tanguay (Armand R. Tanguay, Jr., Ph.D., 1977) were surprised quite late one night by a visit from none other than Richard Chang.

Phil shot Armand a look that screamed "What the heck is *he* doing here this late?", to which Armand's return glance replied "How the heck would I know; he's *your* advisor!"

Phil and Armand had been busy at work for many hours trying to align Phil's optical multichannel analyzer with a pulsed laser, and were using a Polaroid camera mounted to a high speed Tektronix oscilloscope to record each pulse.

Richard, in his inimitable way, quickly got to the point, asking "Is it working yet?" To which Phil replied, "We'll know after this next series, after we reload the camera", which by the way had just run out of film. Richard said "Let me get that", and with one hand unlatched the camera back, grabbed a new roll of Polaroid film, still in its box and light tight aluminum wrapper, tore open the box, tore open the aluminum wrapper, inserted the film in the camera, and slammed the back shut in a fluid motion that lasted no more than three seconds.

When neither Phil nor Armand could speak for a minute, nor stop staring at Richard, he looked up at them and said, somewhat sheepishly, "After 100,000 shots, you get pretty good at it…"

Thank God that digital cameras were invented soon thereafter!

OPTICAL MATERIALS AND DEVICES LABORATORY
520 SEAVER SCIENCE CENTER, MC-0483
(213) 740-4403 FAX (213) 740-8684

3 May, 2008

Professor Richard K. Chang
Department of Applied Physics
School of Engineering and Applied Science
407 Becton Center
Yale University
15 Prospect Street
New Haven, Connecticut 06511

Dear Richard,

Congratulations on a truly amazing career!

I don't even know where to start to tell you how much you mean to me personally, and certainly how much you have meant to my career. I simply cannot imagine making it through those years at Yale without your constant advice, encouragement, inspiration, and insights.

I still look back fondly on the lab course we put together for the Solid State Devices course, in which we developed the notion of "programmed failure" (!), and on the Optical Properties of Semiconductors course that Paul Dressendorfer, Phil Black, Peter Collings, and I co-taught with you to ourselves!

I learned so much about science and scientific style from you. Your scientific abilities and scientific methods profoundly influenced me during my days as a graduate student. I have always been deeply impressed by your ability to articulate, with both clarity and insight, fundamental scientific problems that are worthy of attack and rich in intellectual prospect. You have also shown us great leadership in your ability to focus with laser precision on the key issue, and to pursue it relentlessly until it yielded to the onslaught. You always had an eye for the underlying physics, but an appreciation of the potential applications as well. And no one, simply no one, comes close to you in their ability to be both quick on the uptake and incisive in query. I fondly recall many a Monday evening seminar where the one pivotal question was asked by you (often to the dismay of the seminar speaker, who as often as not just wasn't prepared to respond that quickly!).

With regard to scientific style, who would imagine that these days I give invited lectures to Ph.D. students and postdocs at both USC and Caltech on "The Art of Scientific Presentation" ("The Scientist as World-Class Communicator")? Where do you think I learned all of those presentation tricks from? I still recall sitting in one of your own invited presentations at an OSA Meeting (Rochester, I think, in 1972 or so, on Raman scattering in the microspheres) and being stunned by the clarity, insight, preparation, presentation coherence, style, class, elegance, and unity of the whole effort. I took as many notes then on how to give an effective presentation as on the science, and I have passed it along to well in excess of a thousand students and postdoctoral fellows over the past 31 years since Yale.

I also learned a great deal about family values from you, values that helped form and inform my own family. I long admired the close, warm, loving relationship you had with your beautiful and amazing wife Sunny, and the wonderful way in which the love you shared with each other extended to your three children. As I have watched each of them grow up to become remarkable young people in their own right, I have marveled at the strong positive influence you have had on each one of them. I have often been deeply touched by how fondly you referred to everyone in your family in conversation.

Thank you a thousand times over for this beautiful gift!

Suffice it to say that I owe you so much that the debt is irrepayable. You influenced me profoundly, then and now, and made it possible for me to succeed in whatever modest way that I have. If my students have benefited from my own teaching and scientific style, then they too owe a debt of gratitude to Richard K. Chang. Certainly everyone in my family does!

With warmest regards to both you and Sunny,

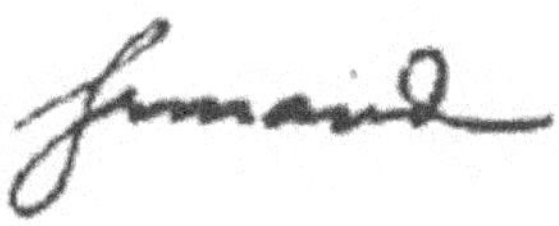

Armand R. Tanguay, Jr. Professor
Electrical Engineering-Electrophysics,
Chemical Engineering and Materials Science, And
Biomedical Engineering;
Neuroscience Graduate Program Ph.D.,

Yale University, 1977

UNIVERSITY OF SOUTHERN CALIFORNIA, UNIVERSITY PARK, LOS ANGELES, CALIFORNIA 90089-0483

Armand R. Tanguay, Jr., Ph.D., Yale University, Engineering and Applied Science, 1977.

Jim Moore, Jayne Miller, and Armand R. Tanguay, Jr., at a restaurant in Vermont following Richard's 60th Birthday Celebration.

Jim Moore and Armand R. Tanguay, Jr., at a restaurant in Vermont following Richard's 60th Birthday Celebration.

Jim Moore, Phil Black, and Armand Tanguay at the Chang's home in Brattleboro, Vermont. Only one was smart enough to locate the wine...

Armand R. Tanguay, Jr. and Richard K. Chang, at Richard's 60th Birthday Celebration, Yale University.

Armand R. Tanguay, Jr., giving a presentation at Richard's 60th Birthday Celebration, Yale University.

Armand R. Tanguay, Jr., making a point during an After-Dinner Presentation at Richard's 60th Birthday Celebration, Yale University.

Philip C. Black, Armand R. Tanguay, Jr., James Moore, and others of Richard's former Ph.D. students at the Chang's home in Brattleboro, Vermont following Richard's 60th Birthday Celebration.

Richard K. Chang at Yale University, on the occasion of his 60th Birthday Celebration, October 20th, 2000.

Richard and Sunny Chang, at a Yale University banquet honoring his 60th birthday, October 20th, 2000.

Letter to Richard on the Occasion of His Retirement in June 2008

Robert Benner, Ph.D.[†]
Vice President of Research and Development
Process Instruments Inc.,
Salt Lake City, Utah 84108 USA

April 18, 2008

Dear Richard,

As you approach retirement from the full-time faculty at Yale, I would like to thank you again for your collaborations and friendship with me over the years. The time I spent at Yale from the Fall of 1977 to the end of 1979 was one of the most enjoyable and productive periods of my career. I find it hard to believe that it has been thirty years since those exciting days. In just over two years we accomplished many things.

I have always been especially proud of our Physical Review Letter on the observation of structure resonances in the fluorescence spectra from microspheres. This paper seemed to mark the beginning of your extensive research accomplishments that were recognized by the David Sinclair Award in 2005.

The experience I gained with gas-phase Raman spectroscopy in your lab enabled me to be first author on my first patent filed in 1985 for an intracavity Raman gas analyzer. This patent provided the basic technology for a Utah company that produced respiratory and anesthetic gas monitors that were widely sold to hospitals throughout the world. The company was sold in 1990 for more than 20 million dollars.

Our use of multichannel Raman spectroscopy to follow the time evolution of SERS correlated to cyclic voltammograms foreshadowed the approach our company is using today for chemical process monitoring. Our company continues to do well. One U.S. refinery is about to purchase its fourth multiplexed Raman system from us for monitoring various refinery processes. We had our first sale in Europe last quarter, and we are expecting a sale in China later in the year.

In addition to the research activities during my days at Yale, the interactions with students, faculty members, and visitors were all very rewarding. The spirit of comradery and positive professional attitudes that characterized your research group are qualities that I have rarely seen in other working environments.

I hope your health is good and that you have an enjoyable retirement.

Sincerely,
Robert E. Benner

†Passed away

Remembrance of Richard K. Chang

Peter W. Barber
Emeritus Research Professor
Desert Research Institute
Reno, NV 89512 USA

I met Richard in the summer of 1977 or 1978 at an annual conference sponsored by the Army Research Laboratory at the Aberdeen Proving Ground north of Baltimore. The Laboratory had an active program in aerosol research and organized the conference to keep up with advances in the field by inviting researchers from across the country. The meetings lasted a few days, and it was a good place to meet other researchers as well as to meet potential research sponsors.

As an outgrowth of these conferences, in December 1978 Richard organized a workshop at Yale focusing on particle resonances. It was at this workshop that we both realized that his experimental work and my computational work complemented each other. I asked Richard about the possibility of a visiting appointment to work with him for a year. He was enthusiastic and we started working on a plan for me to spend the 1979–1980 academic year at Yale.

At the time I was an Assistant Professor at the University of Utah and was not yet eligible for a sabbatical leave that would cover my salary for the year. However, I did have a grant that would fund part of my salary, and Richard volunteered to fund part of my salary from his grants. Finally, Richard arranged for me to teach an electromagnetics class for a semester, so Yale covered part of my salary, too.

My family and I left Salt Lake City in August 1979 and drove to New Haven. We rented an apartment in Hamden, near where Richard and his family lived. Richard planned to take his family with him to China on a sabbatical leave for the second half of the academic year, so he invited us to live in their home in Hamden beginning in January 1980. There we lived rent free, only having to pay for the utilities and take care of their two dogs!

I could not have been happier working with Richard and his colleagues. Although I had an office, I spent most of my time in the basement where Richard's optics laboratories were located. In fact, his laboratories covered almost the entire basement. The environment and the collaborative work were nothing like I had ever experienced. It was the first time that I had ever worked with experimentalists. We were making light scattering measurements and numerical calculations at the same time and comparing results.

There was another aspect of the experience that added to the excitement of working with Richard. In addition to other Yale faculty, postdocs, and graduate students working in his laboratories, there was a steady stream of researchers from other places coming and going. We even had an MIT undergraduate spend a semester working on her senior thesis — she ended up coauthoring three journal articles! Richard had created an exciting and synergistic research environment.

There is a story associated with the first paper that I coauthored with Richard and two other colleagues. We had measured and calculated structure resonances in the fluorescence spectra from microspheres and Richard asked me if I thought that the work was appropriate for submission to *Physical Review Letters*. I said yes. Only a month or so after we submitted the paper, Richard came by one day to tell me that the paper had been accepted and that I sure had chutzpah (he may have used another word) in suggesting that we submit the work. First, as an engineer, I had no idea that *Physical Review Letters* was one of the more selective scientific journals, so my agreeing to submit the paper to the journal was a decision made from blissful ignorance. More importantly, it was Richard's idea to submit the paper, but he was giving me credit for making that bold decision. This was so characteristic of Richard.

I had another experience with Richard that is a great memory. In the Fall of 1979, we flew from New Haven to Baltimore for a meeting. When we arrived at the Baltimore airport for our return flight, we were told that because there were not enough passengers to justify using the regular airliner, we would be flying in a small six-passenger twin-engine airplane. When we walked out on the tarmac, I noticed that there was only one pilot. I was a private pilot, and I asked the captain if I could sit in the copilot seat. He agreed! Our route of flight took us over Long Island where we could see the Statue of Liberty, Manhattan, and the entire New York metropolitan area. I doubt that in today's world a passenger on a scheduled airline flight will ever again have the experience of sitting in the copilot seat. It was great to share this experience with Richard.

After my year at Yale, Richard and I met many times over the next decade at scientific conferences. He visited and gave a seminar when I was at the University of Utah and later visited Clarkson University when I moved there in the 1980s. My year at Yale initiated a collaboration with Richard that resulted in 18 journal articles and 15 conference papers. We also co-edited a book on optical effects in small particles.*

Our joint work only wound down when my career shifted to administration rather than research. However, my faculty colleague Steve Hill, whom I introduced to Richard, continued to work with Richard for many years. In addition to their joint work, one of Steve's Ph.D. students completed a postdoc with Richard and one of his master's students completed a Ph.D. with Richard.**

Richard was instrumental in supporting my career in many ways. He wrote a letter in support of my nomination as a Fellow of the Optical Society of America. And he provided a reference when I applied for my last job as Director of the Atmospheric Sciences Division at the Desert Research Institute in Reno.

Richard had a great influence on me both professionally and personally during my career. He was a brilliant scientist, but a down-to-earth human being. I am still surprised at how much credit he gave me for our joint research. On more than one occasion I heard him tell other people how much he had learned from me when in fact, I felt that just the opposite was the case. His generosity in supporting me and others with his time and talent was much appreciated. Knowing and working with Richard was one of the best experiences of my life.

Editors' note:
*Optical Effects with Small Particles (Advanced Series in Applied Physics) by Peter W. Barber, Ed. and Richard K. Chang, Ed., World Scientific Press, Singapore (1988).
**Dipak Chowdhury and Mohiuddin Mazumder, respectively.

Letter to Richard Chang on the Occasion of His Retirement from Yale University, June 2008

April 18, 2008

Dear Richard,

While we will not be able to attend your retirement party in June, the occasion has been a wonderful opportunity to reminisce about being at Yale in 1979–1980 — this marked the beginning of our scientific collaboration and our friendship (almost 30 years ago!).

My family and I drove from Salt Lake City to Hamden in the summer of 1979. We had shipped ahead about 20 boxes via UPS that were stacked in your garage. We arrived on a weekend out of cash and ATM's had not yet been invented. You may not recall lending me $20 in those first few days, or putting all five of us up in your home for a couple of days when our apartment wasn't ready, but we remember.

I had no idea that working one year at Yale with you and your colleagues would have such an influence on the rest of my career. You have helped many people during your career, but I want you to know that working in your laboratory that year and collaborating with you and our mutual colleagues in the years to follow was the most exciting time of my working life.

I'll never forget the thrill of holding a plot of particle fluorescence measurements from your lab in my hands as a graph of my numerical calculations came rolling out of the printer and seeing that they were just about identical. This and many other experiences gave greater meaning to all of the previous and subsequent years of computer calculations that I was involved in. All of this work led to the light scattering book that Steve Hill and I wrote a dozen years later.*

Beyond the scientific collaboration, our family greatly enjoyed living in your beautiful home while you were all away on sabbatical for the first six months of 1980. Our children enjoyed the Beatle records and they remember to this day chasing your dog down the street or across the golf course when she escaped — Karen and I even participated in dog chasing occasionally ourselves. Karen and I also remember both furnaces running almost continuously that entire cold winter and the stove oil deliveries coming way too often.

Karen and I will never forget the kindness shown to our family by you and Sunny. More than that, the achievements of you and your family have been an inspiration to our family.

We're sorry we were away when you came to Reno last fall, but we were glad that you were able to visit with our daughter April and her children. We hope to visit you in the near future after you have eased into retirement.

Warmest regards,

Peter Karen

Editors' note:
*Light Scattering by Particles: Computational Methods (Advanced Series in Applied Physics) by Peter W. Barber and Steve C. Hill, World Scientific Press, Singapore (1990).

Memories of Richard Chang

Roger L. Farrow, Ph.D. Yale '79
Distinguished Laser Engineer
nLIGHT Corporation
Camas, WA 98607 USA

I first met Richard early in my graduate studies at Yale Engineering in the fall of 1974. I had applied to the Department without any previous contact with any of the professors or staff and I was a little shocked to be accepted. As far as I know, no one from Sewanee (a small liberal arts college in Tennessee associated with an Episcopal seminary) had ever attended Yale Engineering. I suspected but never confirmed that Richard had something to do with my acceptance. He was familiar with details of my application that only review committee members would know, such as a description of a novel experiment dreamed up by my physics professor at Sewanee called a "black box" experiment where the student must determine the contents of a sealed box through non-destructive tests of his or her design. I had managed to figure out what was inside mine and included this information in my application as evidence of experimental skill. Little did I know that Richard prized good experiments and that this detail in my application may have had a decisive effect!

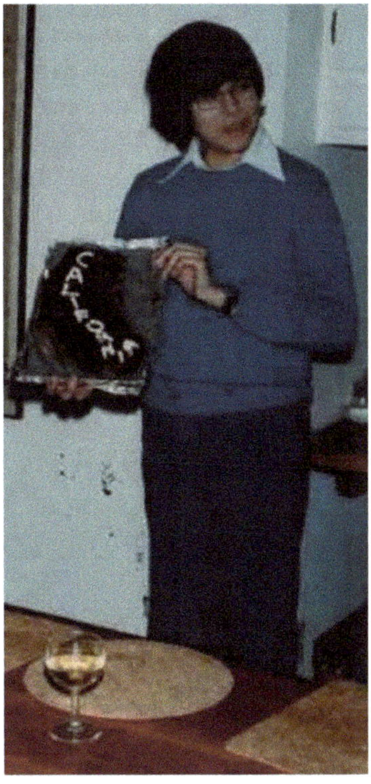

At the beginning of my first year at Yale, I visited with several professors in Engineering, including Richard, to discuss working with one of them on a student project. Sometimes these early interactions evolved into student-advisor relationship for a Masters or Ph.D. degree, so they could be fateful. As it happened, I ended up working with Richard although I don't remember a conscious decision to choose his project. After talking with Richard I left his office with my first assignment in hand and a desk assigned to me in the basement of Becton Center. Little did I know this was the beginning of my career as a student of the legendary Richard Chang!

I'll never forget the wonderful dinners that Richard and Sunny were always giving for his students. It wasn't often that we residents of the Becton Center catacombs had the opportunity to dine on a sumptuous meal and be entertained by a classical music recital. Not to mention the lively dinner conversations and beautiful slide shows of Richard's travels to Machu Picchu and Spain.

One last thing I'll always be grateful for: if not for Richard I would probably still be at Becton working on my thesis. I'm not someone who enjoys writing papers, much less theses, and the deadline kept looming. Even though I had a job offer from Sandia Labs and had agreed to report for work by a specified date, I kept falling behind schedule. Finally I asked Richard if I could take another few months to work on and (possibly) improve the thesis. Typically, he was having none of this. He pointed out that the extra time wouldn't make the thesis materially better but would only delay the start of my new career as I was changing fields from semiconductor physics. This was the best career advice I ever received and I am still grateful to Richard for seemingly putting my interests ahead of his research sponsor. As a testament to Richard's judgment, the dissertation turned out fine (in fact, undeservedly winning a Becton prize) and I ended up starting my new job on time. Of course, I left Yale on a Friday to report to Sandia in California on Monday, and still had to get the thesis typed and submitted later that year. Richard is famous for encouraging his students not to waste time (Marshall Long is a great example) and in my case it was crucial.

I'm very glad to have this opportunity to talk about my memories of Richard. It was an amazing stroke of luck for me that he decided to take a chance on a "kid from Sewanee." Not to mention the good fortune (years later) of Richard sending me a brochure from Sandia that had come through his office. The booklet described the Combustion Research Facility at Sandia Labs, Livermore, CA. This organization emphasizes the use of state-of-the-art lasers to conduct basic combustion research. Again, Richard judged correctly and I was immediately intrigued. This happened well before I was to graduate and Richard recommended that I visit Sandia without delay. As it turned out, I liked the Livermore job so much I took only one other interview (Sandia in Albuquerque, NM). After touring both facilities I accepted the CA job. I worked in the CRF for nearly 30 years in a dream job doing laser diagnostics development for combustion. I can't overstate what a major role Richard played in the direction of my career and I'll never be able to thank him enough.

Letter to Richard Chang on the Occasion of His Retirement from Yale University, June 2008

Sandia National Laboratories

Operated for the U.S. Department of Energy by
Sandia Corporation

Dr. Roger L. Farrow
Distinguished Member of Technical Staff

P.O. Box 969, MS9056
Livermore, CA 94551-0969

Phone: (925) 294-6136
Fax: (925) 294-2595
Internet: farrow@sandia.gov

Professor Richard K. Chang
Engineering and Applied Science Department
Becton Center
Yale University

Dear Richard,

This is just a quick note for your memory book to wish you many years of happiness and contentment in your retirement. Your students and colleagues will certainly miss your contributions to enriching our professional lives, however.

I also wanted to let you know I really appreciated that you gave me chance to enter the Engineering Department at Yale and to work for you. You took quite a chance on a "kid from Sewanee," a small school in Tennessee not known for science. I will never forget how much this changed my life. If not for you, I would have ended up at Vanderbilt University and most likely would have never made it to Sandia, where I've had a long and rewarding career thus far.

I'll never forget the really nice dinners that you and Sunny were always giving for your students. It wasn't often that we dormitory-dwellers had the opportunity to have a delicious meal followed by a classical music recital! Looking at the photos on your retirement web page really brought back pleasant memories of those days.

Looking forward to seeing you again at your retirement party this June.

Best Regards,

Roger

Roger L. Farrow

Exceptional Service in the National Interest

My Encounters with Richard K. Chang

Roman Kuc
Professor of Electrical Engineering
Yale School of Engineering & Applied Science
New Haven, CT 06520 USA
roman.kuc@yale.edu

In 1979 I joined the Engineering faculty in Yale University as an eager new Assistant Professor. A few days after arriving, I met Richard by the Engineering copy machine in the Business Office, where he introduced himself and said with that twinkle in his eye, "So, you are here to shake up Electrical Engineering?" I took this to be my mission at Yale.

Our backgrounds and approach to problem solving were similar. Richard applied his physical optics background to laser sensing, and I applied my physical acoustics to biomimetic sonar sensing.

I was excited when Richard asked me to participate in his research projects along with Marshall Long so I could add my signal processing tools for characterizing turbulent seeded flows with laser imaging. It was a nice problem and the technical discussions I had with him, and his group always focused on the critical issues. But I always felt that our discussions furthered the research. Pedagogically, I applied Richard's approach to teaching by urging students to ask probing questions and make suggestions. On more than a few occasions their questions gave me fresh insights that led to new ideas.

My time at Yale taught me more about Richard's fine qualities as a person. He was always friendly, never dismissive, always considering suggestions. I try to emulate my wonderful encounters with Richard when I meet with my students

As Ukrainians say, *May memory of him last forever.*

Letter to Richard on the Occasion of His Retirement from Yale, June 2008

Marshall Long, Ph.D. Yale '80
Professor of Mechanical Engineering
Yale University, New Haven, CT 06520 USA
marshall.long@yale.edu

Yale University

Department of Mechanical Engineering
P.O. Box 208284
New Haven, Connecticut 06520-8284
Phone: (203) 432-4229
FAX: (203) 432-6775
email: marshall.long@yale.edu

April 21, 2008

Dear Richard,

In trying to write a few words on the occasion of your retirement, I realize this is a very difficult task. It is difficult because, as hard as it is to for me to believe, our history now goes back more than thirty years. It is impossible in a few paragraphs to capture all of the ways you have influenced me, both personally and professionally.

Our history starts, of course, with my graduate school career. It was in the fall of 1977, my second year of graduate school, when I joined your group to work with Bruce Webber (and Phil Black as a consultant) on the low-light-level camera project. Roger Farrow held court in Becton 004 as "Senior Graduate Student," even though he looked like a teenager. Dan Murphy was the strong, silent type in 006, and Bruce and I worked in 002. It was a very busy time. I remember the usual graduate student day — I would try to get there early, stay until dinner, and then go back to the lab after dinner. We all did it. Saturdays were workdays too (for you as well). I remember Susie running the details of the lab with complete authority, her voice calling down on the intercom to the next one in line to run to grants and contracts or walk something through to purchasing. With your guidance, it was one of the most rewarding times of my scientific life. Never again would I have the luxury of devoting all of my energies to a single project, without all of the other responsibilities that tend to accumulate as one moves on in a career.

It wasn't all work of course. Dinners at your house or lab parties were always a big deal. These were not infrequent and were always memorable occasions because of the interesting mix of people, the delicious food, but most of all, because of the hospitality shown by you and Sunny.

Our relationship did not stop when I completed my degree. With your encouragement and support, I had a chance to stay on at Yale. Your generosity in helping me get started in my scientific career truly went beyond anything I could have hoped for. Your willingness to share your equipment and your lab space allowed me to get a running start. Your insistence that I establish a research direction of my own was just one example of the unselfish and wise council that you so freely gave. You helped me navigate the dangerous waters of the Yale tenure process with sound advice and constant encouragement. The cliché that 'I would not be where I am today without your help' somehow does not adequately convey the magnitude of your guidance in my case.

In addition to your influence on my scientific career, your examples and advice have also had a great impact on our family. In 1984 when I was eligible for my first triennial leave, you encouraged me to take the opportunity to travel. Since at that point it was not clear to me that I would ever get another leave, we took Shelby, then two years old, and set off on our most ambitious around-the-world trip ever. Since then, whenever we have had the opportunity to travel as a family we have packed up the kids and gone. Traveling together has been one of the most important things that we have done as a family and all three of our kids continue to have the travel bug. Shelby is leaving for Nepal tomorrow for a trek to Everest Base Camp.

As the years have passed, you have remained a supporter and friend, never once expecting any kind of repayment for all that I certainly owe you. You have had a strong role in helping me along in all phases of my career and, for that matter, my life. For this I am truly grateful. Even thirty years on, I must admit I still often feel like the student, hoping that I can live up to your high standards.

With all best wishes,

What I know about Uncle Richard

Xiaobao Huang, Ph.D.
Senior Facilities Engineer for General Motors
Material Scientist for General Motors Research Lab
xiaobao.huang@gm.com

I first met Uncle Richard in May 1980 when I was in my 2nd year college in Guangzhou, China. That year I got a surprise phone call from Pang-Hua Kevin (Uncle Richard's son) from a hotel near downtown Guangzhou saying his family was there to visit. I rushed to the hotel by bus and met with all of them: Uncle Richard, Auntie Sung-Wen, Pang-Hua, Pang-Mei (daughter 1) and Pang-Yuan (daughter 2). My first impression was that Uncle Richard didn't talk much but observed a lot. Whenever he talked about anything, it made you think, and you would have no choice but to follow his thinking process. He was a true scholar type and seemed very sensitive to just about everything.

When we talked about where to go and see in downtown Guangzhou, I had no idea at all. As a student (still being rough on all edges), not much of any travel experience then, I mentioned some movies or zoo type of visit (who would fly half the world to see a movie or zoo?). I was stubbornly arguing about movies when Uncle Richard slowly said: how about the "Guangzhou Import and Export Fair" or "Sha Mian" the historical strip near the north side of Zhu River? These two items were remarkable icons of the city of Guangzhou. I thought wow, how smart that was. There was no more argument. What have I learned? The best way to win an argument: present a better idea, simple.

The second time I met with Uncle Richard was in the summer of 1986 when he lectured at Shanghai's Fudan University graduate school about his research. Prior to giving his lecture, his family and I went sightseeing and I was able to further my learning experience with Uncle Richard. While we were touring the city of Suzhou by a minivan, Uncle Richard reminded his family, whenever you see a western-style toilet, use it. I didn't realize how important it was since I grew up in China; I thought it was no big deal. I now know how important a western-style toilet can be for others when traveling. Along the way, humor constantly flew out of him with fast thinking, I don't remember all the details but one that always came back to me: when I was being dropped off at a local market, he said "Xiaobao, don't pick up too many women when we pick you up later". Of course Suzhou is the famous place to produce a lot of pretty women since ancient times, the way Uncle Richard linked the history and current event with impromptu made it never boring to be with him around.

Sensitivity for others: In 1986, while I was applying for graduate school in U.S., I met with Uncle Richard and Pang-Mei when they were visiting Shanghai. My father and I took the train from Beijing to meet with them. When my father and I met with Pang-Mei in the hotel, we discussed how to fill out the paperwork. My father was left out of the conversation. Feeling that he could not help, my father decided to walk outside. After a short while Uncle Richard came in and asked where my father was. As soon as he found out that my father went to take a walk, he immediately told us "You cannot do this, don't leave him alone, we need to find him now." I realized then how sensitive Uncle Richard was; this sensibility was part of him naturally and really impacted me for the rest of my life.

Willing to help others, not counting cost: My first interview with the US visa office was December 8th, 1986 during the graduate school admission process. During the interview, the officer pointed out that although I had partial sponsorship by the school with tuition waiver, I would still need someone in the US to provide a guarantee for living expenses. I called "collect" (back then, there was no cell phone nor social media means, everything was landline) in the middle of night to ask Uncle Richard and Auntie Sung-Wen for help. They answered the call and assured me they would help. Sure enough, in about a week, an express mail package arrived to my home in Beijing. In it, there was one piece of paper with Yale letterhead, and Uncle Richard's handwriting stating "We will sponsor Xiaobao Huang from Beijing for his study in the United States." Uncle Richard and Auntie Sung-Wen's extended family had arranged to financially sponsor me. The rest is history. I was able to obtain the visa and came to the U.S. for study. My first semester was completely supported financially by Uncle Richard and Auntie Sung-Wen's extended family. After graduate school, I found a job in Chicago and met my wife. I have worked in the U.S. ever since. We had three children in the U.S. Without Uncle Richard and Auntie Sung-Wen, there wouldn't be our family! I always appreciate what Uncle Richard and Auntie Sung-Wen did for me.

Singapore, 2023.

Memories of Richard K. Chang

Jia-Biao Zheng
Professor of Physics (Ret.),
Fudan University, Shanghai, 200433 China

Richard visiting me at International Student Center when I first arrived at Yale.

I first met Richard in 1980 when Richard attended the International Laser Conference in China. I had the occasion of hosting his visit to the Shanghai Institute of Optical and Fine Mechanics when we conversed and learned about each other's academic focus and research. I was already granted a two-year state-sponsored working-abroad opportunity at the time, and Richard said to me "hope to see you at Yale in the near future" before he left. Soon after he returned to the US, I received his formal invitation letter and IAP-66, inviting me to participate in his research lab as a visiting scholar. Thus started our 40-yearlong professional relationship and our deep friendship.

Richard and I attending a Yale graduation ceremony, David Leach is looking on.

During my years at Yale, I received tremendous help from Richard both inside and outside of his lab. Richard arranged and paid for me to attend Yale's summer school to improve my English, he also invited me to his Hamden home to spend time with his family on numerous occasions, giving me a sense of "home" in a foreign country. After learning of my wife's chronic ailment, Richard sent an invitation to her to visit the States and arranged for her to see doctors at the Yale Medical Center to get some comprehensive diagnosis and treatment suggestions. Richard and his wife Sunny even helped our daughter's college application, hosted her during the summer before she started at Smith College, and attended her graduation 4 years later. Over the years, Richard and I grew from research collaborators to personal friends, I have tremendous respect for Richard's professional achievements and academic integrity, but it was also his heart and his selfless acts to help others that made him forever a friend, a mentor, and a role model.

Richard and Sunny visiting us in Shanghai.

In Memory of Richard

Tom T. Chen
Professor of Physics (Ret.)
National Tsing Hua University
Hsin Chu, Taiwan, 30013 ROC
ttchen@phys.nthu.edu.tw

Prof. Tom Chen, Huey-Ming Tzeng, and Jia-Biao Zheng in the RKC Lab in the 1980s.

My wife and I met Richard and his family when they visited National Tsing Hua University in Taiwan in 1980. We have been friends ever since. Every time they came to Taiwan, they would make an effort to meet with us to spend memorable times together.

Richard was the one who brought up the field of surface enhanced Raman scattering to our research team. By his invitation, I went to Yale University as a visiting scientist in 1981. I had a fruitful year there. Because of our harmonious cooperation, I went back to his laboratory in the following two consecutive summers. I thank him deeply for his inspiration and guidance. He was kind, enthusiastic and energetic. He never let his health condition get in his way of research. Now he has left us physically, but his passion for science will be remembered for a long, long time.

Dorothy Chen, wife of Prof. Tom Chen, and Sunny Chang.

Life at Chang's Lab (1980–1984)

Huey-Ming Tzeng, Ph.D. Yale '85
Principal Systems Engineer at KLA-Tencor
San Jose, CA 95120 USA

I first met Professor Richard K. Chang one afternoon in March of 1980 in the Power Mechanical department building of Tsing Hua University, Taiwan, where I was a graduate student in Power Mechanical Engineering with a focus on flow and heat transfer. Richard was on sabbatical leave at the time from Yale University and was a guest speaker in our graduate school course on Atmospheric Diffusion, taught by Professor B.T. Yang. Richard's presentation of the results from a 2D visualization of laminar air jets based upon elastic scattering of aerosol particles captured with stunning detail edge vortices — at the time — was state-of-the-art. I had questions about the adequacy of flow visualization approach that Richard had presented to quantify turbulent mixing, and after an extended discussion that subsequently continued in Professor Yang's office, Richard asked if I wanted to work as a research assistant in his lab at Yale University in the U.S. I was so drawn to the beauty of the study he presented that I didn't think twice before agreeing to his offer.

Before leaving for Yale, one of my classmates at Tsing Hua warned me of what he had heard about Richard — a tough disciplinarian, demanding and high standards for his students and mentees. A young graduate student with a newlywed wife, newborn son, moving to a foreign country to be a research assistant sounded like a tough journey ahead. But perhaps I was too young at the time to comprehend those challenges. Ultimately, he was not wrong; Richard was tough and demanding, and my first year at Yale was a struggle, but Richard was also a caring mentor who pushed me to be better. I think because I fully believed in Richard's vision that I had the skills to pursue the graduate work that was interesting to me, that I was able to find the confidence to overcome the adversities I faced, and to take on new challenges without fear.

Richard's mentoring for me began well before I stepped foot in his lab at Yale. It was evident, though we had just met, that he cared for my life and my family outside of academics. After our discussion in Professor Yang's office, he immediately began arrangements with his colleagues, including Professor Marshall B. Long, at Yale, to arrange for my admissions and transfer to the U.S. When he learned that I had a wife, Helen, and a newborn son, Tony, he asked to meet Helen to get to know her. When he learned that I could not afford the tuition for an Ivy League graduate school in the U.S., he arranged for a scholarship for my tuition and a living stipend, also offering me teaching assistant jobs on the side to help me make ends meet. Alongside Richard was his wife, Sunny, whose warmth complemented Richard's disciplinarian demeanor. Later, when Helen and Tony arrived in

Marshall Long, Huey-Ming Tzeng, and Richard Chang, circa 1980s.

Marshall Long, Leslie Long, Shelby Long, Huey-Ming Tzeng, Helen Tzeng, Tony Tzeng, and Richard Chang, circa 1980s.

New Haven to meet me in 1981, Richard and Sunny would invite my family to graduate student and post-doctorate researcher parties held at their home and introduced us to other visiting professors and scholars. Sunny was very kind to Helen, helping her adjust to the cultural differences in the U.S. Looking back, we felt privileged to have Richard's and Sunny's mentoring, not only in the lab, but in life.

I arrived at the Yale campus in July 1980 and spent my first summer in the U.S. in Richard's laboratory, located in the basement of Becton Center. The labs were protected by locked gates, a remaining vestige from the protests and violence on campus from the Vietnam War. Before leaving Taiwan, Richard had advised me on which courses to take to prepare myself for work in his lab, knowing that I had little practical exposure to the area that I would be focusing on. His philosophy, however, was that hands-on work in the lab was the best way to learn. My first "test" was to convert a Ruby laser to operate in an unstable mode. Richard gave me some background reading and introduced me to Dan Good, a senior post-doctorate researcher, who would be my coach on laser and laboratory safety and operating of the Ruby laser system. Then, I was left to deal with the laser system myself. Although I had heard about the laser system, this was my first time handling a real big laser system with a six-foot tall power supply and control. Yet, Richard's confidence in me paid off; my first success in the lab gave me the confidence later to modify and set up other laser systems for different projects. This would prove to be one of the most important lessons I learned from Richard's mentoring throughout the years, that only students willing to try and take a risk have a chance to succeed in the RKC Lab!

The first summer passed and fall 1980 came. My initial confidence from my early success at the Ruby test began to waver as my lack of experience began to affect my ability to meet the impending project deadline. I had no hands-on training in digital electronic circuits and was struggling with the digital circuit design/electronic lab. With the project deadline looming, I was in a desperate state of panic. My initial approach was to throw 100% of my time in that electronic lab, in hopes that the time spent trying would manifest results. That was when Richard pulled me aside to share his advice on time management between work and study from when he was a student. He taught me how to set up a plan, to work systematically to efficiently achieve results. His advice worked. I finished the project just in time.

Throughout my time at the lab, Richard would continue to help me build confidence in my own abilities by presenting me with more difficult tests and challenges, in the form of different projects. The success of these projects was important to me, and each success gave me a confidence boost I needed to continue pursuing more challenges. This is the hallmark of Richard's training: take on a challenge, overcome adversities, and then take on a new challenge. I learned from him that the process of investigation follows the pattern: model, test, review results, more questions on the model, and more testing until all questions were answered. As I completed more projects, Richard continued to provide me with more opportunities to explore more complex topics and areas. Later in a series of experiments that ultimately constituted my thesis work on exploring using lasing droplets for size determination, mass transport, and interactions with laser radiations. These experiments were made possible because of a stable droplet generator system I developed. The results we saw surprised us, and perhaps more surprising was that we had fun doing it. I could enjoy the fun of physics really because of the foundation built in the first year of working in RKC lab, under Richard's tutelage and training.

I would be the first student from Taiwan who graduated with a Ph.D. degree under Richard. After me, there was another student from Taiwan, Professor Wen-Feng Hsieh, who also graduated with a Ph.D. degree. Back in 1984, the job market was just starting to thaw after a long slump from the war. Richard understood I had a family to support and that I needed to progress in my career. When the time was right, he helped me explore several job offers, providing strong recommendations on my behalf. On one of his recommendations, I was able to secure a job offer at IBM San Jose Research Center in California, where I moved with my family.

To say that Richard influenced my life would be an understatement — my life today, my career and my journey here exists because of that one afternoon in Taiwan, when I met Richard. That was the day I met a mentor and lifelong friend.

Fudan University, Shanghai, China, 1986

Chang Family Album

Richard, Sunny, Jia-Biao Zheng with Fudan University professors.

Zhongyi Hua, vice president of Fudan University, Richard, and Xide Xie, president of Fudan University, where Richard is made a Consulting Professor.

Prof. Y.F. Li, Jian-Zhi Zhang's mother, Sunny, Richard, with Fudan University professors and students.

Professors from Fudan University, S. X. Qian, Zhi-Ming Zhang, Sunny, Jia Biao Zheng, 1986.

Richard Chang — A Magnificent Man who Made a Difference in My Life

Wen-Feng Hsieh, Ph.D. Yale '88
Professor (Ret.), Department of Photonics,
National Chia-Tung University (formerly), National Chiatung Yangming University,
1001 University Road, Hsinchu, Taiwan, ROC
wfhsieh@mail.nctu.edu.tw

An incident that changed the course of my career was a letter from Professor Shou-Yi Wang, my master's degree advisor from Taiwan's National Tsing-Hua University, asking me to meet a Professor Richard Chang from Yale University. It was 1980 and Richard was spending a sabbatical at the National Tsing-Hua University (NTHU) Physics Department in Hsinchu, Taiwan. My first impression of Richard was that his eyes sparkled when he spoke about physics and his enthusiasm and love for physics were overpowering. Not being very fluent in spoken English at that time, I remember remaining mostly silent, just staring at the sparkle in his eyes. Surprisingly, Richard welcomed me to join his group after my military service, contingent, of course, on my acceptance by Yale as a graduate student.

Unfortunately, being born in a remote area of Taiwan and with farmer parents, who had never gone to school, much of my learning was self-taught. Although, I passed the TOEFL criterion set by the Taiwan Ministry of Education to study abroad, my score was far below what was acceptable for Yale. Thus, in 1981, instead of going to Yale, I accepted a job with the Telecommunication Laboratories (TL) of China Telecommunication Company, a government research institute under the Taiwan Ministry of Communication, to study fabrication and characterization of optical fibers.

In 1983, while attending a conference on atomic and molecular science, I ran into Professor Tom T. Chen, the co-advisor for my master's degree from NTHU. He told me that Professor Chang still asked about my interest in going to Yale. Scientifically, I was just at the point where I needed to go beyond linear optics and was trying to teach myself nonlinear optics, which constantly occur inside optical fibers. I knew that studying under Richard would be a great opportunity to learn from one of the best scientists in nonlinear optics and quantum optics. However, the decision was more difficult personally, as I had to leave my wife and seven months old son behind in Taiwan when I arrived at Yale's Becton Center, the Council of Engineering and Applied Science (CEAS, now the School of Engineering) in July 1984.

Before I went to Yale, several former classmates of mine from the Physics Department of National Cheng-Kung University (NCKU) in Tainan had expressed deep concerns about my pursuing a Ph.D. from Yale. One of them asked a Yale alumnus who had just received his Ph.D. from Applied Physics how he felt about Richard. His suggestion to me was "Don't join the Chang Group. It is the toughest group in Applied Physics and Professor Chang is a harsh professor. Several students have been expelled from his group already!" I understood what he meant because I had met Richard and knew that despite his warm and enthusiastic demeanor, he would not tolerate mediocrity in science from his students. I was worried about that too, but then I had already committed myself to working in Richard's group.

Richard and I agreed that it would be helpful for me to attend Yale English Summer School before the fall semester began in September. Coincidently, at that time, Professor Tom Chen was taking a sabbatical from NTHU and working in Richard's laboratory. Therefore, after my English classes, I worked with him part time in the lab to learn how to operate the Optical Multichannel Analyzer (OMA) system. The OMA system included a software running in a PDP-11 minicomputer of the Digital Corporation and hardware of a Silicon Intensify Target (SIT) detector connected to a Spex-1877C spectrograph for performing Raman spectroscopy. Since I had the TL experience of setting up and writing computer codes for taking absorption spectroscopy and temporal response and baseband frequency measurements of optical fibers, I was quickly able to operate the OMA system. However, I did not get involved much in the scientific discussions. Perhaps that was due to my lack of fluency in spoken English at that time.

Six Lessons I Learned from Richard:

I learned many things about being a professional from Richard, but 6 lessons helped me in my career as a professor: *(1) be open-minded; (2) be strict about safety; (3) be willing to give students the freedom and knowledge to make an informed choice; (4) learn from experts; (5) be able to face problems directly; and (6) not be ashamed to have someone edit your writing or teach you something.*

(1) Be open-minded: In my first semester at Yale, I was assigned to work with a visiting professor from Fudan University in Shanghai, Professor Shi-Xiong Qian. Prof. Qian and I were to study the stimulated Raman scattering (SRS) from micrometer-sized water and ethanol droplets. Prof. Qian wanted me to run the OMA program to take spectral data for him, while he did all the optics alignment. I had read the OMA computer code written by Prof. Marshal B. Long, so I knew it was possible to write a "Macro" command to perform a sequence of command lines that could do all the jobs of exposure SIT camera with a pre-set

exposure time, subtracting background, saving data, and setting a certain delay time, etc. Due to some miscommunication, the meeting time to discuss science with Richard occurred during my class time. Thus, Richard thought I had no contribution to the study of SRS from micrometer-sized water and ethanol droplets and he was unhappy about my performance in the laboratory. At the end of the semester, I explained to Richard about how my class had conflicted with the meetings. Then he assigned a new independent research project for me to do in the second semester: measuring SRS from concentration dependent sodium sulfate aqueous droplets. Unfortunately, I failed to observe any of the Raman signal related to the sodium sulfate, although I did detect water Raman signal at the high frequency region. Due to my apparent lack of progress in the lab, Richard asked me to find a new advisor.

Dr. Huey-Ming Tseng, Richard's first Taiwanese student who came in 1980, left Yale to work at IBM Almaden San Jose, California in late 1984. He returned to Yale for his graduation in May 1985. Huey-Ming and his wife Hao-Lan told me that Huey-Ming had encountered a similar situation, probably even worse than mine, so he persuaded Richard to reconsider my case. Richard was open-minded and decided to give me another chance.

I started my new study in the middle of July 1985 with an almost empty laboratory. With the help of Mr. Joseph Lam (a technician in Richard's group, the middle person in photo below), I rebuilt a new PDP-11 microcomputer to start my OMA system to study micro-droplet lasing pumping SRS from R6G in ethanol droplets. Simultaneously, I cooperated with Dr. Carol F. Wood (a new postdoc, Chemistry PhD from Columbia University) on taking fluorescence spectra of pyrene droplet excited by a nitrogen laser at 337 nm. I helped Carol to solve the problem of electromagnetic interference (EMI) from nitrogen laser firing that were causing a shutdown of computer and optics alignment. Simultaneously, Carol helped me to gather all the optical components needed for my experiment. Richard asked Carol to evaluate my performance. Fortunately, Carol gave me a positive recommendation, and thus I survived in Richard's group. I soon had an opportunity to show Richard that his decision to keep me on was not in vain.

My 1st year at Yale: Taiwanese Yale students gathering (from left to right) Dr. Huey-Ming Tseng (Richard's first Taiwanese student, who came in 1979), Dr. Tze-Jia Wei (now the CEO of Taiwan Semiconductor Manufacture Company, tsmc), Mr. Huang (1984 Master student of Chemical Engineering), and Dr. Kevin Lee (now with Intel, my same year Applied Physics classmate), and myself.

(2) Be strict about safety: Although Richard was open-minded, he was not flexible about safety in the lab. I heard he had once asked a post-doc to leave his lab when he could

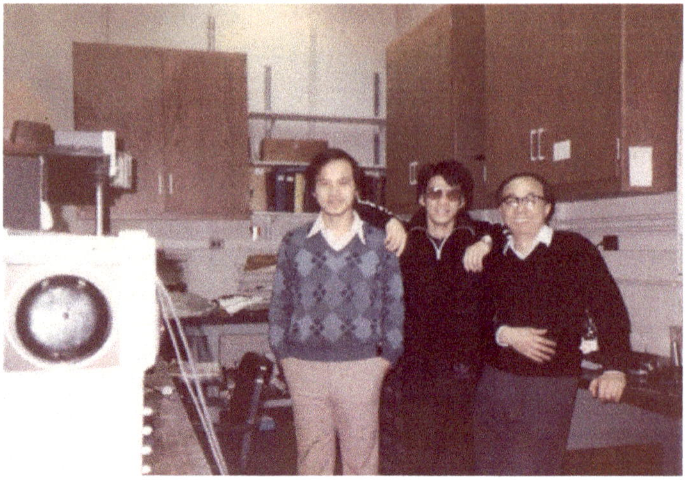

Prof. Jia-Biao Zheng, Joseph Lam and myself (from right to left) in the laboratory.

not get the post-doc to stop wearing slippers or walk barefoot in the lab. My experience with his strict lab regulation was with Charlie. Charlie was a senior technician in the Becton Center basement machine shop where I went quite often. In 1987, he delivered a speech at his retirement party hosted by Richard in the conference room of Becton Center. He said the reason why he was so strict with the students, who used the machines in his shop, was because one of Richard's former students, someone from Russia, had cut off his finger as he was doing machine work in an exhausted condition. Thus, he would never permit anyone, not using a vise to clamp a workpiece on the machinery, in his machine shop. Richard seemed not to have remembered this story. In fact, I later heard from Sunny that Richard did ask the student to leave because he had not taken lab safety seriously and had almost lost his finger...he severed it partially, not completely.

(3) Be willing to give students freedom to seek knowledge to make an informed choice: In addition to doing the microdroplet laser pumping SRS from R6G in ethanol droplets, Richard also gave me five research topics to consider for my dissertation. I immediately chose the study of stimulated Brillouin scattering (SBS) from micro droplets, because I had already surveyed and purchased a Quantronix mode-locked picosecond (ps) Nd:YAG laser with 82 MHz repetition rate for studying the nonlinear optical effects, e.g., second harmonic generation (SHG), SRS and SBS, inside the optical fibers, when I was at TL Taiwan. I made an appointment with Richard a week later to present my thoughts on SBS. However, Richard did not listen to my thoughts on SBS, but rather gave me a hard time for about 30 minutes. He said that he had purposely given me 5 topics to have me think thoroughly over all of them, since human beings were parallel thinking creatures. He wanted me to make an informed choice based on knowledge of all the five topics. I then spent a week of my spare time in the Kline Biology Tower library studying all five topics. Richard's philosophy of "informed choice" inspired me to give my graduate students (especially my PhD students) freedom to pursue their own research topics in my 30 more years of teaching at National Chiao Tung University (NCTU) Taiwan.

(4) Learn from experts: Richard always said, "Learning from the experts is one of the fastest and the best ways to gain new knowledge." To learn more about how plasma was generated

and how it developed into the detonation wave when the interaction of high intensity laser beam with transparent liquid droplets occurred and to understand its possible applications, Richard and I visited Dr. David C. Smith, a scientist and owner of Golden Wings Aviation at Sikorsky, an aircraft manufacturer in Stratford, Connecticut. To know how the time-resolved spectroscopy was performed, Richard took me to Prof. Robert R. Alfano's lab in City College of New York (CCNY). Prof. Alfano had a homemade ps Nd:YAG laser running at one minute a pulse. The few ps pulse-width laser pulse was used to excite a sample, whose emission through a bandpass filter was then detected by a Hamamatsu single sweep streak camera with ps temporal resolution. Later, Richard bought a Hadland streak camera with both single sweep and synchronous scan units with ns/mm temporal resolution and a framing unit at Yale to observe temporal evolution of laser-induced bulge and breakdown spectroscopy of liquid droplets.

One day, Richard got a short-term grant from the Air Force to study laser-induced breakdown of floating carbon micro-particles in the air and to compare laser-drilling holes on tungsten wires both in air and in vacuum. It had to be completed within a week or two, so we needed quickly acquire an electric dipole trap for levitating carbon particles. We visited Prof. Stephen Arnold at New York University and received his help in providing an electric dipole trap and technique to operate it. Inwardly, I had marveled at how helpful all the "experts" we had contacted were, but in 2017 when I had the opportunity to help Richard clean out his files in his VT home, I realized that scientists helping each other was a two-way street because I saw the help Richard had given to them on different occasions.

(5) Be able to face problems directly: Another lesson I learned from Richard was to face problems directly. When my dissertation topic was confirmed, the 10-year old Molectron Q-switched Nd:YAG laser was broken. Richard had me call the technical engineer of the company for help. Unfortunately, the manufacturer was sold to a new company and no one was able to come for technical service. That led Richard to say one should be able to take all the equipment used in one's experiments apart and re-assemble them, not just know how to operate them and to understand their operational principles. He suggested that I should try to repair it myself. Luckily, I was able to track down an engineer who belonged to the original technical branch in the new company to gather the stocked blueprints of the electronic circuitries and optical layouts for our laser. From those blueprints, I contacted the original suppliers for repairing optical parts, e.g., having Crystal Technology Inc. re-coat the mirrors of the laser resonator, the laser crystal and the flash lamp's cavity of the pumping unit and finding replacements for the charging capacitors and the necessary electronic parts. With Joe Lam's help, we finally fixed the laser so that it was like new. This experience of rebuilding the laser provided me the necessary knowledge to design a timing circuit to separately externally trigger the pumping flash lamps and the Q-switching unit with a proper time delay for the time-resolved spectroscopy by using the streak camera. This experience also gave me the courage to study and to build a femto-second (fs) soft-aperture self-mode locking Ti:sapphire starting from theoretical foundation for my own nonlinear optics research lab under limited budget in 1993 at National Chiao Tung University in Taiwan.

The attitude of directly confronting problem was also reflected in the choice of Prof. Hans R. Griem of the University of Maryland as the peer reviewer for my PhD dissertation, which was required by the Yale graduate school, since I learned most of the plasma spectroscopy from his book "Plasma Spectroscopy" published in 1964.

(6) Not be ashamed to have someone edit your writing or teach you something: The sixth lesson I learned from Richard was not to be embarrassed or ashamed when someone is trying to help you make something better, such as your writing style or your experiment. Once, I went up to Richard's fourth floor office for us to revise my dissertation. I saw Richard sitting on a stool against the desk of Susie (his administrative assistant) in Susie's office, which was next door to his. Susie had a red pen in her hand marking over almost all of Richard's manuscript and was saying, "I told you, this verb cannot be used here in this circumstance." Later, Richard said to me upon entering his office that it was not shameful to have someone critically revise your paper, as he busily revised the draft of my dissertation. No doubt, every page of my dissertation was covered with red marks and almost no sentence was untouched. He further told me it is very difficult for one to revise one's own article repeatedly, because one has already gotten used to the logic of the article one is writing. He always read his last version of the manuscript, before submission, sentence-by-sentence, backward by paragraphs, and proceeding to the next paragraph throughout the article. In this way, he avoided taking words for granted in revising the article. I followed his advice throughout my teaching career at NCTU and even passed it on to my PhD students.

A few facts about Susie Engelman should be mentioned here. She was extremely organized, intelligent, competent, and well informed. She had an advanced degree in English literature and was absolutely devoted to Richard and his students. When it came to accounting matters, she was relentless. Once, she spent a whole weekend trying to reconcile the numbers in one of Richard's contracts. She had been off by just **one** cent! It was common knowledge that Richard's contract monitors often called Susie to get the correct numbers, when they had to write final reports. Susie worked with Richard for 19 years and was truly a bedrock of the RKC Lab family.

Some Interesting Tales from the RKC Lab:

The Intercom: In the 1980s, Richard's and Marshal Long's labs and offices in the basement of Becton Center had intercom speakers connected to a telephone line. Susie Engelman (Richard's administrative assistant) and Richard had fourth floor offices from which they could dial a telephone number, then speak through the intercom to call anyone in the basement labs and offices. Although this system was primarily set up to facilitate communication, it was also because Susie had a back problem which made her frequent trips to the Grants and Contracts Office or the Business Office (in another building) very tiring, so sometimes students were asked to take documents or purchase orders to the appropriate office. When anyone left the basement, he or she had to tell his or her office mate that, they would call back within 5 minutes should he or she be paged, otherwise, he or she would be admonished. There was an occasion in 1987 when I went to the office around

9:00 a.m. Michael Winters, my officemate, greeted me with "Oh! Oh! You are in trouble!" because a note from Richard said, "Come see me immediately!" had sat on my chair in the office instead of intercom paging. Michael thought that it would be a more serious situation than being paged on the intercom. I was interested to note that when I visited Yale again in 1993, the graduate students I spoke to (Alfred Kwok, Ali Serpengüzel, and Mohiuddin Mazumder) had no knowledge about the intercom system or the biweekly written reports.

***Biweekly written reports*:** In 1986 when Richard was on his sabbatical leave, he told every one of us, David H. Leach, William P. Acker, Jian-Zhi Zhang and me, to submit two sheets of (8.5 by 11 inches) paper report biweekly. The reports had to be written or typed and figures were excluded. Susie would send them by airmail to him. He then made comments and instructions on all our reports and send them back for us to follow his instructions. We always hoped that Richard would not have to travel too often because that would mean more reports and more instructions. With the advent of cheaper international telephone calls, fax, emails and later the internet, Richard's later students were spared the biweekly 2-page written reports, when he was overseas.

***International phone calls*:** Richard sometimes called past midnight from international conferences he was attending to tell me what experiments should be done for a certain topic or to report on some newly discovered results. He would first say hello in Chinese then switch to English. It was hard for me to switch from Chinese to English quickly in the middle of the night. Therefore, after the first one or two calls, I decided to reply only in English to get around this awkward situation. Huey-Ming and his wife were incredulous when I told them this story sometime in early 1990 at San Jose while I attended a CLEO (Conference for Laser and Electro-Optics) conference. Huey-Ming said Richard had never called him at midnight nor spoken to him in Chinese when Huey-Ming was a student at Yale.

***Feng Shui Seats*:** Richard might or might not have known that his students believed in the so-called "Feng Shui" or the lucky seats. I heard from Mohiuddin in my 1993 visit to Yale, that he finally knew the person who turned a bad-luck seat to a lucky seat in Room 009. He told me that when a student was expelled, his seat became the bad-luck seat; on the other hand, when one graduated, his seat became a lucky seat. My seat had been considered a bad-luck seat, since the student using that seat before me had been expelled. Now I understood why my officemate Michael, who took Huey-Ming's seat, looked at me weirdly when I first settled in Room 009. In 1984, when I first joined Richard's group, I did not know about the existence of lucky and unlucky seats, but I believe most of the other students did and were concerned about Feng Shui. I knew that was why Bill Acker took Kevin Wall's seat immediately after Kevin's graduation and Gong Chen moved to my seat after my graduation in 1988.

***Napkin Science*:** Richard enjoyed napkin science with his students at lunchtime. I remember well the fish filet sandwiches at a cafe on Wall Street that we frequented. While

Photo of Mohiuddin and me, taken in Room 009. During my 1993 visit, I stayed in a guest room of Mohiuddin's rented apartment.

Photo of me, Jia-Biao and Richard (in the background) probably taken by Sunny on the Thanksgiving in 1987 at Richard's home (Hamden). Richard taught me how to carve and to serve turkey.

we were eating our sandwiches, he would write on a napkin with a pen to discuss our research findings. This way he could quickly update our research progress, and he called it "Napkin Science." Sometimes, Napkin Science would be conducted at Richard's Hamden home parties or during lunchtime at Clark's Restaurant on Temple Street across from Helen Hadley Hall where post docs and graduate students lived. Unfortunately, I did not keep the napkins, on which Richard wrote.

***CLEO*:** I was surprised to learn that Richard had never attended a CLEO conference before 1987. During a lunch on Wall Street, he told me the reason why. In his early years as an assistant professor, his first CLEO submission from Yale was rejected by the CLEO review committee. He decided then he would not submit another paper unless he was invited by CLEO to give an invited talk. In 1987, he was asked to give his first invited talk and to organize a special session on microdroplet photonics for the CLEO held at Baltimore. He had everyone in the group attend the meeting.

Because of Richard's 1980 sabbatical in Taiwan, I had no difficulty getting offers for faculty positions from the three major national universities of Taiwan after my dissertation defense in early December of 1987. Because my teaching duties at NCTU did not begin until August 1988, I returned to Yale for a post-doc position with Richard in March 1988. Before I left Yale in July 1988 to begin my career in Taiwan, Richard gave me two pieces of advice: (1) Family comes first, always reserving at least one day of the weekend for family and (2) I should work harder from the beginning, since one has only 15 years to reach the top of one's career in academia. I have kept these words in mind, and I believe I have done well towards my family and studied hard following his advice to reach the top of my academic career in less than 15 years and maintained it until my retirement.

Except for discussing our research progress, I did not have much personal contact with Richard before my thesis defense. Because I still had unsolved research problems to explore, I talked to Richard about returning to Yale during my summer recess at NCTU. Richard promised to find funding for my visits, so I did have two short (2 months) visits to Yale in 1990 and 1993 to continue with my research, which resulted in several scientific publications.

In 1998, Richard accepted my invitation to give an invited speech in the first International Photonics Conference (IPC1998) held at NCTU Hsinchu, Taiwan by the Taiwanese Photonics Society. This was Richard's first trip to Taiwan after his 1980 visit. In March 1999 at the Centennial Conference of American Physics Society (APS) held at Atlanta, Georgia, I was surprised to learn from Richard that he had Parkinson's disease. After a short conversation, Richard was very happy to accept my invitation to visit Taiwan again to give an invited speech for promoting and celebrating the 20 years commemoration of founding the Institute of Electro-Optical Engineering at NCTU in 2000. Richard's movements seemed slower in 2000 than in 1999.

The pictures below were taken on October 21st to 23rd, 2000 for the celebration of Richard's 60th birthday. As I remembered, the ceremony was held at the Yale Graduate Club. Bill Acker, standing at the middle of Picture (a) was the Master of Ceremony. He first introduced Sunny to give a welcome speech in Picture (a) and Richard is on the most right. (b) Richard then gave his birthday testimonial and blew out the candles of the birthday cake. All the attendees (more than 60 persons) gave Richard a big applause. From these pictures, one can see that, Richard not only was free from trembling and stiffness, but he was also in good form. He seemed to have his Parkinson's under control.

(a)

(b)

On the next day after the ceremony, some guests continued to party at the Changs' Brattleboro, Vermont summer home. Richard taught me how to grill a big chunk of beef that was later served to guests. The photo below is of Richard and me sitting on a bench on their deck, with the Connecticut River and Wantastiquet Mountain in New Hampshire behind us.

In the summer of 2001, during our family eastern American-Canadian trip, my daughter Shirley, son Ross, wife May and I, visited Richard in Brattleboro. Sunny was in

Paris at that time helping Nicole with their newborn grandson. In the left picture Richard is canoeing on West River and on the right picture Shirley and May are resting on peak of Black Mountain, VT.

We arrived at Brattleboro, VT on July 13th and stayed for 4 days during which time Richard led us bike riding in the Black Mountain nature area and canoeing in the West River, a branch of the Connecticut River. In those days, Richard was exercising normally, and his movements seemed not to be affected by his Parkinson's. We also spent time talking about science and his experiences before leaving China in 1949 and his recent travels in China.

In 2004, the Physics Department of NTHU at Hsinchu invited Richard to give a talk. During that trip, Sunny and Richard were able to visit old friends at NTHU and to revisit marvelous views of Taroko valley at Hualien, where they had first visited in 1980. We took a train to National Dong-Hua University (NDHU) Hualien one day before Richard's talk there. Richard enjoyed a comfortable massage at their hotel, the Mei-Lun. After he gave his talk the next morning, we had a short trip to Taroko Gorge by taxi. Although Richard was still able to carry his own suitcase then, his movements seemed slower.

After teaching at Yale for 42 years, Richard retired at age 68 in 2008. At his retirement party, there were some scientific talks in a conference room prior to the dinner celebration at the New Haven Lawn Club. I was very happy to attend and to listen to the original talk of the evolution and potential of nonlinear optics given by Prof. Nicolaas Bloembergen. It is well known that Prof. Bloembergen was the person who fully developed the theory and laid the foundation for nonlinear optics. Prof. Bloembergen used old-fashioned transparencies on the overhead projector to present his incisive speech on the germination of nonlinear optics and foresight of the strong field physics — the interactions of atoms with ultra-high laser field.

The retirement celebration continued the next day at the Changs' vacation home in Brattleboro, VT. After a garden barbecue, we went for a short walk on the trail along the banks of the Connecticut River at the foot of New Hampshire's Wantastiquet Mountain. Since Richard was escorted by my classmates, I was walking with Prof. Bloembergen at the end of the group. On our way back, I asked Prof. Bloembergen about his arrangements post his retirement from Harvard University — the reasons why he chose the Optics Center at University of Arizona rather than the Center of Research on Electro-Optics and Lasers (CREOL) at Central Florida University, his original first consideration. He replied that their first consideration was finding a warmer place than Boston. Both Florida and Arizona were good candidates. However, Arizona's drier climate compared with Florida's humid one, made Arizona a better choice. I had similar thoughts about my own retirement plans in Taiwan: warm southern Tainan versus colder northern Taoyuan. In 2014, I encouraged Sunny and Richard to retire in Singapore where their son and his family lived, as Singapore was warmer and had affordable help.

During Richard's first week of retirement in July 2008, he had a near drowning episode in a private neighborhood swimming pool. He spent many months in the hospital and nursing/rehabilitation homes receiving different therapies. With the help of acupuncture, therapies (both Eastern and Western) and medication, he slowly recovered sufficiently to enjoy life again. After his accident, Richard's children (Kevin, Natasha, and Nicole) usually accompanied Sunny and him on many different trips abroad. In 2014, May and I had the pleasure of accompanying them on a week-long-trip visiting old friends and scenic spots in Taiwan. I was happy to be treated like one of their children.

We began our journey first with a local outing in Hsinchu (see photo with statue of Buddha) and then with a delicious Chinese meal attended by NTHU Professors Tom Chen

Photo (3/10/2014, from left to right): Joann and Prof. Edward Yen, Prof. Richard and Sunny Chang, Dorothy and Prof. Tom T. Chen, and Prof. Wen-Feng and May Hsieh at Hsinchu, Taiwan.

and Edward Yen and their families. The next day, March 11th, we headed south to Chu-chi, a remote area on the base of Ali Mountain, to visit my hometown, where they met my elder brother and his wife. On the way to visit NCKU, the college from which I graduated, we had a short visit to a famous photographic spot to see wood cotton flowers, where Richard enjoyed photo shooting.

Our destination before returning to Taipei was to visit physics Professor S.-Y. Wang who had retired from NTHU and to meet Richard's cousin Diana, who was doing consultation work in Taiwan. Although Richard was very happy to see them, his busy day-of-sight-seeing and traveling proved to be too tiring as he asked me whether we were in San Francisco. San Francisco was his cousin Diana's hometown and, of course, where he had always seen her previously. Sunny told me that after Richard's near drowning episode, adequate rest was necessary to prevent occasional confusion.

With me driving and May navigating, Richard and Sunny, in the back seats, we had a memorable time traveling to discover the highlights of Ali Mountain, Jade Mountain, Sun-Moon Lake, the Wen-Wu Miao temple of Yuefei and seeing two aboriginal people singing Taiwanese songs. Sunny even bought two CDs from them. (See photos below.)

The above two photos (taken on 3/14/2014, from left to right) are Prof. S.-Y. Wang, Richard and Sunny at Changhua Taiwan. Photo (taken on the same afternoon): Richard and Sunny met up with Richard's cousin Diana Chang (daughter of philosopher Carsun Chang), who was working in Taiwan.

Our sightseeing included much walking, pausing to rest, photographing and good conversation. Most of the hotels we stayed in were in scenic spots and the accompanying breakfasts were generally plentiful and good. I tried to introduce local Taiwanese foods to them whenever possible.

The next time I saw Richard was a little over one year later in June 2015 when I attended the 8th International Conference on Materials for Advanced Technologies of the Materials Research Society of Singapore (ICMAT2015). Unfortunately, in November 2014, Richard had a terrible fall where he hit his head hard. That accident resulted in a severe concussion and a 7-week hospital stay where doctors worked to adjust his medication to prevent future seizures, which were a common result of such head injuries.

It was my first visit to Singapore and Richard and Sunny wanted to show my wife May and me some of Singapore's tourist attractions. We met at the famous Gardens by the Bay. (See photos below.)

Although Richard sat in a wheelchair, he could walk by holding on the wheelchair for a short distance. We very much enjoyed taking photographs in the Gardens by the Bay and

the Marina Bay area. We also spent an afternoon visiting Yale-NUS College, a liberal arts college in Asia, established by Yale and the National University of Singapore (NUS).

Richard and Sunny's 2015 annual trip home to Hamden, CT proved to be a difficult two and half month's journey for Richard. They were happy to see old friends and family, but the cold November New England weather and the long trip (almost 30 hours) back to Singapore from Hamden took a toll on Richard. In late December and early January, he was hospitalized with a high temperature and very low blood pressure.

I accepted an invitation to give a talk in the World Congress of Smart Materials held in Singapore on March 4th to 6th, 2016. Therefore, May and I were able to see Richard again in Singapore for a week before my retirement in August. Although Richard had returned home from the hospital, he was weak and slept almost all the time during our visit. Occasionally, he would wake up and join us. Sunny took May and me to visit Singapore's Botanic Gardens, an UN World Heritage site and a wonderful place to picnic and to photograph.

Picture (a) below is the photo of Sunny and May taken at the Symphony Lake with the "birds of paradise" flowers, Richard and Sunny's favorite flower. A close-up is shown in (b). On our last day in Singapore, we went to the East Coast Park, while Richard was sleeping at home. The East Coast Park had trails for bicycling and hiking and there were two-person bicycles for rent. Sunny said we could rent them to bike around the trails next time when Richard got better. Picture (c) is the photo of Sunny taken at a snack shop and the background is the harbor lining up with plenty of cargo ships. To end this trip, we enjoyed a dinner with Sunny and Richard, and then took a picture with Richard after the meal, shown as Picture (d). I believe that Richard was so happy to have us visit and did his best to be awake for a short moment with us. This picture is the last photo we have with Richard.

(a)

(b)

(c)

(d)

After Richard and Sunny decided to stay in Singapore, they arranged to move Richard's documents, books, and files that contain lecture notes, papers, project contracts and correspondence, into cartons from their Hamden home to their country home in Brattleboro, Vermont. In 2017, Sunny bought airplane tickets for May and me to have a 2-week vacation in the foliage season of Vermont and to help sort out Richard's Yale documents.

I am grateful that Richard brought me into the world of academic research and that Sunny and Richard always showed concern about both my work and family life. As in the Chinese saying goes "Be a teacher for one day and be a father for life," they are like parents to me. Sunny spoke for Richard under the banyan tree on their 2014 Taiwan visit that I am like a son to them. This may be the reason why Sunny trusted me to sort out Richard's documents and collect Richard's lecture notes.

On this 2017 trip, I brought back three digital photocopies of notes from three lectures, including "Special Topics in Optics", "Optical Properties of Semiconductors 2004", and "Quantum Electronics 2005." Although almost all of those lecture notes are incomplete, I started to work on "Special Topics in Optics" and "Optics and Practical Optical Devices" with an outline of their content sent to Sunny in November 2017. This was about 60% completed by the end of 2019 and I had planned to visit them to discuss the contents in April or May 2020. Sadly, I heard from Sunny the sad news that Richard had passed away on the afternoon of February 16, 2020, a day after I had written them of my concern about the Covid-19 virus in Singapore. Richard is deeply missed, but my memory of him is very much alive.

My Uncle Richard

Li-Ming Pu
Head of People Development
Google, Alpharetta, GA 30022 USA
liming@google.com

My Uncle Richard was someone very special to me. As I was entering high school, Richard started spending much more time in Brattleboro, Vermont and we developed a closer relationship. I very much enjoyed playing in his house and having family gatherings — which were quite often — all eating in the kitchen as one big happy extended family across multiple generations. Richard seemed at ease in Brattleboro, often slowing down to enjoy the beauty of the river and the foliage.

When it came time for me to apply to college, both my Uncle Richard and Auntie Sung-Wen played supporting roles in the process — providing guidance ranging from practical pearls of wisdom to the occasional reality check. During my senior year in high school, right as I was deep in the college application process, I took Advanced Physics. The stress of college applications side-by-side with what seemed to be gnarly topics seemed insurmountable, but my uncle Richard was kind enough to take me as one of his students and diligently tutored me throughout the semester so that I would not only round out advanced physics with a high mark, but that I would also develop an appreciation for the science.

Together, Uncle Richard taught me about vectors, laws of motion, mass, force, and weight. We would bounce balls together to illustrate different concepts. He was patient but he also challenged me to go beyond the textbook (often creating different scenarios so that I would truly grasp the concepts). In reflecting, the time and bond were special, memorable, and showed the loving character of my uncle. I do miss him.

Reflections on Richard Chang

Robert Welch, M.D.
Cambridge, MA 02140 USA

November 20, 2022

I have had the good fortune to have known the Chang family for over 40 years. Kevin and I met as freshmen at Harvard, and we became close friends. I first met Richard and Sunny at their beautiful home on Paradise Avenue in Hamden, Connecticut in November 1979. Richard and Sunny graciously welcomed me into their home, their lives, and their family, and we became lifelong friends.

Over the years, I was lucky to spend much time with Richard, Sunny, Nicole, Natasha, and Kevin in Connecticut, at their lovely home in Brattleboro, Vermont, and in Singapore. I have so many happy memories of weekends at their homes, including visiting New Haven for the Harvard/Yale game, special family celebrations and birthdays, and so many significant life events. While all of us remember Richard for his brilliant intellect and academic achievements, what I remember most of him is the keen interest he showed in me and my life, always talking with me about what I was learning and experiencing, and supporting my own personal and professional development. He was generous with his time and curiosity and I will always be grateful for that.

As many of you know, Richard faced severe challenges in his later life, dealing with severe medical issues and much adversity. Throughout these difficulties, Sunny and his family were a constant presence, and some of my strongest memories are of the tremendous commitment and love which Richard and Sunny shared throughout their lives. I fondly remember seeing Richard in Vermont and Singapore on so many occasions, always surrounded by family and love. In spite of his difficulties, he remained bright and optimistic, always greeting me warmly with his kind and open smile. His perseverance and courage in the face of so many challenges were a gift and a lesson to us all, and the enduring love and commitment which he and Sunny shared served as a rare but remarkable model of caring and dedication.

I miss Richard greatly, but I am always happy when I think of his warm smile and the many wonderful times we shared. I am extremely fortunate to have known him, to have had him, Sunny, and his family as part of my life.

Working in the Richard K. Chang Laboratory

Alfred Leipertz
Professor Emeritus
Lehrstuhl Für Technische Thermodynamik
Friedrich Alexander Universitat
D-91058 Erlangen, Germany
alfred.leipertz@ltt.uni-erlangen.de

Working team in Richard Chang's group at Yale University during the 6 months stay of Alfred Leipertz in 1982: Jiabao Zheng (Guest Professor on leave from Fudan University, Shanghai, PR China), Richard K. Chang (Host and Group Leader), Alfred Leipertz, and Judith B. Snow (Physicist in Richard Chang's group) — from left to right.

Alfred Leipertz in the laboratory at Yale University, 1982.

Letter to Richard Chang on the Occasion of His Retirement from Yale University, June 2008

Professor Richard K. Chang
Department of Applied Physics
Yale University
P.O. Box 208284
New Haven, CT 06520-8284
USA

Erlangen, April 16, 2008

Retirement Celebration

Dear Richard,

Unfortunately I shall not be able to come to your retirement celebration on June 12th, 2008 which I strongly wished to do. So I will use this way to take a little bit differently part in your festivity, to wish you all the best for the future and to thank you again for all that you have done for me and for my career.

When I came to your lab in March 1982 I had a research grant of the German National Science Foundation (DFG) for doing some work on two-dimensional imaging. On the first day of my stay you invited me for lunch and during lunch time you convinced me to shift the focus of my work away from imaging to pure rotational CARS and its possible use for flame temperature measurements. You had used this technique just recently before for room temperature determination (PhD thesis of D.V. Murphy). And this was an excellent idea with a long lasting effect on the scientific orientation of my work. We not only succeeded during my stay in the very first application of pure rotational CARS to combustion research, pure rotational CARS started at that time to stay in the focus of my research activities up to now. In my group this technique has been further developed to become a powerful tool for the investigation of technical processes, e.g., temperature measurements inside vaporizing fuel sprays for the simultaneous determination of temperature and of the fuel-air ratio in homogeneous charge-compression IC engines.

The second topic of my research activities during all the years since then became multi-dimensional imaging using different light scattering and fluorescence techniques. Looking into my publication list you may realize that more than fifty percent of my publications are related to these two topics I started during my stay with you at Yale. And, of course, the success of these activities formed the basis for my professional career. So you might be proud having acted as a stimulant and somewhat like a midwife for this broad research field which has successfully been developed during the last 25 years.

Lehrstuhl für Technische Thermodynamik, Am Weichselgarten 8, D-91058 Erlangen, Germany,
Telefon +49-9131-852-9900, Telefax +49-9131-852-9901, e-mail: sek@ltt.uni-erlangen.de, internet: http://www.ltt.uni-erlangen.de

A short story told by the way. In the time period when I was visiting you, several young "scientists" in the US tried to reach scientific and professional success by manipulating the results of their research work. So, everybody at US universities was very cautious with the results obtained. During my stay I did computer calculations of the pure rotational CARS spectra and for temperature evaluations out of the experimental spectra by comparing them with the pre-calculated ones. Finally, this worked out quite well. In August 1982 I had a two-week trip to California with my family. Coming back, the atmosphere in the lab had cooled down tremendously. Why? During my stay in California other people of the group had tried to use my computer code and failed to reproduce the results I had obtained before. Manipulated results?? Fortunately it turned out very fast that the processor card of the computer had been exchanged during my absence using a different way of processing (I forgot the details — it was something related to floating point operation). Using the "old" processor card again, the same results came out again. Wow! I never will forget the feelings I had during the suspicion of manipulating.

But I also will never forget the great time I had in total during my stay at Yale. Thanks again for giving me this opportunity. I am wishing you all the best and good luck for your future activities in your private and professional life.

With my warmest regards

Prof. Dr.-Ing. Alfred Leipertz

Lehrstuhl für Technische Thermodynamik, Am Weichselgarten 8, D-91058 Erlangen, Germany,
Telefon +49-9131-852-9900, Telefax +49-9131-852-9901, e-mail: sek@ltt.uni-erlangen.de, internet: http://www.ltt.uni-erlangen.de

Letter to Sunny on Richard's Passing

Gwen Kao
Chair, Charles Kao Foundation for Alzheimer's Disease
Hong Kong, China
gwenkao@charleskaofoundation.org

February 19, 2020

I am very sad that yet another of Charles' friends has left us.

You and I will miss our partners, even now over one year since Charles left, I still weep for all the times we no longer share together. It was so kind and thoughtful that you came over especially for the funeral.

Alas, this time I cannot do the same for you. I will find myself in quarantine in Singapore on arrival and returning to Hong Kong, experience the same in Hong Kong!

This virus crisis has affected all our lives.

My thoughts will be with you at this sad time. Will you stay in Singapore to live or return to Connecticut?

Condolences and tears.
Gwen

A Letter to Richard

K. Ulrich von Raben, Ph.D. Yale '82 & Rea von Raben, M.D.
EichBichl Castle
D-83553 Frauenneuharting, Germany

Dear Richard,

I learnt recently that you passed away last year.

I cannot express my sadness and the emotions that I've gone through.

It's so sad. It's simply very sad.

And I tell you why.

As a young German student with a master's degree in physics I got the chance to join Yale Graduate School, thanks to your support. At that time, I didn't know you, nor your labs, nor the Yale community, nor New Haven. Surely, I was used to working long hours to achieve my goals.

When I arrived, I entered your group as a total unknown, and everything was new and fascinating. Happily, I realized quickly, that expectations are to do a good job to survive at Yale. You never told me that, but I could feel and sense it when you were talking to me. You didn't talk to me as the famous Yale professor, rather as a friend who iş willing to make his students successful and happy. Every time when we talked together, I was left recharged, motivated, eager to go to the lab, strengthen my work and dig deeper.

I cannot express the sadness when I learnt about your decease, nor my gratitude and thankfulness to you. You gave me the opportunity to work with you, and that gave me the possibility to explore a new world and to review and readjust my personal focus. You showed me how to work hard, how to concentrate on the final goal and not get lost with all the details and obstacles, and how to handle the balance between work and family. And

never to forget one's own origin and destination.

Writing this letter to you I intend to include three photographs. When I worked at Yale, students didn't take too many photos, for several reasons. But I selected three, and because I don't know if they can be included in the final print, I will mention my selection. The first is a picture of you, and of Stanley Mroczkowski and Victor Heinrich both part of the Yale community, during a trip to DuPont, Delaware, to initiate a joint project dedicated to catalytic processes. It's the picture that I have of you from 45 years ago: full of life, motivated, and motivating others. The second is a picture of you and your wife Sunny. For me the everlasting example of a sovereign couple both deeply dedicated to their profession while keeping the mental balance and caring for others. And the third is a picture taken at your house in spring 1981, when Rea and I had just returned from Germany after getting married. After dinner, you offered us champagne to celebrate our marriage. This year, we celebrated our 40th wedding anniversary. I think and feel, that what we learnt and experienced at the time at Yale, has deeply contributed to building up our marriage and life. And even more, the values, the dedication, the growth and the continuity of one's own personality, the focus on the important goals in life, that's what you taught me. Just by expressing and living it every single day. And I learnt it. For the rest of my life.

Thanks to you, Richard!

In deepest respect for you and Sunny.

Yours,

Ulrich & Rea

Visiting Richard's Labs

Kevin F. Wall, Ph.D. Yale '86
Senior Scientist
Q-Peak Inc.
Elmore, VT 05657 USA
kwall@qpeak.com

I first met Professor Richard Chang in the summer of 1981, shortly after being accepted into the School of Engineering and Applied Science at Yale University. I was looking for a professor that had openings for students in their groups and possible research opportunities.

It was a hot summer day, but Richard took time out of his schedule to bring me to the basement of the building to show me the labs. There I saw rooms filled with state-of-the-art lasers, spectrometers, detectors, and other equipment whose purpose I could only guess at. Not only were there labs filled with equipment but there were students, post docs, and visiting professors bustling about. That was it for me. I knew that was where I wanted to work.

But it wasn't all about the science and great research that was going on in Richard's group that made my years at Yale memorable. It was the way that Richard treated people as equals. Just the fact that he preferred to be addressed as Richard rather than Professor Chang put one at ease immediately. As a student, you felt you were part of his family.

This was reinforced by the numerous occasions when as a student, I joined his family for dinner on occasions to welcome a visiting professor or new graduate students. It was not only Richard that was so gracious but his whole family including his wife, Sunny. The delicious meals were a welcome break from meager diet of a graduate student.

I also remember the occasions when Richard brought myself and other students with him to visit other research groups. A trip to Bell labs was particularly memorable. Richard was always inclusive giving his students the chance to meet and talk with top experts in the field.

One event that stands out in my memory was when I had lunch at one of the colleges on an early winter day. I had hung my winter coat at the entrance to the dining hall and when I returned, it was gone. On hearing that I had lost my coat and with winter approaching, Richard reached for his wallet and gave me money towards a new coat.

Working in Richard's group was a rewarding experience, and this was amplified by the many acts of kindness that Richard showed towards me, my fellow graduate students, post docs, and others.

Letter to Sunny on Richard's Passing and Some Notes on Richard K. Chang

Steven C. Hill, Ph.D.
Research Scientist
Army Research Laboratory
Adelphi, MD 20783 USA

Dear Sunny,

We are very sorry to hear of your loss of Richard. The two of you were such a great team, and now you have to continue on without him. We know he was a great husband, father, son and brother. We cannot think of anyone who took better care of their spouse than you did. We would like to send some words of comfort, to bring you a bit of a smile as you reflect on the marvelous life you and Richard have had together. We hope you can feel our love at this difficult time.

When Steve told me last night that Richard had died, I said, "The world has lost a good man." I can't think of higher praise than that, and it was the first thing that came to my mind. Richard was a good, kind man, and he made tremendous contributions enriching our family's life. In many situations we ask ourselves, "What would Richard do?" Thinking of your dear husband Richard, I find Oliver Sacks words as he neared the end of his life fitting: "I have loved and been loved; I have been given much and I have given something in return; I have read and traveled and thought and written. I have had an intercourse with the world, the special intercourse of readers and writers. Above all, I have been a sentient being, a thinking animal, on this beautiful planet, and that in itself has been an enormous privilege and adventure." Richard had the special family of scientists and the intense interest in the physics of the world. But more than that, he loved and was loved, and touched all who knew him with his bright mind and kind respect. He will be sorely missed. — Libby

Richard was such a great person, mentor and friend. For many years Libby used to tell me how it was so good when I worked with Richard because he got me interested in working on exciting problems and I was happier when working with him. I noticed that also, but it was interesting to have Libby notice and remind me of that. Even in our earliest discussions on surface plasmons killing resonances of spheres on surfaces and of single droplet lasing it was so fun talk with him and learn of such interesting physics. Also, he was so generous. He helped me so much on several early papers, providing insights and suggesting interesting problems to solve and ideas on how to make a paper of interest and understandable. Also, he seemed to have a sixth sense, a social sense. I don't know how else to say that. It is partly that he cared about people and they could recognize that. His and your strength in dealing with Parkinson's and a near drowning were so remarkable. — Steve

Some Notes on Richard K. Chang

Richard was a great person, mentor and friend to me and to many students, post-docs, researchers and others whose lives and careers he enhanced. I first met Richard when he visited the University of Utah in 1983, where I was a graduate student working with Bob Benner. At Utah we had measured "structural" resonances in the fluorescence emission from dye-doped spheres on substrates. Half the resonances were markedly decreased in amplitude on metal surfaces but not on non-metallic surfaces. These resonances were especially small on polished silver surfaces. I was thinking the TE modes would be decreased because of the boundary condition on tangential fields at the surface of a conductor, but we had not yet determined which modes had decreased. When we showed the spectra to Richard, he surprised me by saying something like: "Wait. Don't tell me which modes go away. Let me guess." He thought about 3 seconds and said it was the modes with their electric field normal to the surface. That was the opposite of what I was expecting. He explained this in terms of propagating surface plasmons — which I didn't understand. It turned out he was right. I was amazed. Even in our earliest discussions on surface plasmons killing resonances and of single droplet lasing it was so fun to talk with him and learn of such interesting physics.

I thought Richard should be a coauthor on the paper describing this work because he explained the key results (TM modes, surface plasmons, why resonances are more decreased on silver than on gold or copper), and the explanation was not one I'd have thought of (the only thing I knew about surface plasmons was the name). But Richard declined. He suggested he submit it to Applied Optics as a Rapid Communication, which he could do because of his position in OSA.

Richard was so generous, gracious and kind. He helped me so much on several early papers, providing insights and suggesting interesting problems to solve and ideas on how to make a paper understandable and of interest. He also declined to be a coauthor on a paper in JOSA B in 1986 on which resonances could support nonlinear processes in spherical resonators and the density of optical resonances in spheres. That paper was worked out largely as a dialogue with Richard. He was so generous with his time in drawing out what I was thinking, making sure the arguments carried weight, and helping to get it into a form that readers could understand. He took pleasure in helping launch the careers of researchers.

For many years, especially when I was teaching at Clarkson, my wife Libby used to tell me how it was so good when I worked with Richard because he got me interested in working on exciting problems, and I was happier when working with him. I noticed that also, but it was interesting to have Libby notice and remind me of it.

As I got to know Richard better I also became more impressed by how his brilliance in physics was accompanied by wonderful social skills (not such a common combination). How he cared so much about his students, family, coworkers, and others. He had a wonderful sense of humor. He had outstanding abilities to assemble and motivate teams. Richard seemed to have a sixth sense, a social sense. I don't know how else to say that. Richard cared about people and they could recognize that. Richard and Sunny used to have get-togethers at their home in Hamden. At the first one I attended there were probably 20 to 25 or so scientists on his deck. Richard moved from group to group, making sure everyone was doing well, and introducing persons who might not know each other but had interests in common. At an early one of these Richard introduced me to mangos and showed me how to eat them.

Richard asked questions that led to insights and interesting work. For example, in the later 1990's I explained to Richard how interesting it was that the excitation and emission problems for a fluorophore in a particle were linked by an approximate reciprocity, especially when the particle is larger than the wavelength. This is why the fluorescence tends to go back toward the source of the illuminating beam. The reciprocity is only approximate because the excitation and emission wavelengths are not identical. I was thinking this near-reciprocity was intrinsically lovely. Richard said something like, "There has to be a use for that," and asked, what else can you do with it? His question flipped me out of a rut and we realized quickly that two-photon emission in a sphere would tend to go backward, more strongly toward the source; and three-photon emission would tend to go backward even more. Many months later, after I'd calculated theoretical angular distributions of multiphoton-excited fluorescence, Richard setup a discussion with Jean-Pierre Wolf, whose lab in Lyon was setup to make such measurements with the needed femtosecond lasers. Jean-Pierre invited Veronique Boutou in his lab to measure the angular distributions of fluorescence, which she and Jin Yu did within a week. These measurements overall agreed well with the calculations. We had further excellent collaborations with Jean-Pierre, Veronique and others in Lyon. That illustrates how Richard's broader perspectives, his deep knowledge and tendency to ask good questions, and his excellent social skills which could be used to set up collaborations with world experts, all combined to make interesting research move forward, and made working with him so rewarding. It is clear why Libby noticed my enjoyment in working with Richard.

Richard had a great sense of humor, but specific examples are difficult to remember. One evening at Yale we wanted to see how an instrument would respond to tobacco smoke. Richard and I went to a little shop a couple of blocks from the lab and asked for cigarettes. The clerk asked what kind and I said "the cheapest." Richard thought that was so funny, the health conscious Yale professor being part of a purchase of cigarettes with "cheapest" as the key criterion. The next day Richard brought his dog, a very big dog, to the lab so we could see if the instrument could detect his dog. It sniffed the dog very well and so was christened "The Dog Sniffer."

Working at the Army Research Lab, I've been thankful that Richard was interested in and willing to work on problems in detecting biological and other particles in the atmosphere, not only in fundamental science issues around the problems, but also in more direct work to detect particles, understand the issues and phenomena, and develop better solutions. For example, in 1998 I gave up on measuring good laser-induced fluorescence spectra of individual particles as they flowed rapidly through our system. The quality of spectra was too variable. I thought we had failed at that enough times, and should go to a fallback position: acoustic levitation and/or focusing. Richard wasn't against exploring levitation, but wasn't ready to give up on the flowing-particle measurements either. He and Dr. Yong-Le Pan then worked out how to do it and did it. That example in particular stands out because it has had such a large impact on our work.

Richard and Sunny's strength in dealing with Parkinson's and a near drowning were so remarkable. Richard had a tremendous ability to do what he thought was best. He could ride his bike on West Rock fast and long. Even when his Parkinson's made his life difficult he could do many more pushups than the average person his age. He exercised and was very strong. I don't see how he could have done that if he did not have the great benefit of Sunny looking out for him, looking for any medicine or exercise that could help.

Remembrances of Interaction with Richard Chang

Kenneth Young
Professor of Physics
Chinese University of Hong Kong
Hong Kong, 999077 China
kyoung@cuhk.edu.hk

It was serendipity and my good fortune to have gotten to know Richard Chang. We met through Charles Kao (Nobel Prize in Physics, 2009) who in the late 1980s to early 1990s was the President of the Chinese University of Hong Kong. Charles would spend a part of each summer in New York, in a consulting role with International Telephone and Telegraph Corporation (ITT), thinking about the next challenges in optical communication, including issues such as terahertz technology. He would also spend time at Yale, where he interacted with Richard, and the two became good friends. Thus, it was that Richard came to give a talk, when he visited Hong Kong, and that was how our interactions began.

Having met and chatted, we discovered quite a few antecedent connections, in a typically "small world" story. It turns out that, one of my cousins had been a good friend of Richard's at MIT around 1960, and my uncle had known Richard's father back in Shanghai, presumably in the 1930s. I also discovered that, Richard and I had lived on the same short street in Hong Kong, though at different times.

Richard's talk was about optics in microdroplets, and the exquisite morphology-dependent resonances (MDRs) that could be probed with precision optics. One of the applications (I am afraid the only one that I can recall) was that this opened a path to monitoring, in real time, what was happening to the fuel droplets in an internal combustion engine — no doubt the sort of thing appealed to funding agencies. My interest was rather different: here was an interesting and probably tractable problem in mathematical physics, rooted in real life. We looked at several physical processes, where the MDRs and the resultant resonant concentration of energy might play a significant role, for example Brillouin scattering and enhanced chemical energy transfer.

Just to illustrate how much things have changed in just a few decades, it was a nontrivial task in those days to evaluate spherical Bessel functions $j_l(x)$, when l and x are both large, say about 500, as would be needed, when dealing with the MDRs. Nontrivial in the sense that, it took a fair amount of time on a mainframe. Moreover, we discovered that, the standard International Mathematics and Statistics Library (IMSL) subroutines (in FORTRAN, if there are still people, who know what that is) had a bug with its Bessel functions, and we had to rewrite our own. In the course of that, we learnt something well known to those who do this sort of thing professionally: the standard forward recursion formulas that we all learnt as undergraduates (but alas, no more for the current generation) are mathematically correct, but practically unstable, and instead one has to resort to reverse recursion, going down in l, which at first glance seems counter-intuitive. In any case, we thought it might be useful, if approximate asymptotic formulas for the positions and widths of the MDRs could be developed for practical use, so that for many purposes one did not have to struggle with the Bessel functions. That was a fairly straightforward piece of work [1], but people seemed to have found it useful. Incidentally, Charles Kao was pleased, when I told him about these things, because fields in optical fibers are described by cylindrical Bessel functions $j_l(x)$ and the associated Neumann and Hankel functions, which all appeared in his famous paper [2].

However, another problem turned out to be more interesting, and had a major impact on the direction of research for me and several of my colleagues. The droplets, that Richard and others were investigating were falling under gravity, in the presence of some air, and would thus be slightly flattened by air resistance. The question then arises: how would the MDR frequencies be shifted (indeed the spherical degeneracy lifted) by the small shape distortion? The forward problem of calculating the frequency shifts would occupy mathematical physicists, while Richard and colleagues were more interested in the inverse problem: how the measurable frequency shifts could be used to diagnose the shape distortion. At the time, Richard was collaborating with several experts in computational electromagnetism (EM) theory, including Peter Barber and Steve Hill, for example. They used brute-force T-matrix methods to find the scattering resonances for various spheroidal shapes. When spherical symmetry is broken, all the modes with $j_l(r)Y_{l,m}(\theta, \varphi)$ become coupled and large matrices have to be handled. The tedious work could be carried out on a suitably powerful computer, and indeed the (numerical) answers were all found.

To us, the shift in eigenvalue (even though now complex) upon a small change in the system should be a problem amenable to perturbation theory. There is the standard formula known to any undergraduate, in obvious notation and in units with $\hbar = 1$:

$$\Delta\omega = \frac{\langle \psi \mid \Delta V \mid \psi \rangle}{\langle \psi \mid \psi \rangle} \tag{1}$$

where ΔV represents the perturbation and ψ is the unperturbed wavefunction. Porting this formula from the context of Schrödinger's equation to the context of Maxwell's equations was straightforward.

The difficulty lies in the fact, that the MDRs are quasinormal modes (QNMS) and not normal modes (NMs). Physically, the internal reflection at the droplet boundary that confines the MDR is not total, because the interface is not a plane, so that there is a tiny bit of leakage. Thus, there is some outgoing wave outside the droplet (like the output coupling in any other optical cavity, but much more readily computable because of the spherical geometry). Far away, ψ must look like

$$\psi \sim r^{-1} \exp[-i\omega(t - r / c)] \tag{2}$$

The frequency ω has a small imaginary part that makes ψ decrease with time (energy leaks out, so the stored energy decays). But since time and distance enter as $(t\text{-}r/c)$, the radial dependence is exponentially increasing at infinity, which makes the normalizing integral in the denominator of [4] diverge. (There is no problem with the numerator, because ΔV is non-zero only in the finite parts of space.) Not only is this a mathematical difficulty, but it defies physical sense: the frequency shift cannot depend on what happens at spatial infinity; indeed, it cannot depend on anything outside the droplet.

When couched in these terms, the problem becomes general and interesting to mathematical physicists. For we all have learnt, from our first class in quantum mechanics, the canonical formalism for dealing with conservative systems described by hermitian Hamiltonians, with real eigenvalues and eigenfunctions (NMs) that are normalizable, orthogonal, and complete, and for which various tools such as perturbation theory have been developed. These are not only calculational tools, but more importantly the conceptual framework and the language, in terms of which we think and talk about the physics. The issue then is: *To what extent can we build an analogous formalism for non-conservative systems, especially those that lose energy by "leakage" to the outside?* Droplets are the "hydrogen atoms," that provide the motivation and the testing ground for this wide class of problems; other optical cavities with output coupling also come to mind, but there are other interesting applications, as we shall see.

Zeldovich [3] had taken an important step in the context of one-dimensional (1D) quantum mechanics. He showed, for the normalizing integral:

$$\langle \psi \mid \psi \rangle = \int \psi(x)^2 dx \tag{3}$$

(a) you use ψ^2 and not $|\psi|^2$ — in the presence of a complex eigenvalue you cannot blithely go around conjugating your equations; and (b) you can insert a regulating factor $\exp(-\varepsilon x^2)$ and take $\varepsilon \downarrow 0$ at the end, giving not only a finite answer, but the correct answer to put into the denominator in [4]. Note that $\psi^2(x)$ oscillates about zero (whereas $|\psi|^2$ does not), which is the technical reason, why the regulator works. This recipe, though correct, is unwieldy. After playing with the equations for some time, we arrived at an equivalent formalism [4], that (a) involves only the "inside" up to the boundary and therefore (b) does not require any regularization, making it conceptually much neater, and opening the way to practical applications. We later generalized this to inner products [5,6], leading to the concept of QNMs being mutually orthogonal in a certain sense, thus allowing projections to be carried out, and for example higher-order perturbation theory to be developed [7]. It was also shown that, provided that the boundary demarcating the system from the "outside" is sufficiently sharp (e.g., the dielectric discontinuity for a droplet), the QNMs would be complete. In that case, the decaying signals would be a sum of exponentials, each like a damped harmonic oscillator. Interestingly, when the boundary is not sharp, there is in addition a long-time tail [8], typically decaying with a power in t rather than an exponential.

So, these questions led us on a path that occupied me, several colleagues, and students for many years, with a summary in a review article [9]. There are still important open questions to which we might return — if we can overcome an obstacle to further generalizations. But here, it would be appropriate to briefly point to two applications, the first in collaboration with Richard, the second in an apparently distant field.

By 1991, we had developed a simple and explicit perturbation formula for the frequency shift of the MDRs [4], in terms of the eccentricity of the spheroidal droplet. That summer, Richard had invited me for a short visit to Yale, and we were discussing, in a small group meeting, one of his experimental findings. The photons in the MDR hug the rim of the droplet, their orbit in the intersection between some plane and the droplet surface. The angular momentum **L** of the photon points along the normal to that plane. When the droplet is slightly oblate, spherical symmetry is lost, so L_x and L_y are not conserved and **L** (and hence the plane in which the photon travels) precesses. If the observing telescope is focused at one spot on the droplet surface, at some latitude, the photon direction will hit the telescope once every precession period, much as in a pulsar. So, the observed intensity goes up and down with a frequency Ω equal to that of precession. That much is obvious, as is the fact that the effect is proportional (to first order) to the eccentricity e in the droplet shape. But what is the proportionality constant in $\Omega \; \alpha \; e$?

It is well known that (in obvious notation), $d\omega/dk$ gives the group velocity in the case of linear motion, where k is the linear momentum. So likewise, $d\omega/dm$ gives the group angular velocity, indeed nothing but Ω, where m, the azimuthal quantum number, is just L_z. Since we had an explicit formula expressing the shift in ω as a function of m, it was but a moment's work, right there on the blackboard, for Richard and I to solve the problem in closed form — an "aha" moment, that is very vivid in my mind, as it was in Richard's. The result was written up shortly thereafter [10].

Interestingly, something very similar happens with black holes (BHs). If a BH vibrates around its static state, as for example when it is first formed in the merger of two smaller BHs, then the vibrations are also dominated by QNMs. (In this case, the QNMs are not complete, but are expected to be dominant except for very late times [8]. The QNMs are also much broader.) The QNM frequencies are determined by the spacetime metric outside the event horizon, and if precisely observed, would be a defining signature to prove that one is indeed seeing a BH — just as the spectrum of radiation from a broad-band source placed inside a Fabry–Perot cavity shows sharp resonances, characteristic of the cavity. So, there is much interest, and indeed a substantial literature, on the QNMs of BHs. It is believed that, in the gravitational waves seen in the last few years [11] (Nobel Prize in Physics, 2017), the ringdown signal after the point of merger reveals the first QNM — though one needs better observational precision as well as detection of the next few QNMs to make a definitive statement. The framework for discussing the QNMs in this leaky system is almost the same as the one for dealing with MDRs in microdroplets. For example, a minor adaptation leads to a formula for the shifts in the QNM frequencies when BHs are slightly perturbed, e.g., by accreted mass outside the event horizon, i.e., a "dirty" BH. The subject of dirty black holes has become a minor industry (just look up the key words in Google Scholar) and our paper of 1997 [12] was one of the earliest on this subject — inspired by droplets. I must have told Richard about all this, but only on email and therefore only briefly. I much regret that I did not have the chance to tell him in person; he would have been pleased: "To see a World in a Grain of Sand" [13]; "To see a Blackhole in a Drop of Liquid."

In remembering my interactions with Richard, what stands out is not just the science and the fruitful exchanges, but more vividly the kindness and the human touch. Richard was always gentle and quiet-spoken, curious about a range of subjects beyond his immediate concerns, generous and kind to younger scientists and students. That lofty but understated spirit has certainly rubbed off on all those, who have been fortunate to come into contact with him, and while we much regret his passing, we know that he has left a mark that lives on.

References

1. C.C. Lam, P.T. Leung and K. Young, Explicit asymptotic formulas for the positions, widths, and strengths of resonances in Mie scattering. *Journal of the Optical Society of America* **B9**, 1585–1592 (1992).
2. K.C. Kao and G.A. Hockham, Dielectric-fibre surface waveguides for optical frequencies. *Proc. IEE* **113**(7), 1151–1158 (1966).
3. Ya. B. Zeldovich, *Sov. Phys. JETP* **12**, 542 (1961).
4. H.M. Lai, P.T. Leung, K. Young, P. Barber and S. Hill, Time-independent perturbation for leaking electromagnetic modes in open systems with application to resonances in microdroplets, *Physical Review* **A41**, 5187–5198 (1990).
5. P.T. Leung, S.S. Tong and K. Young, Two-component eigenfunction expansion for open systems described by the wave equation I: Completeness of expansion, *Journal of Physics* **A 30**, 2139–2151 (1997).
6. P.T. Leung, S.S. Tong and K. Young, Two-component eigenfunction expansion for open systems described by the wave equation II: Linear space structure, *Journal of Physics* **A 30**, 2153–2162 (1997).
7. H.M. Lai, C.C. Lam, P.T. Leung and K. Young, Effect of perturbations on the widths of narrow morphology-dependent resonances in Mie scattering, *Journal of the Optical Society of America* **B8**, 1962–1973 (1991).
8. E.S.C. Ching, P.T. Leung, W.M. Suen and K. Young, Late-time tail of wave propagation on curved spacetime, *Physical Review Letters* **74**, 2414–2417 (1995).
9. E.S.C. Ching, P.T. Leung, A. Maassen van den Brink, W.M. Suen, S.S. Tong, and K. Young, Quasinormal-mode expansion for waves in open systems, R*eviews of Modern Physics* **70**, 1545–1554 (1998).
10. J.C. Swindal, D.H. Leach, R.K. Chang and K. Young, Precession of morphology-dependent resonances in nonspherical droplets, *Optics Letters* **18**, 191–193 (1993).
11. B.P. Abbott *et al.* (LIGO Scientific Collaboration and Virgo Collaboration), Observation of Gravitational Waves from a Binary Black Hole Merger, *Physical Review Letters* **116**, 061102 (2016).
12. P.T. Leung, Y.T. Liu, W.M. Suen, C.Y. Tam and K. Young, Quasinormal modes of dirty black holes, *Physical Review Letters* **78**, 2894–2897 (1997).
13. W. Blake, Auguries of Innocence (1803).

Letter to Richard on the Occasion of His Retirement on June 2008

Joseph Lam
Automation Project Engineer
Misumi USA
Bay Area, California USA
jlam@misumiusa.com

Dear Richard,

I still recollect how you proposed that I join you at Center for Laser Diagnostics (CLD) at Yale back in 1983 and remember you telling me that there is no guarantee that it will work out between myself and Marshall Long, your former student and a recent Assistant Professor in Mechanical Engineering at Yale. I was surprised to see that Marshall looked like a hippie or someone out of the Woodstock era. I thought "Is this the guy that Richard had in mind for me?" As it turned out, you could not have found a better person to guide me in those years. Although I am no longer with the CLD at Yale, I will always remember how it turned my life around.

You always looked after me even after I left Yale. Maybe because you knew that I grew

Joe and Merina Lam and their 3 children in the Bay Area, California 2020.

up in a foster home. Like a big brother, you strived to check on me to see if I am doing OK whenever you were in the Bay Area. You have given me more advice on both the personal and professional level since I left Yale. I remember you telling me about the value of having a family and family should always come first. And if it was not for your family, life would have been very different for you. Little did I understand what you were telling me at that time.

After I left Yale, I got a job and became a world traveler, hanging around company executives and shooting for the corporate office. All I wanted was to help take a Silicon Valley company public. Fast forward to today, Merina and I have three children and we enjoy them all. Working for the corporate office is less important to me than spending time with my family.

I am very fortunate to have found someone like Merina, my wife, and a big brother like you.

Letter to Richard Chang on the Occasion of His Retirement, June 2008

Burt V. Bronk, Ph.D.
Research Scientist,
Air Force Research Laboratory
Wright-Patterson Air Force Base, OH 45433 USA

2008

Dear Richard,

I have had the chance to interact with you on and off for more than two decades, ever since I left academia and went to work for the Department of Defense at Edgewood in 1985. Everyone at the Army Labs always looked forward to hearing each year about remarkable new discoveries by you and your students about the properties of micron-sized droplets.

Not only did I learn some science from you, but I enjoyed a warm relationship with you and your wife Sunny. Whenever I tried to trace you down by calling you at home, Sunny always made me feel like a family friend. You once gave me a copy of a wonderful memoir ("*Bound Feet and Western Dress*") by your daughter Pang-Mei in which she told the story of her great aunt. According to the book, your aunt survived a disastrous marriage to your uncle, who was a brilliant and famous Chinese poet, but a neglectful and destructive husband. Your aunt surely had great strength to be able to make a new life as a successful business-woman. As I once told you, after reading the book, I felt I knew more about that part of your family, than I know about some secrets of my own family.

Your students and associates at Yale seemed to take on attributes which I observe in you. I had experiences working in the lab with three of them: Yongle Pan, Kevin Aptowicz, and Steve Holler (who got his start in micron particle physics with another remarkable colleague, Professor Steve Arnold of Brooklyn Poly). Each of these three was highly skilled, cordial, respectful, open minded to suggestions, and like you, fun to work with.

Many good wishes to you and your family,
Burt (Bronk)

Editors' note:
Pang-Mei Chang, *Bound Feet and Western Dress: A Memoir*, 215 pages, Anchor (1997).

Remembrances of Richard Chang

Alan C. Eckbreth, Ph.D.
Director Emeritus, United Technologies Research Center
Hartford, CT 06108 USA

I don't recall precisely when I first met Richard. However, in researching the only paper we ever co-published (D.V. Murphy, M.B. Long, R.K. Chang, and A.C. Eckbreth: Spatially resolved coherent anti-Stokes Raman spectroscopy from a line across a CH4 jet. *Optics Letters* **6**, 167–169, June 1979), Richard cites Office of Naval Research sponsorship under Project SQUID. So I suspect we first met at one of the periodic Project SQUID meetings in the mid 1970s. The Office of Naval Research was an early supporter of research into laser diagnostics of combustion. Richard was instantly likable with an inquisitive sparkle in his eyes and we quickly became friends. The linear imaging described in the paper was Richard's idea and he suggested we jointly pursue it. I agreed and we performed the experiments in my laboratory at UTRC. Then graduate students Dan and Marshall spent a long weekend with me doing the experiments. I had SQUID funding after I had discovered laser modulated soot incandescence as a major interference to Raman scattering diagnostics in practical flames. My SQUID funding demonstrated no viable approach to suppress this interference for the very weak Raman process, so my research then turned to coherent anti-Stokes Raman spectroscopy (CARS) which indeed turned out to be viable for practical combustion systems. Incidentally, laser induced incandescence emerged as a viable diagnostic for soot particle sizing and there are commercial systems available today but we did not pursue it at UTRC although Lynn Melton on sabbatical leave at UTRC from UT-Dallas suggested that possibility.

At Richard's invitation, I occasionally gave seminars on my CARS work to his classes at Yale. We saw each other at the biennial Gordon Conferences on Laser Diagnostics in Combustion which commenced in 1981 thanks to David Crosley and John Daily who were respectively the first and second Conference chairs. I was the third Chair in 1985, Richard the fifth in 1989. Marshall Long was Chair in 1997. At the 1985 conference pictured below, Richard and Marshall are seated in the front row, left. On Tuesday afternoons of our Gordon Conferences, we had a Chairs meeting which, in 1985 and for quite a few subsequent conferences when the conference was in NH, was held on top of my boathouse on Stocker Pond in Grantham, NH. My wife Barbara prepared a wonderful buffet lunch and after our "business" meeting kept purposely short, swimming and water skiing followed.

Our relationship was more than collegial. Richard shared with me the history of his and his parents lives and their arrival in America, very eye opening for someone who lived a sheltered life growing up in NJ. My wife Barbara and I spent several evenings over the years with Richard and Sunny at dinners and plays. Most notable for us were the dinners in Richard and Sunny's house when we had the opportunity to meet their children. Barbara and I also have very fond memories of reading Natasha's book *Bound Feet and Western Dress* which Sunny recommended to us.

1985 Physics and Chemistry of Laser Diagnostics in Combustion Gordon Research Conference group photo (the photo is the property of Gordon Research Conferences and is used with their permission).

Memories with Richard Chang on the Occasion of His Retirement in June 2008

1985 — Yehiam Prior, Richard Chang, and Alan Eckbreth in Negev Desert, Israel.

Richard working hard at another Gordon Conference Chairs' Meeting on Alan's Boathouse. L-R: Ron Hanson, Richard Chang, David Crosley. Guess 1991.

Gordon Conference Chairs' Meeting at Windsong on Stocker Pond Grantham, NH, July 1993. L-R: Larry Rahn, Jean-Pierre Taran, Richard Chang, Alan Eckbreth — host, Ron Hanson, Kermit Smyth, Dave Crosley, and John Daily.

Some Flashes from New Haven 1985

Marcia Temperini
Professor of Chemistry
Instituto de Química
Universidade de São Paulo
São Paulo, Brasil
mlatempe@iq.usp.br

There are many things that I remember that portray Richard's concern and goodness towards people.

The first image is of the day of my arrival when Prof. Richard waited for me at the New Haven airport despite my two-hour delay. Besides that warm reception, he also provided me with all the things that I needed to live in the International Center House (Prospect Street 460??) and he also went with me to this House to check whether I had everything I needed.

During all the research, he always gave me continuous support and encouragement towards my work demonstrating his interest.

I remember him on a barbecue party when he spoke in Portuguese in a sincere demonstration of friendship.

Despite the time and distance, as I got back he wrote some letters showing his concern about my research.

I would like to thank him for his contribution to my scientific life.

Today I am Professor at the Chemistry Department at University of São Paulo, I am trying to give to my students the same good lessons and way of doing the things that I had learned from Prof. Richard.

Thanks once more,
Best regards.
Yours,
Marcia Laudelina Arruda Temperini

Note to Richard and Sunny

Richard H. Brodhead
President Emeritus of Duke University
Former Dean of Yale College
Guilford, CT 06437 USA
richard.brodhead@duke.edu

December 11, 2017

Sunny and Richard, Cindy and I are back in the New Haven area after 13 years at Duke. I stepped down as president at the end of June, and we have moved into a house we built overlooking the Thimble Islands in Guilford. We've been reentering the Yale world very gradually, but I did agree to be a fellow of one of the two new residential colleges, and when we went to the holiday party last night, we ran into Dan Prober. We had been very curious about the two of you since we spent so many wonderful times together with you — I remember once in a Chinese restaurant where they served all things not on the menu because you were there (was it called Shangri Lee?), and another time when we went kayaking in the Connecticut River, and another when you went on a sleigh ride at our Vermont house — and perhaps my favorite memory of all, meeting your daughter at the time of *Bound Feet and Western Dress*! Dan gave us a quick update and shared your email, Sunny. We think of you with great warmth and affection and send you and your family all our best wishes. At Duke, I used to go to China two or three times a year (but never Hong Kong any more) … do you ever make it back to these parts? We would welcome any news. All best wishes, Dick

Letter to Sunny on Richard's Passing

Dear Sunny,

Cindy and I regret to loss of such a wonderful friend. You know it far better than we: Richard was so deeply intelligent in the academic sense but also so wise, courteous, and kind. We will look with pleasure to so many things: the time you took us to the Chinese restaurant where they served special dishes just for you, which had become invisible when we went back on our own; your sharing of your daughter's excellent book *Bound Feet and Western Dress*; the sleigh ride we took when you visited us in Vermont; Richard's willingness to take on any and all duties at Yale — and maybe rarest of all, his and your willingness to welcome us as friends in spite of our differences of field and age. What a splendid person. All our love to you and your family, Best, Dick

Editors' note:
Pang-Mei Chang, *Bound Feet and Western Dress: A Memoir*, 215 pages, Anchor (1997).

Letter to Sunny on Richard's Passing

Gorden W. Videen, Ph.D.
Senior Research Scientist
Army Research Laboratories
Adelphi, MD 20783 USA

Dear Sunny,

For some reason, I have thought of you and Richard many times the last few weeks. I have thought of the times we have had together, too few, and the influence you and Richard had on my life and my outlook. I am saddened by Richard's passing, but I know that we will meet again.

In 1982, my grandmother died of a heart attack. At the time, I knew it somehow, although we were 2000 miles apart. My grandfather was very ill and in the hospital at the time, but when my mother called, the first thing I did was to ask her what happened to grandma. She was revived in the hospital and the first thing she did was curse out the doctor for his hard work! She described how she visited with her mother. Her mother told her it was not her time yet, and that she must go back for a little while. That little while became 25 years. She was never afraid of death after that and I found the entire episode incomprehensible, especially my somehow knowing, but I am comforted also.

I am saddened by your loss, but I know that you and Richard will always be united and that he is smiling and thinking of the wonderful life he lived with you by his side. I know that this is a busy and difficult time, but remember that you have many friends who love you deeply,

Gorden

My Cousin Richard

Hortensia Meimei Chang-Blanco
Wife of Former Argentinian Ambassador to Singapore and China
Buenos Aires, Argentina

Although close cousins, fate and family situations made Richard and me grow up in different parts of the world and only meet later in life. Even though the late encounter, I have always felt a strong bond to Richard, who was the son of my father's eighth brother, but the first Chang relative I met when he came to Italy to visit us during Easter (which is why we called him zio Pasqualino) and so finally, meeting a member of my Chinese family, unknown to me until then. Before my meeting Richard, I had only known the Italian side of my family. Afterwards, "diplomatic" life and propitious situations brought us together and new feelings added to the normal affection between cousins: firstly pride and respect for his exceptional academic achievements in his outstanding career as a Yale professor of Applied Physics; followed by great appreciation for him as a human being for his gentle soul, kindness, family love, generosity, cheerfulness, travel & sport quasi addiction, joie de vivre; and later on, deep admiration for his extraordinary acceptance of his illness and resilience, through which he proved to be the quintessential example of human strength and sacrifice.

The way Richard lived his life with courage and humbleness, and navigated his suffering as a stoic will always be truly exemplary to me: he taught all of us the power of patience and love. I remember his eyes very fondly, always filled with hope and gratefulness, especially for his wife and guardian angel Sunny. Tears come to me as my heart bursts in pride, admiration, and love for him. Yes, these are the three words that would define Richard Chang for me: a rich, inspiring example of a Great Man whom I have had the privilege to know and consider my mentor in life — and happy to boast about it!

I am sure I am not the only one to experience these feelings for Richard. THANK YOU DEAR RICHARD for the precious lesson of life you have taught me, you're my top professor!

Remembrances of Richard Chang

Daniel Chernoff, M.D. Ph.D. FACR
Albany Medical Center
Albany, NY 12208 USA

I was a callow New Haven, Connecticut public high school junior in 1975 or 1976, good at math and interested in electrical engineering, when my independent studies teacher, Sunny Chang, suggested I get some practical experience in science research. Somehow, she had prevailed upon her husband, Professor Richard Chang, to allow a high school student with few skills into his lab. If there was more to this story, I wasn't aware of it or have forgotten, but that's the way I recall it now.

I remember that the Chang laboratory was in the Becton building on the Yale campus, in the basement with room numbers 001, 002, ... I met Professor Chang and was given a small amount of space in room 007, which I thought was very cool from a James Bond perspective. At first, I believe my job was to set up samples of crystals of I-know-not-what for X-ray crystallography, in which the sample is exposed to an X-ray beam and an X-ray diffraction pattern forms on photographic film placed behind the sample. I never did learn much about the significance of this work, but I think it got me introduced to the process of scientific inquiry.

Later Dr. Chang asked me to construct some kind of electronic circuit to be used in other research. Knowing relatively little about electronics other than an enthusiasm for taking things apart and attempting to put them back together, this was a bit too much for me to get right, but I did spend a lot of time soldering, learning to use oscilloscopes and other electronic test equipment. The "stars of the show" for this circuit were large Burr-Brown operational amplifiers, or "op-amps" which were precision analog integrated circuits of the day and I believe quite expensive. I was always nervous I would kill one of these with an overvoltage, short circuit or the like. Even though I don't think I got the circuit fully working, Sunny induced me to write it up for the Westinghouse Science Fair that year, where I believe I got an "honorable mention" for my trouble.

Sunny Chang had a huge influence on my academic performance in high school and set me up for acceptance and academic success at the Massachusetts Institute of Technology (MIT), my dream school. Dr. Richard Chang's generosity in tolerating a high school student in his lab of mostly graduate students was a big help to me. I eventually earned 4 degrees at MIT and Harvard University: Bachelor of Science, Master of Science, PhD and MD, although I left academics early for private practice medicine once I had a family. I am so grateful to them both.

Paper Dedicated to Richard on the Occasion of His Retirement and Memories for Richard

Andreas Otto[†]
Professor of Physics
Institut für Physik der kondensierten Materie
Heinrich-Heine-Universität
Düsseldorf 40225, Germany

J. Phys. Chem. C 2008, 112, 30, 11075–11077

SERS of CO_2 on cold-deposited Cu: An electronic effect at a minority of surface sites*

M. Lust[a], A. Pucci[a], W. Akemann[b]
A.Otto[b]
[a]*Kirchhoff Institut für Physik, Universität Heidelberg, Im Neuenheimer Feld 227, 69120 Heidelberg, Germany*
[b]*Institut für Physik der kondensierten Materie, Heinrich-Heine-Universität, 40225 Düsseldorf, Germany*

*Dedicated to R.K. Chang on the Occasion of His Retirement

Richard and Marcia Temperini on the Thimble Islands, CT during the summer of 1985.

Andreas Otto, Richard, and Marcia Temperini on the Thimble Islands, CT during the summer of 1985.

[†]Passed away

My Interactions with Richard

John R. White, Ph.D.
Chief, Physics Division (Ret.)
Edgewood Chemical Biological Center,
U.S. Army, Darlington, MD 21034 USA

I first met Richard in 1984 or 1985 at a Smoke and Obscuration Conference held in Edgewood, MD. From the mid-1980s until 2015, I worked at the Edgewood Chemical Biological Center, first as a principal investigator and later as a Branch and then Division Chief. My investigations in the first part of my career involved the U.S. Army's use of smoke as an obscurant. My work was specifically oriented toward the theoretical studies of which properties of a material made it effective/not effective as an obscurant, and how to best disseminate a material to make an effective smoke screen. It was, during this time, from about 1986 to 1994, that it was my privilege to work with Prof. Richard Chang of Yale University. At that time Richard had been funded by Dr. Ed Stuebing, at Edgewood, to characterize single particle interactions with high energy laser beams and the quality of Richard's work was just outstanding. Furthermore, the work Richard was doing blended nicely with my interests, and we began to work together.

This was the period when the Strategic Defense Initiative Office (SDIO) was formed and heavily funded and the proposal of work I submitted to the SDIO was accepted. My program was classified, and the tasks were immediate and demanding. Richard had the classification requirements, the expertise, and a recognized standing in the physics community that would have made him an asset to any program he worked on, and I was able to get a contract in place with Yale in a very short time. Since this work was classified, I cannot discuss it, but I can say his work was always met with anticipation and enthusiasm at SDI program reviews. Although I cannot specifically talk about Richard's contribution to this program, it was innovative, well executed and well presented.

My SDIO program was the last substantial work I could do with Richard. Starting in the mid-1990s my interests had shifted to distributed computing technology I was invited to present some of my work at a meeting of the Russian Academy of Science and a group of my colleagues, including Richard, and I attended this meeting. I mention this because it was interesting and inspiring to watch Richard's interaction with the Russian scientists, as well as their reaction to him! Richard knew several of the Russians personally and many more of them by reputation. It was obvious they were well acquainted with Richard's work. I benefited greatly by watching Richard's interactions with our hosts as well as their reaction to him. The Russian scientists were very interested in Richard's outstanding work, not so much in mine.

Over the years Richard and I became good friends. I never failed to be impressed with the integrity and the quality of his work, with the patience he showed with difficult individuals and with the poise and kindness he always demonstrated to everyone he interacted with. In my professional capacity I was occasionally asked to write letters of commendation for some of my contract performers. In several cases I was specifically asked to mention that their work compared favorably with Richard's. I could never do that, but it is a testament to the high regard others had for his work.

How Richard Chang Changed my Scientific Career: Remembering a Great Collaborator and a Special Man

A. Douglas Stone
Carl A Morse Professor of Applied Physics and Physics
Yale University, New Haven, CT 06520 USA
douglas.stone@yale.edu

Yale University

Department of Applied Physics
P.O. Box 208284
New Haven, CT 06520-8284

Campus address:
305 Becton Center
15 Prospect Street

A Douglas Stone
Carl A Morse Professor of Applied Physics and Physics

email: douglas.stone@yale.edu
Phone: (203) 432-4279
Fax: (203) 432-2207

My PhD and postdoctoral work were in condensed matter (CM) theory, and in 1986 I was hired into the Yale Section of Applied Physics (as it was called then, now it is a department) as part of Yale's build up in CM theory at the time, which initially also included R. Shankar, Subir Sachdev, Nick Read and Karin Rabe. My main research area was micro/nanostructure physics and quantum transport, and I was one of the first theorists to work in the new area of mesoscopic physics, studying quantum fluctuation effects of electrons at low temperatures in metallic and semiconducting micron-scale circuits. I was fortunate to have two active experimenters on the Yale faculty with whom to collaborate, Daniel Prober (metallic and superconducting structures) and Robert Wheeler (2D electron gases on semiconductors), and I published papers with both. However, one fateful day in the spring of 1994 I attended a seminar by one of the current PhD students of my colleague, Richard Chang, that immediately sparked a new research direction, and ultimately led me to do something rare in academic science: change my primary research focus completely.

I am now an optical physicist who "used to work in CM theory", thanks primarily to my scientific collaboration and supportive interactions with Richard. To explain how this happened I must go back a few years and explain a bit about the interdisciplinary field of quantum/wave chaos, toward which my interests had shifted by 1989. This subfield had formed around the general question, originally posed for quantum systems: what are the quantum manifestations of classical chaos in conservative Hamiltonian systems? Most textbook single-body quantum Hamiltonians have high symmetry and in the classical limit correspond to integrable dynamical behavior: the system has a number of constants of motion at least equal to its dimension, and these constants can be used to find the geometry of individual trajectories analytically (e.g. think of standard orbit theory in a 1/r potential). All orbits lie on low-dimensional subspaces of the full phase space. In contrast, when there are no such symmetries, there are regions of phase space, where a single trajectory fills a finite fraction of the whole space, and where many pairs of trajectories with slightly different initial conditions diverge in phase space exponentially, a technical definition of chaotic behavior.

Over long times such trajectories (as well as other types of phase space variables) exhibit pseudo-random behavior. When such a Hamiltonian describes a microscopic system and hence is described by the quantum Schrödinger equation, one finds that its quantized energy levels and wavefunctions behave pseudo-randomly (in space, not in time), and the quantum-classical correspondence of this type of system was being extensively studied at the time. Studying this kind of pseudo-random wave behavior was closely related to the statistical studies, I had done in the field of mesoscopic transport physics and Anderson localization, so it both intrigued me and seemed like reasonably fertile ground for my research. It should be noted that due to the formal similarity between the Schrödinger equations and the Helmholtz equation of electromagnetism, many experimental studies in this field were performed not on quantum systems but on classical electromagnetic systems, primarily microwave cavities.

In 1990 my interest in this field led to a published paper with R. Jalabert and H. U. Baranger, entitled *Conductance fluctuations in the ballistic regime: A probe of quantum chaos?* It showed how differently shaped mesoscopic junctions would exhibit conductance fluctuations with distinct statistical properties due to their different shapes, the first being circular and the second having a hyperbolic shape, known to generate chaos in the classical dynamics. The essential predictions of this paper were confirmed in experiments by Charles Marcus and coworkers in 1992. However, I quickly realized the challenges in trying to study classical-quantum correspondence in these electronic circuits. It required ultralow temperatures to avoid phonon scattering, and the relatively simple Hamiltonians one wished to study were only imperfectly realizable by state-of-the-art fabrication methods. In addition, of course the electrons interact, making the use of a one-body wave equation approximation unreliable.

At the same time, I had become fascinated by a phenomenon in classical dynamics which I had not known about previously, the Kolmogorov–Arnold–Moser (KAM) transition. This occurs in relatively simple systems, when one begins with a highly symmetric and hence integrable system, and smoothly adds a perturbation which breaks the symmetry. As the perturbation parameter increases instead of a sudden transition to chaotic motion, bit by bit the phase space breaks up into chaotic and regular structures, often in a fractal hierarchy, which can be visualized using a Poincaré surface of section plot of the type I will show below. As this transition occurred a single trajectory would begin to fill all of phase space by a kind of diffusive process, aspects of which could be predicted quantitatively. This physics was well-known in various dynamical models. But up to that point it had been was very difficult to measure effects related to the *classical* KAM transition in a real experimental system. I was interested in both the classical behavior, which should be seen in the ray optics limits and the wave manifestations of the transition, about which almost nothing had been studied experimentally in either CM or electromagnetics.

There seemed to be no hope of doing such experiments on electron waves (for the reasons mentioned above). Little did I know that a very appropriate and accessible system was under study in the basement of Becton Center at Yale, in Richard's lab, around that time. Unfortunately, I cannot remember which member of Richard's group gave the solid-state seminar I attended, or exactly when it was in the spring of 1994. However, during that seminar, for the first time I started to understand that his group was studying the resonances and laser properties of small droplets, with an injected dye to provide gain when pumped, leading to lasing. When the droplet was far from the orifice at which it was produced, surface tension made it highly spherical, and it would lase on the well-known high quality factor (Q) whispering gallery modes of a sphere. Whereas, near the orifice, it was strongly deformed, and bit further downstream this deformation led an oscillating, non-spherical droplet, with a well-defined, primarily quadrupolar deformation from a sphere. While in that state its lasing properties could be studied, and highly anisotropic but reproducible field intensity patterns were observed. The droplets were much too deformed to analyze these modes in terms of the standard whispering gallery modes of a sphere.

A comment was made in the seminar to the effect there wasn't even a qualitative understanding of the laser emission from the deformed droplet lasers. The thought struck me suddenly: these highly deformed droplets were, in an essential way, like a 3D version of a "quantum billiard". The light within the droplet, in the ray optics limit, travels in straight lines between scattering specularly from the boundary and a complicated ray dynamics results, due to its non-spherical shape. The fact that the droplet itself was changing shape as it fell was not very important, since the time scale of resonant scattering of light was very short compared to the elastic oscillation time, so that the droplet was essentially static on the time scale of laser dynamics (although the shape of the cavity varied between prolate and oblate as it fell, leading to varying ray dynamics). However, there was something essentially new in the system compared to an ideal quantum billiard: the system was leaky in phase space, since when the angle of incidence of a ray fell below the critical angle for total internal reflection it would be subject to refractive escape according to the Fresnel formula. There was a "trapping" region in phase space where the ray motion was very similar to a closed system, and then a rapid escape near the critical angle.

This was similar to studies of 2D billiards with escape channels in certain spatial regions, which had been studied before (including by me), but interestingly different because the escape channels were defined by the angular momentum of the rays and the "holes" were in phase space. It immediately struck me that it would be possible to implement ray evolution in phase space to find the fate of ray bundles. This could possibly then be used to predict the emission pattern and typical Q values of the higher Q resonances, which would be relevant to lasing of deformed droplets, and might reveal structural property related to the KAM transition. Of course, I realized that we might also have to include wave and interference effects to correctly describe the system. I had found a system which might be controllable enough to study both classical (ray) chaos and certain wave effects associated with the transition to chaos! Immediately I told Richard with excitement that I thought there was some interesting basic physics relating to classical and wave chaos that could be studied in his deformed droplets. He was very interested and encouraging to me, and so the very next day I went to the office of a relatively new graduate student in my group, Jens Nöckel (now professor at the University of Oregon) and said: "I think I have a great idea for a thesis topic".

I then launched into a proposal for modeling these droplet resonators as deformed 3D chaotic billiards. Initially we could study the emission from such systems through Monte Carlo style ray simulations (in phase space, in contrast to the familiar ray tracing); every time a ray hit the boundary we would plot the position on the boundary as the x-coordinate and the sine of the incidence angle as the y-coordinate. Such plots are known as a "surface of section" and extensively used in analyzing nonlinear dynamical systems (see Fig. 1). Eventually we would validate/compare the predictions of such a ray model to full-wave simulations (averaged over resonances) and see if and when the wave nature of light needed to be taken into account. We both agreed that this seemed to be a rich, interesting and novel project. Three years later Jens' thesis, *Resonances in Non-integrable Open Systems*, was approved with distinction, and later he shared the 1998 American Physical Society (APS) Division of Atomic, Molecular and Optical Physics (DAMOP) outstanding PhD thesis award for this work.

We decided not to immediately attempt to use this model to explain and predict properties of lasing droplets, preferring to work with effectively 2D systems, closer to standard dynamical billiards, which could be achieved by studying cylindrical systems with a uniform deformation from circular symmetry. We began using the term "dielectric billiards" for such systems in the ray limit, which were thus subject to refractive escape. We immediately implemented the ray escape model for deformable cylinders and the results indicated that indeed KAM physics controlled the escape of the rays. Richard was all in on this idea, and provided valuable guidance, although this initial model was too crude for comparison to his experiments. The first paper was produced within several months, submitted to *Optics Letters* in June 1994, and published in November [1]. It focused on our finding that the average Q values of the higher Q resonances (relevant to lasing) had a sharp threshold deformation above which Q decreased with a power law characteristic of the KAM transition. However, we also found that despite the emerging ray chaos, the light emission was highly directional (and not pseudo-random, as one might naively guess).

This was a relatively easy property to study experimentally and we immediately set

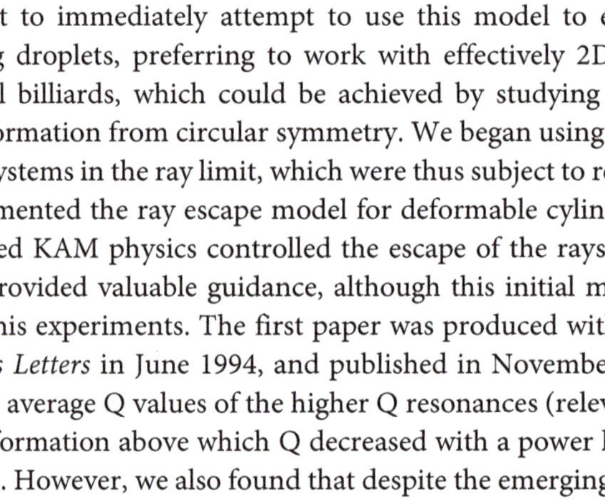

Fig. 1: (a) Surface of section plot showing at what polar angle the chaotic dynamics allows rays of light to reach the critical (Snell) angle for refractive escape; this predicts the angle at which the deformed droplet should look brightest. Blank regions are inaccessible to the rays for this value of L_z, and (b) measured lasing intensity pattern from lasing droplets, showing bright spots as predicted; modified from Ref. [2].

out to apply it to Richard's experiments on lasing droplets. At this point we were lucky to have a talented undergraduate, Attila Mekis, working with my group, focusing on how to extend the framework to deformed droplets which varied in shape in three spatial dimensions, unlike the cylinders. In principle this system has a larger phase space making it harder to analyze and visualize. Fortunately, we were helped by the fact that the deformed droplets still maintained rotational symmetry around the vertical (z) axis, meaning that the z-component of angular momentum, L_z, was conserved, and dividing up the phase space into many slices of different L_z. One could see that when L_z took on its maximum value the ray motion was in the equatorial plane of the deformed droplet, which has a circular cross-section which generated no chaotic motion. This would lead to high-Q whispering gallery lasing modes with little emission. In contrast, when $L_z = 0$, the ray motion was in a plane slicing through the poles of the droplet, leading to highly deformed cross-section and highly chaotic motion, generating low-Q modes which wouldn't lase.

The lasing modes producing the most output power, hence dominating the emission pattern observed, would be at some intermediate value of L_z, which translated into two bright spots in each hemisphere at the points of maximum curvature for the prolate deformation. For oblate deformations the theory predicted only one bright spot in a region around the equator. Both predictions were confirmed by experiments (see Fig. 1(b) for prolate case) [2]. This success of the theory was very exciting to both Richard and me; our outside-the-box theoretical model had explained the emission patterns that were previously a mystery, and which had started the whole project in the first place. Our paper comparing the theory to experiments was published in *Physical Review Letters* in 1995 [2], quite a coup since it wasn't easy at that time to get optics papers into that journal. We benefited from the fact that we had brought together ideas in nonlinear classical dynamics to explain properties of laser emission, something of general interest by connecting different areas of modern physics.

For the next three years Richard continued related experiments, switching from lasing droplets to studying a continuous stream of ethanol containing a lasing dye from an orifice, with cross-section approximating the deformed cylindrical dielectric resonators we had studied first theoretically. We found results consistent with the ray theory, including an anomalous splitting of the bright spots into two, which can be explained by the presence of stable ray orbits at the critical angle for certain shapes and deformations [3]. Shortly thereafter (in 1996) Jens, Richard and I applied for a patent for "Asymmetric Resonant Optical Cavities", which was eventually granted in 1998 [4]. Further observations of droplet emission, performed by S.-C. Chang from Richard's group, were also explained from this theory in a paper published in 2000 [5].

However, a major missing piece on the theory side was validation of our ray model with exact solutions of the Maxwell's equations for the resonances of such deformed cavities. This was also achieved around the same time, and to our great satisfaction the resulting paper, with significant extensions of the theory, was published in *Nature* [6], the cover of which was graced with a beautiful graphic produced by Jens Nöckel.

The last phase of my collaboration with Richard was in the period 2001–2004, after Jens Nöckel had gotten his PhD, and two new talented PhD students were working in my group on chaotic lasers, Hakan Türeci (now professor of Electrical Engineering at Princeton), and Harald Schwefel (now professor of Physics at the University of Otago). By this time new experiments were being done with semiconductor lasers in Richard's lab. His PhD student, Nathan Rex, was producing interesting data about lasing from deformed cylindrical microcavities which were not easily explained by our previous ray model. This led to an intense period of close collaboration, in which we discovered that wave effects really mattered for these smaller lasers and we were able to include them in the theory. In particular, the high emission directions were not always well predicted by simply launching rays and collecting them in the far field; instead, certain unstable periodic ray orbits seemed to dominate the emission pattern, knows as "scars" in the quantum chaos terminology. Moreover, the light emitted from these scars didn't go in the direction predicted by Snell's Law, due to an effect we called "Fresnel Filtering". This work also made it into *Physical Review Letters* [7].

A year later the same team began working on more highly chaotic dye laser provided by J. Zyss from École Normale Supérieure (ENS) Cachan. Here we discovered that the high emission directions from such lower index cavities could be predicted by looking at the flow of rays near a few short unstable periodic orbits. This turned out to provide the best insight for how to implement the ray model in a fully chaotic setting, and led to one of our most cited papers (with over 220 citations) in *JOSA B* [8].

Just around the same time another of Richard's PhD students, Grace Chern (now Grace Metcalfe) was working on InGaN semiconductor lasers with Richard along with a team from industry. They implemented a different cavity design, proposed by Richard, in which the radius of a nearly cylindrical cavity spiraled outwards, and then abruptly jumped back to the initial radius at a sharp notch. The idea of this design was to get primarily a single output beam from the notch in this cavity, whereas the chaotic cavities we had proposed all had produced several beams. This worked quite well and led to something of a breakthrough in this type of microlaser. The theory of this laser was also interesting and different from the chaotic cavity resonators.

This became our second most cited paper from the collaboration [9], after the *Nature* paper, and led to a second patent, granted in 2006 [10]. Sadly, this work, published in 2003, was my last joint paper with Richard. Our collaboration had produced 8 publications, multiple PhD theses and two patents, which up to this time have received over 2200 citations. Richard then turned his research direction to the detection of toxic particles from nonlinear scattering, partially as a reaction to 9/11 (I believe) and his desire to do something for his country. Just a few years afterwards his health began to deteriorate, leading to his retirement from research in 2007, a major loss to Yale, and to the field of microcavity optics in general. His subsequent physical challenges and untimely death is a tragedy for all who knew him.

The events of our collaboration occurred more than twenty years ago as I write this remembrance, so my many detailed conversations and interactions with him are difficult to recall. But I will never forget my admiration and appreciation of Richard as a scientist, a man of humility and great personal integrity, and a friend of graciousness and warmth. I will end by quoting from remarks I gave at his retirement event at Yale in 2008.

[Richard] has given me a fabulous gift, by making it possible for me to enter the new field of optical physics in mid-career. This would have been difficult, if not impossible without his help. Richard is an anomaly even among the many genial collaborators with whom I have worked in my career. With Richard it was always: how could he help me out, how could he deflect credit from himself and onto me and our students.

At various times he has gotten me invitations to speak, given me more than my share of joint grants, allowed my students to work in his lab, and even supported them when I was short of grant funding. I don't feel that I have been able to return these favors in equal measure, although I have tried. But at least I can thank him publicly and let him know how much I have appreciated his support as well as his scientific contributions.

I am a theorist, who knows basically nothing about what goes on in the lab except to try not to look directly into the laser.

Richard always approached me with great respect, but he reminded me of Senator Sam Ervin during the Watergate hearings, "I'm just an old country lawyer, but ...". Richard would pretend he didn't understand much of the theory, but then would provide a key insight, suggestion or correction that would make all the difference.

I have also admired how late in his career, and faced with personal challenges, he committed himself to projects to help the country addressing the energy problem, biodefense, medical and security technology, not to mention helping the economy with all the successful startups his mentees have created. I must say that Richard is perhaps the most beloved advisor/mentor I have known in my career. At his 60th birthday party his former mentees just wouldn't stop repeating how wonderful he is — I said to myself, "wow, these folks really love this man". The community that Richard has created through his life and science is not a matter of luck, it is a matter of character. Richard has brought out the best in all of us who have been touched by him.

References

[1] "Q-spoiling and Directionality in Deformed Ring Cavities," J. U. Nöckel, A. D. Stone and R. K. Chang, *Optics Letters* **19**, 1693 (1994).

[2] "Ray Chaos and Q-spoiling in Lasing Droplets," A. Mekis, J. U. Nöckel, G. Chen, A. D. Stone and R. K. Chang, *Physical Review Letters* **75**, 2682 (1995).

[3] "Directional emission from asymmetric resonant cavities," J. U. Nöckel, A. D. Stone, G. Chen, H. L. Grossman and R. K. Chang, *Optics Letters* **21**, 1609 (1996).

[4] "Asymmetric resonant optical cavity apparatus," R. K. Chang, J. U. Nöckel, and A. D. Stone, U. S. Patent 5,742,633 (filed October 2, 1996; granted April 21, 1998).

[5] "Observation of emission from chaotic lasing modes in deformed microspheres: displacement by the stable-orbit modes", S.-C. Chang, R. K. Chang, A. Douglas, and J. U. Nöckel, *J. Opt. Soc. Am* **B17**, 1828 (2000).

[6] "Ray and wave chaos in asymmetric resonant optical cavities," J. U. Nöckel and A. D. Stone, *Nature* **385**, 45 (1997).

[7] "Fresnel filtering in lasing emission from scarred modes of wave-chaotic optical resonators," N. B. Rex, H. E. Türeci, H. G. L. Schwefel, R. K. Chang and A. D. Stone, *Physical Review Letters* **88**, 094102 (2002).

[8] "Dramatic Shape Sensitivity of Directional Emission Patterns from Similarly Deformed Cylindrical Polymer Lasers," H. G. L. Schwefel, N. B. Rex, H. E. Türeci, R. K. Chang, A. D. Stone, J. Zyss, *JOSA B* **21**, 923 (2004).

[9] "Uni-directional lasing from InGaN multiple quantum well spiral-shaped micro-pillars," G. D. Chern, H. E. Türeci, A. D. Stone, R. K. Chang, M. Kneissl and N. M. Johnson, *Applied Physics Letters* **83**, 1710 (2003).

[10] "Locally-outcoupled cavity resonator having uni-directional emission," M. Kneissl, N. Johnson, G. Chern, R. K. Chang, H. E. Türeci and A. D. Stone, US patent, 7,031365 (filed 5/2/2003, granted 5/18/06).

Letter to Richard on the Occasion of His Retirement from Yale in June 2008

April 18, 2008

Dear Richard:

At the occasion of your retirement I want to thank you for your friendship and the example you have set for the twenty-two years I have been at Yale. For twenty-two years, when we needed astute advice, we have gone to you (and Sunny). When we were newcomers to New Haven you gave us the wonderful dentist and brilliant (if wayward) plumber that we still use today; when we were newcomers to China this year, you told me not to miss HuangShan (the Yellow Mountain) which turned out to the highlight of our trip. More importantly, your work inspired a new research direction for me, which has changed my scientific career in a rewarding manner. And throughout our collaboration you always bent over backwards to give my students and me a larger share of the credit than we deserved. You would perhaps have received more credit for the several seminal works you pioneered had you been the kind of person who aggressively pursued recognition, but that never was your style. Richard, you are a superb scientist, a loyal friend and mentor, and an inspiration to us all.

With Warmest Regards,

Doug and Mary

Alfred Douglas Stone

Carl A. Morse Professor of Applied Physics and Physics

When I Worked with Richard

Johannes H. Eickmans, Ph.D.
Vice President Business Consulting North America (Ret.)
Bergisch Gladbach 51429, Germany

I would like to share some thoughts from the time and circumstances when I met and worked with Richard. I studied physics in Germany and was very keen to work at an American university for a period after completing my PhD. This came about because my thesis advisor, Prof. Andreas Otto, recommended that I should apply to work in Richard's lab at Yale University.

Jian-Zhi Zhang's mother, Professor Yu-Fen Li of Fudan University, Barbara and Alan Eckbreth, and Richard at the tailgate party in Harvard-Yale American football game, November 1985.

My wife and I arrived in New Haven, CT, in May of 1985 with 4 suitcases, great excitement and great expectations. We had never before been to the USA. Everything was new and uncertain.

We were very warmly welcomed by Richard and his wife Sunny. They both very lovingly assisted us in our first steps, such as finding an apartment, opening a bank account, and dealing with paperwork with the authorities.

Jian-Zhi Zhang's mother, Professor Yu-Fen Li of Fudan University, Barbara and Alan Eckbreth, and I at the tailgate party in Harvard-Yale American football game, November 1985.

Richard inspired my research and gave me a lot of freedom. I admired him for his creativity and learned a lot from him scientifically.

Besides work, he often invited my wife Gaby and me privately to his home. This would have been completely unusual in Germany for a professor to invite his students or post docs privately to his home. Through these private invitations and in many conversations, Richard introduced me to the Asian and American cultures, e.g., I got to know the phrase and the custom "tailgate party," when he invited me to my first American football game.

I think back very fondly and with gratitude to the almost 2 years I worked with him. Richard was certainly one of the most significant people who shaped me. I admired him both scientifically and as a person. He always had a smile on his face, was always very approachable, supported me in every way and seemed to do everything in an easy manner. Thanks to him, my time in the USA was certainly one of the most beautiful and memorable times in my life. I have a lot to thank him for.

Contribution to Richard Chang Memorial Book

William Acker, Ph.D. Yale '89
Executive Director
New York Battery and Energy Storage Technology Consortium Inc.
Albany, NY 12203, USA
acker@nycap.rr.com

July 23, 2021

When discussing this book, my wife commented that "Richard was a genuinely good person." Richard Chang's scientific achievements have been celebrated, it is all too fitting that we also discuss his impact on other people. In addition to being a researcher, Richard was a superb teacher both inside and outside of the classroom leaving a lasting influence on multiple generations.

During his long tenure at Yale, Richard educated and molded numerous students. His positive influence on so many is the result of what he taught us, both through direct education and by example. In thinking about his success and the success of his students, it is clearly built upon a keen intellect and scientific insight, but also involves his approach to problems and the way he interacted with collaborators and students. The latter has been exceptionally important to my career. In addition to teaching us physics, he taught us a management style and philosophy of life. One of respect, being the consummate gentlemen, encouraging excellence in yourself and others, being open-minded and constantly learning, and working and playing hard.

Richard's retirement was a catalyst for me to express to him something I hope he already knew:

"This is an excellent opportunity for me to thank you for all you have done for me over the years. Next to a man's family, his major professor and thesis advisor exerts the greatest influence on his life (at least in my case). You have always been a superb teacher, gentleman, and friend."

I know I am not alone in this sentiment. Richard's influence can be seen across the scientific community and beyond.

Tracy and I have many fond memories of Richard, his family and mutual friends. From gatherings at his house, frequently to welcome visitors from around the world, to discussions of travel or other interests, to the major events in our lives. Richard's love of scuba diving was effused in us resulting in some wonderful experiences. His attitude of fully embracing a full life was contagious.

Abraham Lincoln said "whatever you are, be a good one." Richard carried this to the next level, encouraging sufficient focus to be the best in an area. He personally achieved these high aspirations both scientifically and as a human being.

Right to left: Richard Chang, Sunny Chang, Tracy Acker, and William Acker (1989).

Right to left: Richard Chang and William Acker (1989).

Letter to Richard Chang on the Occasion of His Retirement from Yale University, June 2008

Dear Richard,

This volume, prepared on the eve of your retirement from Yale, is an excellent opportunity for me to thank you for all you have done for me over the years. Next to a man's family, his major professor and thesis advisor exerts the greatest influence on his life (at least in my case). You have always been a superb teacher, gentleman, and friend.

During your tenure at Yale, you educated and molded numerous students and contributed greatly to the advancement of science. I am quite proud of the research we conducted together. Your positive influence on so many is the result of what you have taught us, both through direct education and by example. In thinking about your success and the success of your students, it is clearly built upon a keen intellect and scientific insight, but also involves your approach to problems and the way you interact with collaborators and students. The latter has been exceptionally important to my career. In addition to teaching us physics, you taught us a management style and philosophy of life. One of respect, being the consummate gentlemen, encouraging excellence in yourself and others, being open-minded and constantly learning, and working and playing hard. For this, the years of further collaboration when I was with Texaco, the many more years of friendship, the introduction to the joys of scuba diving and so much more, I thank you.

Best wishes,

Bill

William P. Acker ('89)

My Interactions with Richard

Lei Xu
Professor of Physics
Fudan University
Shanghai 200433 China
lei_xu@fudan.edu.cn

I am a physics professor at Fudan University in Shanghai, an institution with which Richard had formed a long and good relationship. I met him for the first time in 1986 when I was a physics graduate student at Fudan University. He came to receive an Advisory Professorship and gave a lecture on laser spectroscopy. Naturally there were many interesting comments and questions from both faculty and students. Stimulating scientific exchanges later continued through correspondence that resulted in Richard introducing a tool kit to our Laser Laboratory in 1988. During that period, there was no software to prepare scientific viewgraphs and all figures and characters had to be hand-written. The tool kit was a whole set of type matrix with various sizes, that one could use to prepare beautiful figures on transparencies and submitted papers. I used that often when I prepared my Ph.D. thesis in 1990.

In 1997, I met Richard at the International Conference on Laser Spectroscopy in China. I was then the Director of the Fudan Laser Laboratory and felt it was my duty to continue the good relationship with international scholars that the older generation of Fudan had established. Richard was of course a big name. During that meeting we discussed scientific research and areas of possible collaboration between Fudan and Yale. A young professor named Yongle Pan was sent to Yale and did research there until Richard retired more than some ten years later.

In 2004, I was the organizer of the National Conference on Laser Spectroscopy (NCLS). This conference had a long history starting from 1980, right after China reopened its door to the outside world and accelerating scientific research. I invited Richard and his family to Lijiang, a place famous for its unusual and marvelous scenic views. Prof. Y.R. Shen from Berkeley, and Prof. Jun Ye from Joint Institute for Laboratory Astronomy (JILA) also attended the conference. Together with Richard, they gave sharp but very important comments and views to Chinese optical scientists. Different from other scientific conferences, NCLS had a session that institution and lab leaders reported their major accomplishments in the past two years, and the audience were free to give comments and suggestions on the development of the group. Richard very much enjoyed the discussions and told me he was happy to hear my impression on a father/son relationship response to a question that I received during my talk. I believe Richard and his family found the trip to be memorable.

Although my scientific research started from atomic laser spectroscopy, I shifted to integrated optics in 1995. Our group had a solid background on fabricating glass optical waveguides by using organic/inorganic hybrid materials (called the sol-gel technique). High quality thin films with controllable refractive index could be easily prepared on silicon wafers or optical glasses. These films could also be patterned by exposure to ultraviolet (UV) light. Laser dyes could also be doped into the sol-gel precursors, so that the waveguides became excellent laser media. In 1998, triggered by Richard's research, I started to do some work on optical resonators. We fabricated microdisks on dye-doped glass substrates and successfully observed clear periodic resonances from the disks, when pumped by a pulsed 532 nm laser. We then put the device in liquid. When we changed the refractive index of the liquid, we saw movement of the resonances. This was the first time that an optical sensor based on whispering gallery modes (WGMs) was realized. Richard visited my lab in 1999 and I showed him the preliminary results. He was very excited and encouraged me to keep on exploring in this direction. Around 2002, he sent me a paper published in *Optics Letters*, which analyzed theoretically how a whispering gallery mode resonator can work as a sensor to largely lift sensitivity. Unfortunately, research from my group in that field had stopped during the 2 years I stayed at University of California, Berkeley.

Research in that area continued after I went back to Fudan in 2002. Although on-chip microcavities were much closer to final application, to fabricate high quality cavities was very challenging. This was later solved by Prof. Vahala's group from Caltech by using CO_2 laser annealing of silica plate to form microtoroids, which have ultra-smooth cavity boundaries.

I concentrated on optical fibers, because fibers have a very smooth shape naturally, and our organic/inorganic materials can form very good quality thin films on fibers, therefore the films are basically high quality-factor microcavities. I noticed that Andrew Poon, then a Ph.D. student of Richard, published a work on WGMs from a square fiber. I got some square fibers from Richard and asked one of my students to do some work on microcavity lasers by coating dye-doped organic/inorganic films on the square fiber. The results were fascinating, as we saw highly directional emission from the square microcavities. This was very interesting because WGM cavities have very high quality-factors, but very weak and direction-less emission. We submitted the work to *Optics Letters*, and it was quickly accepted. Meanwhile, Richard worked on directional emission from on-chip semiconductor lasers. He developed a series of microcavities that had non-circular shapes, especially the spiral microcavity, with the goal to realize directional laser emission. Richard visited China a few times in the subsequent years; each time he discussed his spiral microcavities with me. Later, I followed Richard's work and fabricated a dye-doped organic/inorganic glass spiral microcavity on silicon chips, and successfully observed clear WGMs with high directionality.

In 2006, I started to think about coupled microcavities. A high quality-factor optical microcavity is just like an atom and has been explored for several decades, but the work on coupled cavities was minimal. The main reasons were that modes in coupled cavities were more complicated and coupled cavities were more difficult to fabricate. However, when I use optical fiber as the framework for high quality-factor cavities, a coupled cavity can be simply formed by attaching two optical fibers together. I remembered the classic textbook example that a coupled laser cavity that can generate a single longitudinal lasing mode. The principle was that a particular mode, which matches the resonant condition of both resonators of the coupled cavity will have the lowest threshold to lase. My idea was to prepare two touching parallel optical fibers that have a very small difference in diameter with both fibers coated with dye doped organic/inorganic film, thus forming a very simple coupled cavity. The two cavities have a little difference in resonant spacing, therefore only the modes that match in frequency lase first. The free spectral range (FSR) of the coupled cavity is reversely proportional to the difference of the individual resonant spacing, so theoretically the FSR of the coupled cavity can be so large that only one whispering gallery mode exists in the gain spectrum.

Our experiment successfully obtained single mode lasing. Additionally, because the cavity breaks circular symmetry, emission from the cavity is highly directional. In the winter of 2008, I flew to Beijing where Richard was visiting his daughter and had an intense discussion with him for about 2 hours. At that time, Richard was not able to control his walking easily, but he insisted on walking by himself. He named the coupled cavity "figure 8 cavity" and said that the principle could be simply concluded as coming from the Vernier Effect. He also recommended to me a Korean group, who worked on calculating modes and light emissions from asymmetrical whispering gallery mode resonators. Our work was eventually published in *Optics Letters*, and I believe that this was among the first publications on coupled whispering gallery mode (WGM) resonators.

Richard was not only a top scientist, but he was also an excellent leader. Since 1980s, he maintained a very good relationship with our lab at Fudan. My colleagues, senior and junior, Professors Jiabiao Zheng, Shixiong Qian, and Yongle Pan worked in Richard's lab, and I believe that their stay in the lab had substantial influence on their future research career. Yongle Pan remained in the US, while Zheng and Qian returned to Fudan. Two Fudan undergraduates, Jianzhi Zhang and Gang Chen, went on to Yale to obtain their Ph.D. degrees. In 2005, I had a Ph.D. candidate Qinghai Song, who was very smart and hard working. He had already published a few high-quality papers in *PRL* and *APL*, when I asked Richard whether he could accept Qinghai to work in his lab for 3 months. I felt strongly that working experience in a top research lab was very important for a young researcher. I believe Qinghai's experience in Richard's lab changed his way of thinking a lot and he also made good connections with top researchers in the field. After 2 years of post doc work with Prof. Hui Cao, he joined Harbin Institute of Technology and subsequently was awarded some top national prize for young scientists.

In summary, to me Richard was not only a top scientist, but he was also a very close friend. His humor, elegance, and fantastic choice of research will be remembered forever. I love you Richard.

Memories of Richard

Jian-Zhi Zhang, Ph.D. Yale '90
Business Development Director
Corning Incorporated, Corning, NY 14831 USA
zhangjj@corning.com

I joined Richard's group the first summer of my graduate life at Yale in 1985. I graduated in 1990 with a Ph.D. I may be one of few who owe Richard a professional life and a happy family.

I remember Thanksgiving of 1984 vividly. This was before I joined Richard's group. I had arrived at Yale in the fall of 1984 from China. Richard invited the new physics graduate students from China to his home for Thanksgiving. He personally picked us up from Helen Hadley Hall. At his home, we met not only his family, but also international students, his son had invited from MIT. I was very excited to meet people from countries like Mexico and Vietnam. The international students took turns to try slicing a cooked turkey using an electric knife. The photo of me slicing the turkey is still highlighted in our family album until this day.

Richard really took care of his students. I remember him taking an extended leave to Asia in 1986. This was before the internet era, and email was not yet broadly used for communications. Each of his graduate students would report technical progress weekly, and his secretary, Susie, would mail him all the reports. Richard would write back in detail his response in the form of long letters, often using hotel letterhead. This form of supervision and care kept all of us going without interruption.

In the summer of 1985, Richard was kind enough to invite my mother* to Yale as a visiting scholar. To make sure her stay was pleasant and convenient, Richard personally spent time verifying the suitability of her housing arrangement I had made. The place turned out to be between Richard's home in Hamden and Yale. He would give her a ride to or from work, whenever his schedule allowed, and on more than a few occasions, Richard and Sunny would invite my mother for weekend breakfast. Richard even took my mother to watch the Yale-Harvard football game at the Yale Bowl. I had nearly finished my PhD thesis writing by late 1988. I was due to return to China after obtaining the degree per original arrangement. I told Richard my desire to see Europe before heading back to China. He immediately wrote to his friends in Germany and Austria about a visiting scientist guest position and received a few positive offers. I ended up choosing University of Konstanz and arrived there at the end of 1988. During my one-year stay in Konstanz, I witnessed the fall of the Berlin Wall, and the June 4th event in Beijing completely changed my plan to return to China.

I also met my wife in Konstanz. We started our family in the US, and raised two boys, who have by now both moved outside the home. We have kept in touch with Richard and Sunny throughout the years. In the summer of 2010, my wife and I visited Richard's home in Vermont and enjoyed delicious lobsters Sunny prepared. The last time I met Richard was in Taiwan, during the time I was working in Taiwan on an expat assignment for Corning.

I feel blessed to have had Richard as an adviser and our family treasures the relationship with Richard and Sunny throughout the years.

Editors' note:
*Professor Yu-Fen Li of Physics Department, Fudan University, ShangHai, China.

Here is a photo taken in May 1990 at the Yale University commencement. In the picture were Professor Zhi-Ming Zhang (left, of Fudan University, my undergraduate alma mater), Richard and me (right). It was a great honor to have Richard attend the ceremony in his Harvard gown.

Brief Encounters with Richard Chang

Nicholas F. Borrelli, Ph.D.[†]
Corporate Fellow, Corning Inc.
Corning, New York 14830 USA

I only met Professor Chang on a few occasions, as I recall twice when he visited our Corning Research Lab and once in New Haven where he invited me to give a talk. I must say I have never met in my career a more gracious and friendly man. His entire demeaner was warm and engaging. In New Haven, he treated me and my wife to his home and the best Chinese dinner I have ever had. We will fondly remember our brief friendship.

[†]Passed away

Knowing Richard Chang

Dipak Chowdhury, Ph.D.
Division VP & Technology Executive, Corning Inc.
Corning, New York 14830 USA
chowdhurdq@corning.com

I believe it was Fall 1989 when I first met Richard Chang at Yale University's Applied Physics Department. I was a graduate student at Clarkson University, Potsdam, New York, working with my PhD advisor Dr. Steven C. Hill on problems of non-linear light scattering from small particles. I had arrived in Potsdam in the fall of 1987 from Bangladesh and was 24 years old. Rakiba, my wife of 36 years now, arrived a month after I did. I still remember the day we experienced snow for the first time in Winter of '87. Growing up in a tropical country, experiencing snow for the first time was miraculous.

I heard of Richard from Steve Hill as a well-respected and well-known professor of Applied Physics at Yale University, who had done some pioneering work on nonlinear optics of laser illuminated liquid droplets. It was clear that Steve considered Richard his mentor. On top of that, when I heard that Richard had done his PhD at Harvard under the tutelage of Nobel Laureate Professor Nicolaas Bloembergen, my mental image of Richard was of an elderly professor with a severe demeanor who would never let a smile pass his lips.

In the fall of 1989, when I first met Richard, all my perceptions of him were shattered — he was energetic, compassionate, and caring. Most of all, I remember his smile when he greeted me for the first time, as if he had known me forever. Later, I realized that he had the unique capability of making anyone feel like they were the only person he was paying attention to in a large crowd. Of course, I did not know that in a couple of years I would be working for him as a post doc, a position I would hold for two years.

I arrived at New Haven the Summer of 1991 to work jointly as a post doc for Steve and Richard. We heard that New Haven was not the safest place to live and did not have any experience in finding suitable housing in the US. Richard and Sunny made this tremendously simple — we rented an apartment just outside of New Haven that they owned. It was a wonderful apartment at a safe location close to the lab. I still remember the first day I was at the lab. Richard took me out before returning home after work. He took me to a local store and recommended that I buy a steering lock for my car since many of the graduate students at the lab had had their cars stolen at least once. I remember when Sunny and he invited us to attend *Phantom of the Opera* at Woolsey Hall, where a world-renowned organist was playing the organ in sync with the original 1925 silent movie. Sunny helped find Rakiba a teaching position at the local public school. Richard and Sunny were doing everything they could to make us feel safe, comfortable, and useful. The experience we had from Rakiba's teaching at the inner-city school was a unique and priceless one that shaped our thinking about poverty, education, and race relations in the US. Richard and Sunny became our surrogate parents in the US.

Richard Holding Lubabah during their visit to our Hamden, CT, apartment, near New Haven, CT.

Our first child was born in New Haven. Of course, Richard and Sunny came to visit us after Lubabah was born. We spent a lot of weekends at their place with Lubabah and shared stories about their parenting. Our parenting skills were influenced tremendously by their experience and advice. During one of these visits, I noticed that Richard was busy working on some type of schedule. I asked if it was for an event. He surprised me by answering that he was planning dates, times, and locations that they planned to visit for a holiday in Thailand in almost a year's time. By then I knew that Richard was very organized, but to plan for visiting a particular location at a particular day and time a year in advance was not something I could have even imagined. As I write this note after multiple reminders from Sunny and at the last possible minute, I only wish that some of his planning skills had rubbed off on me.

After finishing my Post Doc in 1993, I joined Corning Incorporated and moved to Corning, New York. The reason I ended up in Corning had to do with being at Yale when a couple of Corning scientists visited Richard's group to find qualified talents to recruit. They were visiting Yale because their then-CEO, Jamie Houghton, had received some high-tech treatment at Yale University Hospital and had been very pleased with it. He therefore wanted to see if Corning could recruit some Yale students. Richard's guidance resulted in my getting the job at Corning in many ways — a story for another day. While at Corning, I had the good fortune to continue research collaboration with him.

Richard and Sunny with Lubabah during the visit to our home at Hamden, Connecticut.

I eventually left for a three-year Corning assignment in France, from 2007–2010. I heard that Richard was having some trouble with sleeping and restless leg syndrome. However, he seemed the same energetic and caring individual I had come to know over the years. Around this time, he was diagnosed with Parkinson's disease. He had a near fatal drowning incident around 2008. I recall coming to visit him from France during the Winter of 2009. Richard and Sunny were living at their Vermont home at the time. I vividly remember the conversation I had with Richard during that time. As I arrived at their home, he received me with his usual smile and welcome. It was clear that he was very happy to see me. I was, and still am, thankful that I made that visit. That was the last time I visited him while he recognized me. He described how he was forgetting who in the circle was living and who was not. I did not know what to expect, in terms of his physical ability. When he volunteered to show me the house, I was nervous. Sunny assured me that this would be good for him. He walked through the two-story house and showed me each of the rooms and described different things. I remember him standing on the doorway as I was driving out of the driveway.

As I was driving back from Vermont to Corning, I could not help but cry because of the sense of loss I was feeling already. At the same time, I felt very fortunate to have made the decision to come to the US from France just to visit him.

As it turns out, Richard's situation improved after this, and he was able to supervise his last graduate student Hermes Huang who defended his work in the spring of the same year. Richard and Sunny were able to travel around the world. I did see him one more time in New York City to celebrate Richard and Sunny's anniversary in 2012. However, it was a surprise event and Richard was surprised, shocked, and found everything a bit out of place.

I knew that Sunny and Richard had relocated to Singapore in 2014. Their quality of life improved with help that they were able to find in Singapore. Richard had a bad fall in November 2014 that put him in the hospital for about 7 weeks. His last trip to the US for about 3 months was in 2015. After that, traveling was too much for him to handle. Finally, in 2020, I heard from my friend Ali Serpengüzel, one of Richard's students whose time at Yale overlapped with mine, that Richard had passed away.

In life one meets people who are inspiring, caring, hardworking, and organized. Rarely are these qualities all embodied in one person. Moreover, one must be tremendously fortunate to meet such an individual at a time in life when their wisdom can shape one's own life. Rakiba and I were two of those rare fortunate ones. We happened to have met two such couples after our migration to the US. One of them was Richard and Sunny. Who we are today, in terms of values and what matters to us, are so much influenced by Richard and Sunny that we consider them our surrogate

Richard, Sunny, and I during 2009 visit to their Vermont home.

parents who shaped our lives in the US. I have not seen another person as life affirming and positive as Sunny. Her positivity was the only reason, in my opinion, that Richard had the length of life he had after the Parkinson's diagnosis.

It was a privilege to know Richard Chang, a hardworking, highly disciplined, and compassionate leader who changed the course of many lives. We have been in the US for more than 30 years and as we look back at the early years of our lives in the US and compare them with who we are today, signs of Richard's influence are distinct and vast. We feel fortunate that we had the time we had with Richard and Sunny. I was not just his student; he was the father figure of part of our life in the US. While we miss him, we strive to live up to his example every day.

Letter to Richard Chang after His 60th Birthday Celebration in 2000

Allan Bromley†
Professor of Physics and Dean of Engineering and Applied Sciences
Yale University, New Haven CT 06520 USA

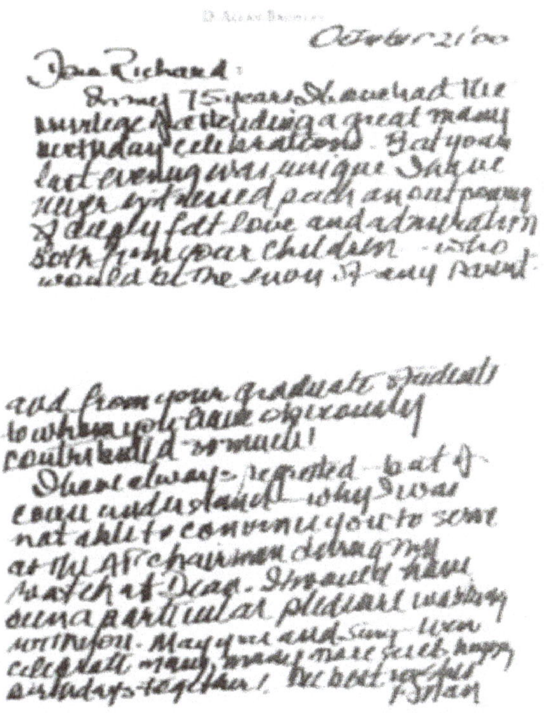

D. Allan Bromley

October 21 '00

Dear Richard:

During 75 years I have had the privilege of attending a great many birthday celebrations. But yours last evening was unique. I have never witnessed such an outpouring of deeply felt love and admiration both from your children – who would be the envy of any parent – and from your graduate students to whom you have obviously contributed so much!

I have always regretted that I could never understand why I was not able to convince you to serve as the [illegible] chairman during my watch as Dean. It would have been a particular pleasure working with you. May you and [illegible] celebrate many, many more such happy birthdays together! All best wishes,

Allan

†Passed away

A Contribution to the Richard K. Chang Memorial

E. James Davis
Professor Emeritus of Chemical Engineering
University of Washington, Seattle, WA 98105 USA
chedavis@uw.edu

As University of Washington Professor Emeritus of Chemical Engineering and Editor-in-Chief Emeritus of the *Journal of Aerosol Science* I am pleased to contribute to the memorial of Richard K. Chang. For many years I followed Richard's work on the application of morphology dependent resonances (MDRs) and bioaerosol characterization, and my own work on MDRs and surface enhanced Raman spectroscopy (SERS) of bioaerosols was influenced by his work. I much appreciate his extensive work on stimulated Raman spectroscopy (SRS).

It is difficult for me to recall my first contact with Richard since our paths crossed many times at scientific conferences for more than thirty years, but I certainly recall our last meeting. In 2004 Richard and I were keynote lecturers at the 7th International Congress on Optical Particle Characterization in Kyoto, Japan. At the main banquet of the conference, I sat with Richard and his wife, Sunny, and we shared travel experiences. Shortly before the conference I had visited Guizhou Province, China, taking a break from lecturing as Adjunct Professor of Chemical Engineering at Sichuan University in Chengdu, China. Sunny informed me that she came from Guizhou where her maternal grandfather had a distillery producing the famous Maotai liquor, called Hua Mao. Hua Mao is the predecessor of the present-day Kweichow Maotai, now owned by the Chinese government. After the founding of the People's Republic of China the government consolidated all of the Maotai distilleries and made Maotai (using the original Hua Mao process) the national liquor and the primary lubricant for the development of Guanxi (Chinese networking). Sunny left China in 1949 with her mother and sisters to join her father who was working at the United Nations in New York. Richard told me of his moves from Hong Kong, Chungking, Shanghai, Tokyo, and São Paulo to the United States for high school, his studies at MIT and Harvard and his long and successful career at Yale.

When I was Associate Editor of the *Aerosol Science and Technology* and later Editor-in-Chief of the *Journal of Aerosol Science* I called on Richard for reviews of manuscripts, and he was most cooperative and diligent. As former President of the American Association for Aerosol Research (AAAR, 1997–1998) I was very pleased to see Richard receive the 2005 David Sinclair award for sustained excellence in aerosol research and technology, an honor well deserved.

In 1991 I was surprised to hear from Richard in his capacity as a member of the awards committees of the Optical Society of America (OSA, now OPTICA) that my doctoral student, Theresa Allen, was given the Newport Research Award. The following year I was able to reciprocate somewhat by serving as the outside reader of the doctoral dissertation of Richard's student, Ali Serpengüzel. Upon Richard's retirement from Yale University, it was my privilege to contribute to the festschrift *Optical Processes in Microparticles and Nanostructures* dedicated to him and edited by Ali Serpengüzel and Andrew W. Poon. Thanks to Richard, Theresa and Ali have academic positions at the University of Idaho and Koç University, Istanbul, respectively.

For the many friends and colleagues who shared Richard's Guanxi I propose a toast to the memory of Professor Richard K. Chang, preferably with a glass of Maotai.

Good Old Days in the RKC Lab

Ali Serpengüzel, Ph.D. Yale '92
Professor of Physics
Koç University, Microphotonics Research Laboratory
Sarıyer, İstanbul 34450 Türkiye
aserpenguzel@ku.edu.tr
microphotonics.ku.edu.tr

Here is what I remember from those happy days in New Haven, Connecticut, USA.

How did I end up in the RKC Lab? During the beginning of PhD studies at Yale University, as new graduate students, we had to find a PhD thesis advisor, specifically for the first summer break. Stress, stress!

Towards the end of the fall term, there were open houses in pretty much in all the research groups, Dr. Carol Wood presented the RKC Lab, as Richard was on a trip, (later on we would learn that, this was quite typical of him) I was very much impressed by the RKC Lab, lasers! lasers! and wanted to join an optics lab, or an electronics lab, and RKC Lab with all its fancy lasers looked wonderful.

I was an engineer and a physicist. I wanted to be an experimental physicist. We did not have many of those in Boğaziçi University, or Bosphorous University, or former Robert College, the first US college outside of US.

When Richard returned from his trip, I must have requested an appointment. At one of the meetings, he said "Call me Richard." First name basis, a friend, that was unusual! That was not usual in the US either, it was him, Richard.

Every couple of weeks, Richard assigned papers for me to read and discuss with him. That's how I learned, what a scientific paper was. I knew of magazines, manuals, books, data books, and had heard of patents, since their numbers were written on products.

Every week of the spring term, we met, I was reading the papers, and explaining, what I understood to Richard. It was like an unofficial reading course.

There was no internet, we had to go to the various Yale libraries such as Physics, Chemistry, and Engineering, and the main Sterling library, find the journal volume, make photocopies (or xeroxes as they say in the US) of the papers. The journal citation indexes were also in form of directories, similar to telephone directories. There was no search engines, we had directories for everything.

I think, the first paper, I studied was on Raman scattering by Michael Raymer and Mostowski entitled "Stimulated Raman Scattering: Unified Treatment of Spontaneous Initiation and Spatial Propagation," *Physical Review* **A 24**, 4 (1981). I remember discussing these papers with Richard on his green glass board with white chalk. I was not accepted to the Chang group yet, so there was some apprehension and excitement in these meetings, I had to impress him.

At the end of the spring term, he was satisfied with my progress and offered me a summer job, by saying "Welcome on board". I copied him, and kept saying this adage to my graduate students later on.

I knew some optics, electronics, mechanics, and chemistry, before I joined Yale. However the rest of the subjects, such as photonics, optoelectronics, piezoelectrics, pneumatics, and hydrolics, I enjoyed learning in the RKC Lab.

Richard as a teacher: I took two courses from Richard, first solid state physics, which he and Victor Heinrich taught together. That was quite exciting, as I never had solid state physics before.

Then, there was a nice introduction to quantum electronics/nonlinear optics course. He invited experts from the industry to his course like Alan Eckbreth from United Technologies, who were using nonlinear optics to measure combustion diagnostics in the development of aircraft engines. That was so exciting: industry was performing applied research to develop their state-of-the-art products. There were also some professional students, so the courses were later in the day and Richard would make sure that, there would be a good supply of coffee and cookies for us.

How were the chemical, pneumatic, hydraulic, piezoelectric, mechanical, electronics, and optical components supplied? We would go to the onsite machine shop, if required. We could have gone to the glass shop in the Chemistry Department. Or we would go to the Physics Department store room to get electronic components. Or we would go to the Chemistry Department store room to get chemicals. The nitrogen tanks were available at the Mechanical Engineering basement in Mason Building. We would charge these to the RKC Lab in Becton Center. We could also buy small items from the market and later get reimbursed as petty cash.

How were the chemical, pneumatic, hydraulic, piezoelectric, mechanical, electronics, and optical components and equipment supplied? For the components not found in the store rooms and for the equipment we had to go to the Purchasing Department to get a purchase order (PO) number, then we would give these to the company, we were ordering from. When the order came, it would be delivered to the group secretary, Susie, and she would call down to the lab from the fourth floor, "There is a package. Can someone from the Chang group please come up and get this package?" Then the item will be picked up and delivered to the right lab and would be included in the experimental setup.

What is expected of a research assistant? Richard was running a tight ship, he specifically said that a research assistant has to do three things: "maintaining and upgrading of the laboratory systems, performing project experiments for applied research, and performing scientific experiments for basic research." He used to say that the projects were our "bread and butter," which are what they are for any experimental research group. We had to adapt the latest scientific equipment to illuminate the way for the next generation technology and science.

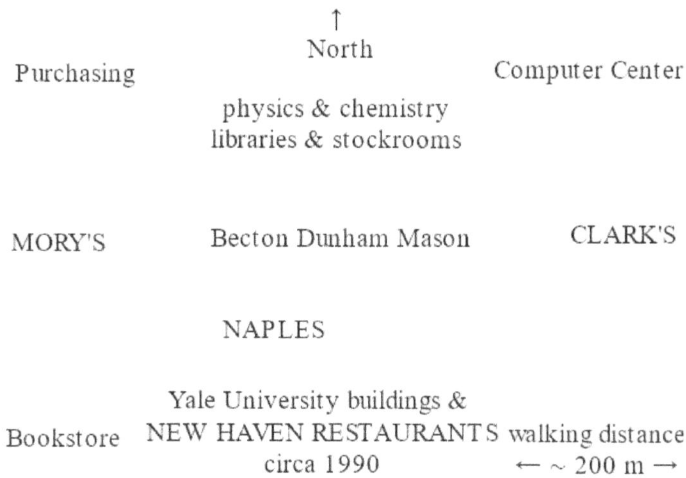

How was life in RKC Laboratory? Looking back, I can say that life in the RKC laboratory was fun. There was a certain routine. We would go to our offices in the basement of Becton Center by bicycle during the day and by car in the evening. We could study in our offices. There was always filtered coffee in the basement. That's where, I started drinking coffee. This might come as a surprise, since I am from Türkiye. Back then, Turkish Coffee was reserved for adults (meaning married couples), not students, there were no coffee shops around every corner, and the most popular drink was, and still is, the tea.

Anyways, we would come by bicycle to our offices in Becton Center in the morning, if we had experiments planned, we would work in the lab till lunch time. Then we would go out for lunch to Naples or Clarks. After lunch, we would continue to work in the lab or our offices. Then we would go to our apartments for dinner and come back to the lab to work until midnight.

Arriving by car to the department during the daytime was the privilege of the professors. The students could only drive in the evening and park in the university parking lot. If you parked on the street during the daytime, you had to feed the mechanical parking meters, and also pay close attention to the street cleaning days and times, not to get your car towed.

Life would go on like this, till some vacation such as Thanksgivings or Christmas. Then the domestic students will leave the Yale campus, and the international students would stay all alone. That would be the time for the famous Thanksgiving celebrations at the Chang's home in Hamden and at Shangri Li in Hamden, where Richard would treat us. Chang family was 100% integrated with the Chang group.

How was the group managed? Richard was the boss, the leader, the advisor, the professor, the teacher, the trainer, and the coach. Susie was the group secretary, the communicator, the accountant, the manager, the English teacher, and the reviewer.

How was the experiment setup designed and setup and run? We setup our experiments on optical tables, used optomechanical holders and positioners for our optics such as mirrors, lenses, filters, polarizers, prisms, waveplates, and detectors. The lasers were either placed on the optical table or next-door tables, the spectrometers were placed on the optical tables, and the linear array detectors were placed in the place of the exit slit. The electronics that was not readily available were homemade. I knew about breadboarding, but I learned wire wrapping at Yale. There was also a body system in the RKC Lab, although we were in different labs, there were always group members running experiments simultaneously, there were also graduate students from Long group, and other neighboring groups in the basement of the Becton Center.

How were the data gathered? We used Digital Equipment Corporation (DEC) PDP-11s and later Apple Macintosh IIs for data acquisition, as well as 8 inch floppy disks and later 3.5 inch floppy disks for data storage. Our generation of graduate students moved the RKC Lab from the PDP11s to Macintosh IIs. Everybody used the optical multichannel analyzer (OMA) program of Marshall Long for data acquisition. The OMA program was in a continuous state of development and was applied to all the detectors of the Long and Chang Labs.

The data were also recorded in an official chemically resistant bound Laboratory Notebook with ink, where nothing is erased, just crossed over, all pages are numbered, no pages are torn. The laboratory notebook was the proof for priority of invention, although, I believe this later (2013) changed in the United States, and patents became first to file, rather than first to invent.

How was the data analyzed? After experimental data collection or numerical simulation, there were many one-on-one meetings with Richard, then group meetings, the spectra or signals were printed on US letter size white paper, later on US letter size transparency slides, to be shown on overhead light projectors. Even slide projectors were used for data taken on 35 mm slides, but this was not very often. Unfortunately the slide projectors disappeared before the overhead projectors, and digital data shows became the standard now. On those days, analog data shows were only available on transatlantic in-flight entertainment.

How were the simulations performed? We were initially using the mainframe at Yale Computer Center (YCC) on Whitney Avenue over terminals, later on we got a group Workstation, and we started using the Apple Macintosh IIs. One could also purchase a Macintosh at an academic discount at Yale Computer Center (YCC) on Whitney Avenue. We were initially using the Apple Macintosh II for word-processing, later on we used the Apple Macintosh II for data acquisition as well.

How was the research written up and published? I did not know how to write a scientific paper, so I was emulating what our group mates were doing. Richard taught me that the paragraphs had to be connected, that the last sentence of one paragraph had to lead to the next paragraph. That was something he learned from one of his professors. I should now reorder my paragraphs, accordingly! He also insisted that we should make our figure captions explanatory enough so our papers would be almost understandable by reading the figure captions. Richard also gave an importance to tables. Tables were used to summarize our results. Then inputs and the outputs of the experiments could be summarized in form of tables. Since then, I have been recommending my students to use tables to summarize, what they know, and what can be deduced in form of knowledge/innovation from, what is missing from these tables.

After many iterations, we would FEDEX our manuscript to the journals, get the reviews, do the revisions, and FEDEX the revised manuscript to the journals again, We would order approximately 50 copies of preprints, mostly in black and white. As color figures cost extra, we had to choose, which figures would be printed in color.

How was the research communicated? Conference on Laser and ElectroOptics (CLEO) collocated Quantum Electronics and Laser Science (QELS) was the conference for us if it was in Baltimore, MD. If the conference was in San Jose, CA, we could only go after postdoc. We would rent the most affordable car from Thrifty, and drive down to Baltimore, MD for the conference. We would leave New Haven in the morning, and stop for lunch and driver exchange in New Jersey. We would arrive in Baltimore close to the evening to stay at an affordable hotel during the conference. Richard would treat us to a very nice local restaurant one night.

Before the conference, we had to review our presentations approximately five times, two to three times to the group without Richard and twice with him. He coached us through our presentations over and over again in the Becton Center meeting rooms next to the elevators to make them logical and effective, and audible. He would often tell me that he could not hear me to make me speak up louder.

How do you keep the group in the excited state? As humans, we need something to look forward to, some excitement to break the routine. We never had a boring day in the lab, there was a constant flow of visitors, in form of project managers, academics, or industrial scientists. We had posters about our research in the corridors outside of the laboratories, and we would explain our posters to the visitors, we would also make presentations and get their feedback, which was very useful, when analyzing our data. We would also go to scientific conferences like CLEO & QELS and Gordon Research Conferences and workshops with other groups such as Steve Hill's group. Richard would also treat us to a Yale-Harvard football game or to a lunch at Mory's, if we were entertaining visitors from the industry. He would treat us at Mory's for each group member's graduation as well.

How was the research so successful? Later observations on the perfect mix of basic and applied research components show the following: Yale University brings excellent faculty from all over the world. In sciences and engineering departments the majority of Yale faculty were experimentalist running their individual research laboratories. Yale science and engineering faculty working on theory and numerical simulations were smaller in numbers. Couple side notes here: US university research laboratories are the brainchildren of their individual professors and are never shutdown or cannibalized. If a professor retires,

the lab is offered to a newly arriving faculty. In case a professor moves to another university the research laboratory and group moves with the professor. One can say that the value of the research laboratory is more than the sum of its components. Yale Graduate School brings excellent graduate students from all over the world. From what I remember from those days, there was approximately 5000 undergraduate students in Yale College, each year graduating about 1000 students. The same was true for the Yale Graduate School, about 5000 graduate students, and each year graduating about 1000 doctorates and masters, with considerably more doctorates than masters. Yale faculty brings excellent project funding from the research offices of different government institutions such as foundations, agencies, and departments, as well as industrial companies. Initially to the graduate student, these projects were just some letters and numbers at the end of the papers in the Acknowledgement section. Later on, it became clear that these funding institutions enable the development of their countries as they convert basic research in form of science, that the universities perform into applied research in form of technology in a span of 20 years through their respective research centers, and pass it along to the domestic industry, that coverts it into technologically advanced products, which enables the next stage of basic research, as well as contributing to the growth of the country's economy. One can say that these countries operate in basic research, with strategic applied research, and the tactical outputs are novel high technology products. Any one of these universities, technology centers, and production facilities can be non-profit such as state institutes or foundation institutes, or private forprofit companies.

Overall, it was great fun to work with Richard and colleagues in RKC Lab on the science and technology of nonlinear optics in droplets and sprays.

Letter to Richard on the Occasion of His Retirement in June 2008

Alfred Kwok[†]**, Ph.D. Yale '92**
Professor of Physics,
Pomona College,
Claremont, CA 91711 USA

Dear Richard,

When the Lambda-Physik dye laser arrived in the basement of Becton in Nov 1991 a few weeks ahead of schedule, you commented that it was an "early Christmas present." The dye laser was indeed a great present because it allowed us to observe how lasing and SRS interacted with each other in droplets. However, the greatest gift you gave me was your model of hospitality. All of us can attest to the Thanksgiving feasts and the graduation parties on Paradise Avenue. I can still remember my first dinner at your place when you invited David Leach and me over only a few days after I arrived at Yale. More importantly, your hospitality was extended to my family as well. After you learned that my parents were visiting me (from Hong Kong) a few months after I arrived at Yale, you invited us to lunch immediately. I have worked for/with three other PI's since I left Yale, and none of them have made me feel so much like family. While I cannot follow your scientific footsteps, I have tried hard to follow your footsteps as a mentor. There are a few dozen Franklin & Marshall and Pomona students who owe you a big "Thank you" for the dinners they enjoyed in my house.

Last, but not least, thank you for giving me the privilege to be your first graduate student from Hong Kong.

Fondly, Alfred

[†]Passed away

Letter to Richard on the Occasion of His Retirement in June 2008

Donald M. Engelman
Eugene Higgins Professor of Molecular Biophysics and Biochemistry
Yale University, New Haven, Connecticut 06520, USA
donald.engelman@yale.edu

Yale University

Donald M. Engelman
Eugene Higgins Professor of Molecular Biophysics and Biochemistry

Department of Molecular Biophysics and Biochemistry
266 Whitney Avenue
P.O. Box 208114
New Haven, Connecticut 06520-8114

Telephone: 203 432-5600
Fax: 203 432-6381
Email: donald.engelman@yale.edu

29 April 2008

Dear Richard,

I remember so many good times we shared when my stepmother, Susie Engelman, was working as your assistant. She always enjoyed her life with you in Becton, and I know you had a warm personal relationship as well. She would wish to celebrate this transition with you, I am sure, and would hope that you have a splendid time in retirement.

I, too, offer my congratulations on your fine career, and my expectation that you will fully exploit the freedom that retirement offers. Your energy and creativity now have no constraints, as you apply them to living life to the fullest extent!

All the best,

Don

Letter to Richard on the Occasion of His Retirement in June 2008

Patricia Brodka
Administrative Associate
Yale University
Department of Applied Physics
New Haven, CT 06520 USA

April, 2008

Dear Richard,

It seems like it was just last week when you started thinking about retiring "someday." How quickly that someday arrived! The last fourteen years have gone by very fast. I've been in Chemistry four years now and have missed you every day.

Always the quintessential teacher, I learned so much from you. Just watching you deal with students and colleagues was an education in itself. But it was what you taught me about myself that was the real lesson and, one of the reasons I moved on. I remember when I first started in Applied Physics, Arlene Bjorkli would say "Oh, Richard is so demanding! Suzie is the ONLY one who can handle his work." Of course, I was terrified of you! Then, during quiet times, I'd sneak into Suzie's office to ask if she needed any help. Little by little I came to understand that it wasn't that you were demanding it was just that there were many tasks to be completed and they needed to be done correctly to make the lab run smoothly. I don't know if you remember, but in between Suzie and Katrina, I would work for you in the afternoons. It was then that I realized that method and organization made your lab function like a well oiled machine.

I will never, ever, be able to thank you enough for rescuing me from the Provost's Office. There were times when I doubted myself in handling the work that you trusted me with. But the fact that you did trust me, and treated me like a colleague, made me rise to the occasion. From there, I gained confidence that I could do more. But, you did that with all of your students, and that is why we love you.

I hope you are able to spend some quiet time in Vermont. When you've rested some, I'm sure your thoughts will turn back to science. But this time, you'll be able to enjoy the subject and have impassioned discussions without worrying about how it will lead to supporting your group!

Richard, Walter and I wish you, and Sunny, the very best that retirement brings. Enjoy.

With love and admiration,

Pat

Letter to Richard on the Occasion of His Retirement in June 2008

Thomas Heldt, M.S. Yale '97, M.Phil. Yale '98
Professor of Electrical and Biomedical Engineering,
Electrical Engineering and Computer Science
Massachusetts Institute of Technology
Cambridge, MA 02139 USA

Dear Richard, April 20, 2008

It is with great pleasure that I write this letter on the occasion of your impending retirement from Yale's faculty, as it gives me the opportunity to express my sincere gratitude for the way in which you have influenced my personal and professional life.

I arrived at Yale in 1995, and naturally gravitated toward the applied side of physics given my dual interests in physics and medicine. I remember meeting with you to secure a summer research position after my first year of course work. I had checked out a bunch of books on laser physics prior to our meeting and had studied them quite thoroughly. I was rather surprised (and caught somewhat off-guard) when during our meeting, you never once talked about physics in any great detail, but rather asked about my personal background and my general interests. I remember leaving your office wondering how you would be able to decide whether or not to hire me on the basis of what had been said. Needless to say that when years later I started to interview young students for research positions, I found it very helpful to first understand a student's background, interests, and goals before assessing the student's level of technical competency.

I also remember my first presentation during our weekly group meetings. I thought it went alright. After the group meeting, however, you asked me to stay behind. You talked me through every single transparency I had prepared, pointing out the important aspects of our work, how I could have made my points more precisely and succinctly, and what I should have left out. Your clarity of thought, both in your presentations and your writing, has always inspired me and has set a high standard to which to aspire in my own work.

Your most significant support of my career came when I decided that I wanted to continue my physics training with my interests in medicine in a more coherent manner than what was possible at Yale at the time. You strongly advised me to apply for a spot at the Harvard University — MIT Division of Health Sciences and Technology. I am convinced that I would never have been admitted to this program had it not been for your strong support of my application and your encouragement to apply in the first place. (It probably didn't hurt that the director of the program at the time, Roger Mark, was a former classmate of yours at MIT!)

As I look back on my time at Yale, your support and guidance in my first steps toward independent research and my attempts to navigate the academic environment are quite apparent. I count myself very fortunate to have benefited from your mentorship, Richard.

I wish you all the best for your retirement, and I hope to stay in touch with you for many years to come. My parents asked me to forward their best wishes, too. They fondly remember the evening celebrations at your home in Hamden on the day of my graduation.

With best wishes for your future,

Truly yours,

Thomas Heldt (M.S. '97, M.Phil. '98)

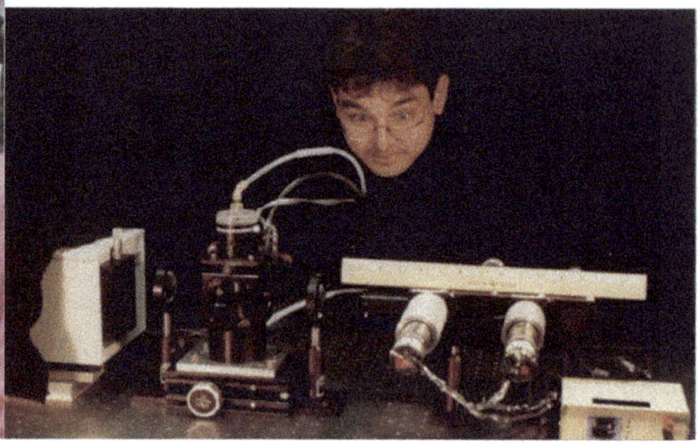

Memories for Richard on the Occasion of His Retirement in June 2008

Paul Gang Chen, Ph.D. Yale '94
Research and Development Manager
Foxconn Interconnect Technology, Ltd.
Irvine, California 92617, USA

Richard and Gang Chen (Ph.D '94) at the 1994 graduation.

Richard, Prof. Jiabiao Zheng (a visiting professor from Fudan University, China) and Gang Chen (Ph.D '94) at the 1994 graduation.

Richard, Chris Swindal (Ph.D '94) and Gang Chen (Ph.D '94) at the 1994 graduation.

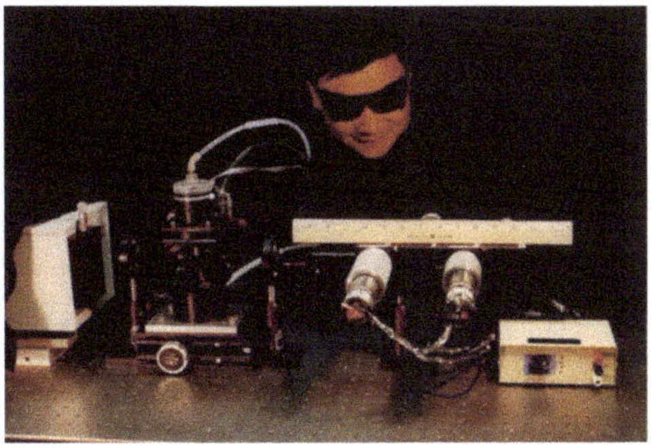

Letter to Richard on the Occasion of His Retirement from Yale in June 2008

Robert J. Schoelkopf
Sterling Professor of Applied Physics and Physics, Yale University, New Haven, CT 06520, USA
robert.schoelkopf@ yale.edu

Yale University
Department of Applied Physics

Robert J. Schoelkopf
Professor of Applied Physics and Physics
Phone: (203) 432-4289
e-mail: Robert.Schoelkopf@ yale.edu

423 Becton Center
15 Prospect Street
Box 208284 Yale Station
New Haven, CT 06520-8284

Tuesday, May 13, 2008

Dear Richard,

I just wanted to write a short note of congratulations and good wishes on the occasion of your retirement.

My strongest recollections of our time together date from my early days in New Haven, when I was still a postdoc. I wanted to get some experience in teaching, and wound up assisting you in teaching junior electromagnetism, AP 322. I remember fondly the couple of years when we "shared" this course, as I tried to help out with some demonstrations and lectures about Smith charts and the mysteries of transmission lines. Since this was essentially my first time doing anything like this, I remember it being a bit rocky, but it was still an invaluable experience when the big transition to teaching a whole course on my own came. I remember well the patience, encouragement, and enthusiasm you showed to the students, which I like to try to emulate myself, with probably varied success. I also remember doing the oral examinations of the students at term's end — a tradition I continue whenever I teach this course today! There are of course numerous and more substantial ways in which you have influenced the course of events at Yale and elsewhere, but that early collaboration in teaching is one example which I will remember in particular.

I have been very happy here (for more than a decade already!) in Becton, and in the small but very special department that is AP. I know how much of the character of the place comes from the examples set by all of my senior colleagues. So I especially want to thank you for your role in establishing the character and ethos of the department. We will try to carry the flag forward in the same way in the future.

All of us in Applied Physics have benefited greatly from having you as a colleague, mentor, and friend. Congratulations on a magnificently successful career, and I wish all the best to you and your family in your retirement!

Your friend,

Truly a One-of-a-Kind Academician, Scholar, and Teacher

Kang-Bin Chua, M.S. Yale '96, Ph.D.
Director, Autonomous Systems
Physical Optics Corporation
a Mercury Systems Company
Torrance, CA 90501 USA

Professor Richard K. Chang was truly a one-of-a-kind academician, scholar, and teacher. Of all the professors that I have met and interacted with, I have not come across anyone else like him. The first time I met him was while attending his solid-state physics class. The graduate course was very condensed and fast-paced, but he never rushed his lectures. He was patient in explaining the physics and effective in using graphical illustrations to explain physical phenomena, and he always emphasized the importance of digesting new technical materials slowly, so that his students could gain a deep intuitive understanding on the subject.

In one of my visits to his office, I learned that Richard loved scuba diving. Before I had attended Yale, I went diving on Sipadan Island (Pulau Sipadan) in the Sabah state of Malaysia. Sabah is on the northern part of Borneo and has many spectacular pristine diving sites. It just so happened that Richard had always dreamed of going diving there. He even named his dog Sabah. Since I was from Sabah and wasn't able to go diving with Richard in the summer of 1995, I arranged for my parents to give him and Sunny short tours of Kota Kinabalu and Tawau during their transits. They tasted local delicacies and tropical fruit such as durian that they had never eaten before. Later, Richard told me it was one of his most memorable trips.

Mataking Island, Sabah, Malaysia.

Although I didn't spend much personal time with Richard, while I was at Yale, we had more interaction after I left New Haven. I was drawn to him because he genuinely had keen interests in the well-being of his students, was generous with them, and cared very much about their welfare and career prospects after graduation. As a result, we often kept in touch after I left Yale. At one point of my Ph.D. thesis research at Dartmouth College, I had difficulty finding an ultra-short pulse laser in the ultraviolet (UV) for electron time-of-flight measurement in Germanium-doped silica. When I asked Richard if he had such a laser, he immediately asked me, when I wanted to use it, so that he would ask his postdoc to configure an ultra-short pulse laser in one of his laboratories to produce 3rd harmonic generation in the UV. I quickly planned my experiments, packed up my instrument, and drove to New Haven. After I arrived at his lab, he tasked his postdoc Yongle Pan to provide me with everything I needed for my experiments amid the busy schedule Yongle already had. This was just one of Richard's many generous deeds.

Richard's beautiful country home in Brattleboro, Vermont, is next to the Connecticut River bordering New Hampshire and Vermont, with foliage on both sides of the river. There were a few large trees in his backyard, including an oak tree. Sitting in the yard, while looking at the foliage and the river, is a relaxing experience. In one of our visits on a beautiful summer afternoon, we talked about a variety of topics ranging from his most recent sailing trip to my interest in precision target shooting and how my job search was progressing. He suggested that I consider applying jobs at small business companies, because they would allow me to have bigger impact in this environment.

Months later, while I was waiting for my job to start at National Institute of Standards and Technology (NIST) in Gaithersburg, Maryland, out of the blue a small business in Torrance, California, that I had not applied to called me for an interview. After the interview, the hiring manager persuaded me to join the company instead of going to NIST. He succeeded. And just as Richard had predicted, working in a small business innovative research company focusing on developing new technologies not only allowed me to have direct contribution to the project success, but also gave me the opportunity to work on the technical areas that I liked to pursue. So I was able to turn my passion for target shooting into several long-term contracts that resulted in a developed technology being transitioned to the government. To this day, these were the most enjoyable contracts, I have ever worked on, and I credit Richard for steering me on that path.

After Richard recovered from his drowning accident, he told me that he had an experience after he lost consciousness. He said he'd seen passed away relatives almost instantly. Whoever he wanted to see, they would appear from a subway-like carriage. He was amazed by the speed and efficiency of this transport process.

Later, he asked me, how I went through my leukemia battle and a bone marrow transplant, which had required finding an unrelated bone marrow donor. Finding matching bone marrow donors for Asians was very challenging in the 1990s because there was a low percentage of Asians in the national bone marrow registry. I told Richard that I was in great depression and fear and that those were the darkest days of my life. The only time I had peace was, when I fell asleep. However, as a Christian, I put my life in God's hands and I prayed to Him every day, asking Him for a perfect matching donor. And a perfect matching donor was miraculously found within about four months after I was diagnosed with the leukemia. Richard listened to my story and was very content with my answer.

Even though Richard and I didn't cross paths as often as others, who have worked alongside with him for many years, his character made a huge impression and a positive impact on me. The little things that he did out of his genuine caring and interest to help others, and his high standard in pursuing scientific research, helped shape my practices both at work and in my personal life. Not only he was a truly one-of-a-kind academician, scholar, and teacher, he was also a wonderful friend to me and to so many others. Richard, we will always be thinking of you.

August 31, 2021

Twilight (light scattering) in Los Angeles, CA, USA.

Memories of Professor Richard Chang

Mohiuddin Mazumder, Ph.D. Yale '96
Senior Principal Engineer
Intel Corporation
Santa Clara, CA 95054 USA
mohiuddin.mazumder@intel.com

I came to US during late December of 1990 and started my Masters program under the supervision of Professors Steve Hill and Peter Barber at Clarkson University in Potsdam, New York. Our team at Clarkson was modeling various phenomena that Richard's team at Yale was investigating experimentally. During the summer of 1991, we drove to New Haven, Connecticut to meet Richard's team for a review of the joint research collaboration work. I was extremely nervous about my first research presentation to an audience outside my group. The way Richard introduced us to his team and led some informal conversations made me feel a little more relaxed, and I was truly excited when he asked probing questions with genuine interest to better understand my work. We were invited for dinner at his home. I was surprised to see his whole team there and from their interactions with Richard and Sunny, it was clear that their relationship with Richard and his family extended well beyond their research boundaries. I was still suffering from home sickness and it meant a lot when Richard took the time to get to know me personally and how I was adjusting to the new US life. That evening was a turning point for my life. I realized that an opportunity to work under Richard's direct supervision would be extremely rewarding, and I started working even harder for that opportunity.

It is hard to imagine that thirty years have passed since my first meeting with Richard. Even though my direct work with Richard spanned only four years from 1992 to 1996, his influence on many aspects of my life remains strong even today. I will share a few stories about my interactions with Richard just to illustrate how they have shaped me to be what I am.

Richard motivated us, inspired us to ask probing questions during our weekly group meeting. He valued the work of each member regardless of the seniority in the team. He used to let everybody present his/her work, listen to all for their inputs, and act as a catalyst to accelerate the progress of each member. These all seemed natural for Richard, but I now realize that building and managing a high-quality team is a challenging task. Richard earned our trust and respect through his compassion and caring for every team member, his painstaking efforts to achieve higher quality, and his relentless pursuit of understanding the physics of various phenomena he studied and finding applications for them.

In one of the group meetings during first year at Yale, Richard informed us about an upcoming trip to United Technologies at Hartford, Connecticut to showcase our team's work. He asked each of his team members to prepare a 15-min presentation. I did not have much experience in giving presentations and I kept working hard to generate more results until the night before the day of our dry run with the team. I could not finish even half of my presentation in the allotted time. I felt very disappointed during the meeting. I was really scared when Richard called me to meet him in his office. He thanked me for all the hard work and paired me with one of his senior team members for working on improving my presentation. That was a great learning opportunity. He trusted me that I would do my best to improve it and congratulated me later for a good presentation at United Technologies. The way Richard handled my disastrous presentation was very graceful and taught me a few lessons on how to strive for high quality team performance without demoralizing any team member for occasional poor performances. I often share this personal experience with our children and my team members when I try to motivate them to achieve better results.

Richard was much more than a renowned scientist to us who were touched by so many of his kind and generous gestures. On January 5, 1995, just a week after our wedding, Rahima suffered concussions from a road accident in Dhaka, Bangladesh. Even though we made it to USA safely, Rahima could not get back to the normal classes due to the treatments she was going through. During that difficult phase of our life, Richard extended his full support enabling us to handle the situation to the best of our capabilities. Richard helped me create a small home office that allowed some flexibility in my work. It was a very difficult semester for us, but Richard had a surprise for me — news of a scholarship to attend a summer workshop in NATO Advanced Study Institute at Les Houches, France. Richard suggested that I take Rahima along with me even though the scholarship would not cover her expenses. We were very excited at the opportunity. A few days before our flight to Paris, Richard came and gave us a $500 cash gift which we did not want to take but he insisted. We were moved by his thoughtfulness and affection for us. When we reached Les Houches, we found out that Richard had called to make sure that they had a special arrangement for us. We did not have enough words to thank Richard. Everything he did to make our life easier and enjoyable was completely beyond our expectations. After that two weeks of restful time in France, Rahima recovered quickly from her ordeals of concussions. We are ever grateful to Richard for treating us with so much love and care and for teaching us how to care for others.

On the surface, my current work seems to be far away from what I did for my Ph.D. work in nonlinear optics. But I use many of the skill sets that I developed working with Richard. The key mission of my current work is to lead development of standards to drive doubling the bandwidth of low-cost electrical I/O interconnects every few years. The sheer optimism that Richard instilled in us gives me the confidence and strength to find root causes of various limitations and push the technology envelopes to find solutions. I still vividly remember the excitement Richard had when the first semiconductor microdisk laser was demonstrated by Bell Labs and he could see the great potential applications of the microcavity physics he was pioneering. The work becomes so much more exciting when we passionately drive ourselves to better understand the fundamentals and to find solutions to various problems as Richard always did.

Richard and Sunny made us a part of the Chang family with their love and care. Every year, we looked forward to Thanksgiving dinner at their home. It truly felt like a big family where everybody cared for each other. Besides such yearly event, Richard made us feel special in various other ways. Once, he ordered a custom-made pottery as a special gift for us which is still a treasure for us. When we went to Gordon Research Conference in Plymouth, New Hampshire, he took me out for canoeing just to give me a new experience. Richard showed us how to be a great leader and a good friend with humility and modesty while setting high expectations on quality.

I have been fortunate to have the opportunity to work under Richard's guidance and learn from him during my four years of stay at Yale. But the influence Richard has left on diverse aspects of my life will always remain. All of us together will carry the torch that Richard has left behind and will make the world a better place to live.

Letter to Richard on the Occasion of His Retirement from Yale in June 2008

Richard has been such a positive influence in my life. While I was at Yale, I enjoyed my research work thoroughly and never felt tired even when I worked long hours. Everyday, Richard would come to my office in the morning with a wonderful smile on his face and we would sometimes discuss various ideas on the board and decide on some strategies. Almost 12 years have passed since my graduation. Many of the physics lessons that I had learned have faded away over the years, but many wonderful life lessons learnt from my Yale life with Richard have taken much deeper root in my life. How Richard led his team could be a leadership model for individuals and corporations. He was a great friend and a great teacher at the same time. He did not tolerate mediocre quality work, but he led us to attain the highest quality through motivation, inspiration, and role modeling. His work ethics have always been an example for me to follow. On one winter Saturday morning, I got a call from Richard. He was stuck in the Chicago airport due to a snow storm. He called me to give me some references that may help explain some of my experimental findings. I was so impressed and excited that I immediately went to the library and looked up the reference. That Saturday morning's idea from Richard was a significant step towards resolving some of the key problems of my Ph.D. thesis work.

Every year, we would get together at Richard's house for Thanksgiving dinner and we all felt so much love and affection that we became part of the Chang family. On so many non-work occasions, Richard was always there with full support for my family. Today, as we try to balance work and family life, Richard and his family as we always knew, remain a source of strength for us.

I have been one of the luckiest persons to have the opportunity to work with Richard and to be blessed with his sincere affection. Now, I strongly feel that I have achieved a lot more than a Ph.D. degree. The intangible lessons that Richard has passed on to us through his life are having more and more influence in my life.

Letter to Richard on the Occasion of His Retirement in June 2008

Mitchell Fields[†], Ph.D. Yale '97
Vice President and General Manager,
Broadcom Inc.,
San Jose, CA 95131 USA

Dear Richard,

It is hard to imagine that you are retiring. Not only does it remind me that I am getting older, but it also means that the next generations of students at Yale will not have the opportunity to learn from you. The learning I describe extends beyond the scientific mentorship that so many of us were fortunate to receive. Through your example, we have all learned integrity, professionalism, and dedication to friends and family.

I would like to remind you of some moments that made particular impressions on me:

- I prepared for about a week to introduce my ideas for injection seeding of lasing in microdroplets and the relevant physics (mode competition, spatial overlap, etc.). After our first discussion, I saw that you were enthusiastic, but you did not provide much feedback. It was a relief to me when at our follow-up meeting the next day you provided me with excellent feedback and direction that sent me eagerly back to the lab.
- You always included your students in important events in your family life. I remember when your daughter became engaged and the party that ensued.
- Your love of simple pleasures. Good times were celebrated with "dipped roast beef sandwiches" and apple pie that had to be pre-ordered to ensure that the dinner didn't run out.
- I was honored when you asked Juergen and me to co-author an invited chapter for *Progress in Optics* with you. The chapter wound up taking a bit longer to write than we all expected, but it was certainly fun!

Thank you, Richard. Laura and I wish you the best.

Sincerely,
Mitchell Fields

[†]Passed away

Richard Chang and His Visits to Moscow

Mikhail Vladislavovitch Jouravlev
Lead scientist, Moscow State Technology University STANKIN,
Associate Professor, Docent, Russian State Agrarian University -
Moscow Timiryazev Agricultural Academy, 127550

It was in March 1994 that I first met Prof. Richard Chang who came to Moscow to attend the International Aerosol Symposium. At that time, I was a junior research scientist at Karpov Institute of Physical Chemistry, Moscow where the laboratory of Aerosol Technology, Department of Aerosols was located. That Department of Aerosols is the oldest established department of aerosols in the world and specialized in the study of the physical chemistry of aerosols. My scientific chief was an academician of Russian Academy of Science, I.V. Petryanov-Sokolov, who was internationally recognized as one of the founders of the twentieth century nuclear age.

I had been asked to meet Prof. Chang, when he arrived at the Sheremetyevo Airport. He asked me to assist him with his large luggage. It was surprising that Professor Chang could lift with one hand, about 10–20 kg weight, which amazed me. He was in excellent athletic form and very strong. I could not have lifted such a heavyweight with one hand, and I am not exaggerating. It turned out that he had a daughter and family in Moscow, and he brought them many gifts from the United States.

The International Aerosol Symposium was held in a historical place, the building of Karpov Institute of Physical Chemistry in the center of Moscow. Professor Chang visited the historical laboratories established for developing of methods of electroformation of Petryanov's non-woven tissues, fiber tissues, pore tissues and filters. He gave lectures and reported on aerosols and their charged particle properties. I showed him our particle lasers, power gas lasers, and aerosol chambers. I accompanied him to the Stars Town, to see the Russian Space Center established for the control of the satellite missions.

He kindly agreed to supervise my Ph.D. thesis devoted to photothermal instabilities in liquid water aerosol droplets. He became one of my scientific advisers to carry out my Ph.D. research and supervised the experimental part of my thesis for "Candidate of Sciences Degree" in Russian, which I defended later in 2000 at Radiophysics Department of Lomonosow Moscow State University. He advised me on how to build the experimental setup for the control of optical parameters for the scattering on the levitated single droplets and the solid particles.

Together, we formulated the experimental problems in the field of optics and spectroscopy of aerosol droplets with chemical contaminants. We made measurements, based on the available laboratory devices and the experimental setup.

We dealt with the formulation, the theoretical problems, and the methods for the calculation of nonlinear scattering effects, Raman, and Mandelstam-Brillouin scattering on the single levitated aerosol droplets. It was interesting to establish the new approaches for the study of the optical discharge on the ultrafine metal and dielectric particles for the detection of chemical impurities in the gas flow in the aerosol chambers.

He established the specific approaches to solve the scattering characteristics of acoustic waves on aerosol droplets, advised me on how to make an installation with an acoustic generator, the excitation of standing acoustic waves in the cavity, "Whispering Gallery Modes" and explained the instability of oscillations in hollow resonators, the injection and the capture of the liquid aerosol droplets, to study the rate of the evaporation of the water droplets with the organic impurities and the laser dyes.

We even discussed the scattering of the sun light, THz, and laser irradiation on cosmosols or metal ultrafine particles in nearspace, optical filters for capturing of the aerosols particles by the polymer nanofibers being in the first production and known as Petryanov's non-woven tissues.

He came to Moscow regularly and I was always delighted to meet him.

It is a privilege to share my memories of Professor Richard Chang. I will never forget him!

Letter to Richard on the Occasion of His Retirement from Yale in June 2008

Justin Hartings, Ph.D. Yale '97
President, Biaera Technologies, LLC
Hagerstown, MD 21740 USA
jhartings@biaera.com

April 2008

Dear Richard and Sunny,

Deciding to work for Richard is the second best decision I made while I was at Yale (asking Nicole to marry me is first!). As a young, scared kid from the Midwest, struggling to fit in and hoping that I was cut out for Yale, I had no idea as I surveyed the faculty that the choice of advisor that I was about to make would have such a lasting influence on me. Among all of the great memories that I have of my time at Yale, there are two that particularly stick with me, each characteristic of common experiences among Richard's students, colleagues, and friends.

Email was in wide use in 1994, when I started working for Richard. But the internet was still a bit of a novelty. One day as my colleague Mitch Fields and I were looking for information on some optics related issue, we thought we would try the internet. At the time, the internet was slow, clumsy, and poorly organized. (A very poor signal-to-noise ratio as Richard might say!) So Mitch and I sat down in 013 and started on what was the first experience that I ever recall with the internet. After searching around, following links, and trying to navigate through (search engines hardly existed), we somehow ended up on the web page for Burger King. Richard had a remarkable ability to show up in the basement of Becton Center at just the right, or wrong, time. Well, just after the Burger King page had loaded, Richard strolled in. With his typical dry wit and without skipping a beat, he said, "Are you looking for some take-out?" As two new students in his lab, still trying to show what we could do, we thought we had blown it. Richard, of course, was very kind about the whole thing, but I think it was another five years before I visited the internet again.

As I started in Richard's lab, each of the first two papers on which I was an author went to the same "unnamed" reviewer. The reviewer obviously had different theories, techniques and approaches than we had, and our manuscripts were all picked apart. They were both accepted, but only after considerable revision and back-and-forth with the editor. The third paper I was quite excited about because it would be my first as the primary author. I worked hard on it, I was sure that it was right, and I was anxious to get past this milestone. The reviews finally came back, and the first reviewer recommended publishing the paper with only minor revisions. The second review was a five page treatise detailing the errors in our technique and the shortcomings in our theory authored by the same "unnamed" reviewer. I was infuriated and discouraged. I marked up the review, detailing what I thought were all of the reviewer's errors, and went to see Richard in his office.

I have, and will, always remember that meeting. In his usual calm, dignified, and professional way, he told me that the best thing we could do would be to use this review to inspire us to do even better science. While we both thought we were right, we should work even harder to strengthen our case, stay true to the science, and let the data drive the answers. I learned more from that one experience than I did in years of toiling away in the basement of Becton Center. It is easy to say that science is an objective pursuit when you are not personally invested in it. It is much more difficult when you have poured months of late nights, failed experiments, and background research into your work. But that is what professionals do. And that is what Richard is: a professional. Richard handled it with class, professionalism, and thoughtfulness, all while being sensitive to the emotions of his young graduate student.

I have since, of course, realized that these same circumstances play out in virtually all aspects of life; science, business, and public service to name a few. Sometimes you are right. Sometimes you have to do more work to show that you are right. And sometimes you have to do more work to convince yourself that you are wrong. Our paper was ultimately published. As my first publication, it hangs in a frame on my office wall. And every time I encounter particularly stiff opposition to a new idea, it reminds me of that meeting with Richard.

When I started at Yale, I thought that I would be taught to be a scientist. I was. I didn't know that I would be trained to be a professional and good citizen. I will always be grateful to Richard for this training, for the example that he was to all of us, and for the effort he made to make me believe that I belonged at Yale. And thanks to both Richard and Sunny for welcoming me into your home, treating me like family, and making me a part of your lives. It was a blessing to work with you both. It is a privilege and an honor to call you both my friends.

Best wishes for a long, happy, and peaceful retirement. Justin Hartings '97

More Than Clients

John Allen
John Allen General Contracting
Brattleboro, VT 05301 USA

I first met Richard and Sunny in the spring of 1997. They had just purchased a house in Brattleboro, Vermont, one street down from Sunny's mom and dad's house. They hired me to do some small renovations to the house, nothing major, just some sprucing up. A few days in we noticed a few more problems that needed some attention. One week had now turned into 3 weeks of work. On one of the final days, I had noticed a leak in the roof. An explanation is needed here. A flat roof in Vermont is not a clever idea. We happen to receive copious amounts of snow. I discussed this small problem with Richard, and he agreed that something should be done about this. I jokingly mentioned that they should put a second floor on the house. I almost fell over when they called back and said could I design something and put a price to it. Well, 24 years later I am still working on the place. That was the beginning of a wonderful relationship.

One day in the beginning of our relationship, Sunny mentioned to me that Richard was diagnosed with Parkinson's disease. This floored me because he was very, very active. They went swimming, kayaking, biking, hikes up Wantastiquet Mountain in New Hampshire. They traveled extensively. One of the great traits of Richard is that he never let the disease define who he was as a person.

Richard was a very intelligent person, who never made you feel uncomfortable when talking to him. He loved talking about what was happening in our lives, what the kids were up to, what other projects were coming up. You always had to be on your toes because his humor was quick and sneaky.

Richard and Sunny moved to Singapore a few years back so that Richard could have better quality of life. We would do occasional video chat, but it was never the same. Cathy, my wife, and I missed both of them daily. Richard will be missed by myself and my family. As I said in the beginning, Richard and Sunny were more than clients, they became one of our truly great friends.

With much love,
John, Cathy, Griffin, and Daniel

Letter to Sunny on Richard's Passing

Juan Fernandez de la Mora
Professor of Mechanical Engineering
Yale University, New Haven, CT 06520 USA
juan.delamora@yale.edu

Dear Sunny:

Many thanks for thinking of me for this memorial book. My personal and scientific interaction with Richard was brief, and probably insignificant at the scale of those of his many other friends and collaborators. It therefore seems to me that my contribution to that book would not be up to the standards of the effort.

Nevertheless, I do owe you a letter of condolences on Richard's passing. I had in fact become quite fond of him, and viewed you also as a most admirable person. I was consequently greatly saddened first by his drowning accident, and then by the news of his death. Therefore, I will use this opportunity to try to fill that gap with some personal remarks. They are written for you and your family. Nevertheless, if for some reason you think that they should be included in the book, we may certainly do so.

Richard and I were colleagues since I joined Yale's faculty in 1981. We both shared an interest in small particles suspended in air, but he was fond of relatively large visible particles, while I liked small invisible ones. As a result we hardly interacted until about 2005, when we supported each other to obtain the 2005 and 2006 Sinclair award of the American Association for Aerosol Research. Richard was then working on the optical detection of bio-threats. Somehow he found that I had done work on a focusing technique that enabled bringing an increasing number of these particles to the focal region of his optical detectors. He thus kindly invited me to collaborate, which I was very pleased to do. My contributions to his project were nevertheless rather modest, because Richard needed a practical device to fit his machine, while what I had at hand was mostly theoretical. I then recommended my industry friend Susanne Hering as the person that might be able to design and build the required hardware. Susanne did indeed develop it, but I am afraid it did not provide the large benefits one would have expected from the funds Richard's grant invested in it. I remained interested in testing and perhaps improving the device, but I was never able to bring it to my lab.

While that brief collaboration produced no scientific publications, it offered an avenue to get to know each other. Richard was in fact unjustifiably kind to me in spite of my poor contribution to his project. He brought me to numerous discussions in his stunning lab, and I even enjoyed a fabulous dinner party at your home. Richard managed to make me feel as if I was a great guru. You may also recall that you, Richard, my wife Mercedes and I sat together for dinner at least once during our famous Engineering parties. Mercedes was in fact most impressed by your personality and your elegance, and sends her warmest regards. There was indeed great potential for a lasting friendship, but building deep ties takes more time than what was then available.

In summary, I am most grateful to Richard and to you for numerous signs of friendship and appreciation during our brief crossing of paths in this life. I Should also say that I admire you for the devotion you showed him during his final years. Finally, I pray that the kindness and generosity Richard showed to others in his noble life has received its reward in heaven.

With best wishes
Juan

Letter to Richard on the Occasion of His Retirement from Yale in June 2008

Dear Richard:

Although the elegance of your optical work has generally been above my grasp, I do have a Chang anecdote to share. I was visiting Ian Kennedy at UC Irvine (June 2004), who was showing me his work on optical resonances in microparticles. When I remarked on its beauty, he said: didn't you know this is the work of your colleague Richard Chang?

Let me conclude with a simple note of gratitude for your generous scientific support of my modest contributions to your bio-threat detection studies. I also very much enjoyed the privilege of joining you, your family and your students for an unforgettable dinner at your home. Thank you also to you and to Dan Rosner for carrying me in the wake of your respective Sinclair awards.

My student Guillermo Vidal is making good progress on the hard subject of aerodynamic focusing at relevant flow rates. I look forward to the blessing of a continued collaboration with you on this or any other subject.

With best wishes
Juan

Memories of Richard K. Chang

Hongxing Jiang[1] and Jingyu Lin[2]
Paul Whitfield Horn Distinguished Professors
Edward[1] & Linda[2] Whitacre Chair in Electrical and Computer Engineering
Texas Tech University, Lubbock, Texas 79430 USA
hx.jiang@ttu.edu & jingyu.lin@ttu.edu

Prologue

We came to know Richard as early as 1998, not by personal contact but through his research work on microcavities. It was almost 3 years later when we finally met each other. We still remember vividly numerous interesting conversations with him. During our interactions with Richard, which lasted for many years, our *perceptions* on academic careers and personal lives were deeply impacted. We are indebted to his valuable advice, generosity, and support.

Professional Interactions with Richard and Career Impact

Our professional interaction with Richard first started through the book he edited entitled "Optical Processes in Microcavities".[1] We started to work in 1994 on III-nitride wide bandgap semiconductors, while we were at Kansas State University (KSU). It was a very exciting time. The III-nitride field just had several breakthroughs, which ultimately led to the *Nobel Prize* in Physics *awarded* jointly to Isamu Akasaki, Hiroshi Amano, and Shuji Nakamura in 2014.

As newcomers, we wanted to work on novel topics which had not been studied in material systems from which we may make an impact to the field. We decided to fabricate and study various microstructured photonic devices. One type of microcavity structures we studied was III-nitride microsize pyramids. We observed very complicated optical modes which we had difficulties to understand. We dug through literature in the field of microcavity. Naturally, we found the book edited by Richard. With what we learned from the materials in this book and in other literature, we eventually understood that a micropyramid can be treated essentially as a triangular microcavity, based on which an optical ray tracing method was developed for calculating the optical resonant modes inside the pyramidal microcavities and the calculation results agreed perfectly with those experimentally observed in GaN micropyramids. We believe this is just one example illustrating how researchers in other disciplines benefited from Prof. Chang's seminal

[1]Optical Processes in Microcavities (Advanced Series in Applied Physics) by Richard K. Chang (Editor) and Anthony J. Campillo (Editor), World Scientific Press, Singapore (1996).

works in microcavities. Likewise, we believe that many other people and *new generation of researchers working in related fields* will benefit from his contributions.

We met Richard for the first time around 2001 during one of the meetings under the program of Semiconductor Ultraviolet Optical Sources (SUVOS) funded by the Defense Advanced Research Projects Agency (DARPA). At that time, several politicians in the US Congress and news media members received anthrax spores through the US mail and this caused a big disturbance in society. The SUVOS program aimed to develop compact UV light emitters emitting around 280 nm, to be deployed to detect possible biological agents in the form of aerosol particles. Through this program, we overlapped and had common interests to develop a biological agent detection system. While Richard was focused on handling optical properties of bio-aerosol particles under a UV light source excitation, our interest was to develop a portable UV light source based on Al-rich AlGaN and study the optical, electrical, and optoelectronic properties of nitride semiconductors and devices. We enjoyed and benefited from many fruitful exchanges with Richard.

During one of these professional meetings, the two of us met together with Richard and Prof. J.J. Song of Oklahoma State University. Prof. Song was very well respected in the nitride semiconductor community. During our conversation, we learned that Richard was her Ph.D. thesis advisor. We mentioned to Richard that Prof. Song's research group was the first to achieve optically pumped lasing in GaN materials, which was very exciting news in the nitride semiconductor community because GaN blue laser diodes had not been demonstrated back then. Richard was very proud of the achievements of his former student and conveyed congratulations to Prof. Song.

After our successful completion of III-nitride microcavity work, we wondered what new features a microsized blue/green LED would possess. This curiosity led to our invention of microLED technology in 2000. Presently, microLED is recognized as the ultimate display technology. The invention of microLEDs laid the foundation for many emerging applications, which currently are under intensive pursuits by almost all the big names in the tech industry for applications in flat-panel displays, smart watches, smart phones, i-glasses, dashboard projectors, and picoprojectors, 3D, augmented reality (AR), and virtual reality (VR) displays. Our invention of microLED and microdisplays has also significantly enhanced our reputation in the field and is our most proud career achievement. Looking back, without our initial pursuits in the microcavity research, the invention of microLED would probably not have been realized.

Professional Support from Richard

We are one generation younger than Richard both in academic career and personal life. Richard was also an endowed professor in a top university. During the period of our interactions, we were young and very energetic, while he had experience and wisdom. Our professional relationship evolved naturally into his guiding and helping both of us grow in all aspects. He was very generous to us and was willing to help us whenever possible. As one example, Hongxing was recommended by his colleagues to be considered for the award of the university distinguished professor (UDP) title by KSU. Since the title of UDP represents the highest honor KSU can bestow on its faculty, the group consists of only a small number of about 25 faculty in the university. Hongxing provided Richard's name as one of the external references when requested by the nomination committee. Hongxing was awarded the title of UDP by KSU in the fall of 2004, thanks to the generous support of pioneers in the field, including Richard.

We also benefited greatly from Richard's valuable advice regarding how to deal with rejections. The career of a university professor is very competitive, particularly in the early stages. One must take responsibilities for teaching, research, publications, and secure funding. Often, many proposal applications submitted by new professors are rejected. One also receives reviewers' reports for one's submitted manuscripts, occasionally comprising quite discouraging comments. The rejections could lead to self-doubt. We were wondering from time to time, if we received more negative comments than others because we were at a less prestigious state university. We asked Richard for his thoughts and advice on how to deal with negative comments/rejections. Richard told us a story regarding one of his papers receiving very negative comments from a reviewer about the understanding of optical processes in microcavities, even though Richard was a pioneer and laid the foundation for the field of microcavity. He advised us only to focus on the most important issues and positive outcomes. His personal experiences helped us gain a better understanding of the academic environment and to establish confidence in our own abilities. His advice also helped us in many ways in choosing and carrying out difficult research projects with genuine impact as well as in the advancement of our careers. He inspired us to focus on what we want and love to do instead of being distracted by some negative comments and criticism.

We are now at the same age as Richard's, when we first met him. We remember all the help and support we received from Richard and other friends and colleagues. We learned from Richard and other generous senior colleagues: we also try to carry on their generosities to next generations — to help young faculty in whatever way we can. We have been promoting and writing many supportive letters for young people seeking jobs and/or career advancements, including promotion and obtaining fellowships in professional societies. Another example is our willingness to share our laboratory facilities to support younger researchers since our laboratories are equipped with some unique capabilities.

Several groups including Columbia University, General Electric (GE), National Air and Space Agency (NASA), Corning Inc., Applied Materials Inc., University of Wisconsin, and Dartmouth College have sent either graduate students or researchers to our laboratory to use our material growth, laser liftoff (LLO), and device fabrication and characterization facilities. In return, we gained satisfaction from the process of helping others. For example, we are so delighted to hear just recently that our Alrich AlGaN deep UV materials grown for NASA are currently used by a NASA Jet Propulsion Laboratory (JPL) team to search for organic matter and minerals on Mars. As has been said "generosity pays off in *the* long run." In this sense, the DNA of generosity from Richard has been passed on.

Personal Interaction and Friendship with Richard

For many years, we also enjoyed a personal friendship with Richard. He provided much wisdom for our personal life throughout this period. Other than research, we found we had quite a lot of other topics in common. One of the reasons, looking back, maybe our backgrounds have certain similarities. We could talk smoothly switching between English and Chinese. He spent his childhood in China and knew traditional Chinese culture better than us in many aspects. We told him stories about our upbrings in China.

We invited Richard, around 2003, to give a colloquium for the students and faculty in the Physics Department at KSU. He visited many labs in the department and delivered a very interesting talk about his work on optical scattering of liquid droplets. He was very excited talking about his research. However, he was not a person, who only likes to talk about himself. He asked many intriguing questions about our research projects. It was probably too much energy drain with all the interactions and lab visits due to his health condition. After lunch, he said that "I need to put down my head just for a few minutes." He stayed in Hongxing's office and took a short break for about 15 minutes and then he appeared refreshed and energetic again.

We remember well one thing that Richard had mentioned that affected our careers later. Richard mentioned that it would be difficult to move one's lab when one gets close to 55 years old. When an opportunity to relocate to Texas Tech University (TTU) in 2007, when we were both nearly 50, was offered to us, Richard's earlier advice resonated with us and helped us make the final decision to move.

Our relationship became more personal after his visit to KSU. He told us that he had been diagnosed of Parkinson's disease a few years ago. It may have been caused by his overuse of dichlorodiphenyltrichloroethane (DDT) to kill mosquitoes, when he was a teenager living in Brazil. He said he had such a good time observing the mosquitoes diving down under the spray of DDT, but never knew back then this could come back to hurt him. We talked about the fact that the Nobel Prize in Physiology or Medicine 1948 was awarded to Paul Hermann Müller "for his discovery of the high efficiency of DDT as a contact poison against several arthropods," and then the US banned DDT in 1972, and it was banned completely in 2001 internationally. We then told ourselves that this is an example illustrating that even a Nobel Prize discovery needs to stand the *test of time*.

During his visit to KSU, he gave us a surprise gift — a book written by his daughter, Pang-Mei Natasha Chang. The title of the book is very interesting to us, "*Bound Feet & Western Dress: A Memoir* (小脚与西服)." The copy he gave to us has been translated into Chinese. As described by the introduction of the book, "In this exquisite memoir, Chang Yu-Yi (張幼儀), the daughter of a distinguished Chinese family, recreates her life for her American-born grandniece, Pang-Mei Natasha Chang, a Harvard student who is conflicted about her identity." Most Chinese of Richard's generation who got advanced education in the US were associated with privileged families in China. The book recounts the facts revolving around the marriage and the divorce between Chang Yu-Yi and her ex-husband, Zhi-Mo Xu (徐志摩). Zhi-Mo Xu is considered to be one of the most important figures of modern Chinese poetry and his divorce from Chang Yu-Yi was the first of its kind in the history of modern China. Almost every Chinese knows the name of Zhi-Mo Xu and considered this divorce event as a courageous rejection to an old feudal Chinese tradition by a talented modern poet. Moreover, as we grew up in China, our brains were filled with the stories of Zhi-Mo Xu's romantic love affairs with Huiyin Lin (林徽音) and Xiaoman Lu (陆小曼) and never bothered to ask what had happened to Chang Yu-Yi at the end. Richard told us that the main character in the book (Chang Yu-Yi) is his aunt. As we read through the book, we were so shocked to learn that his aunt continued to take care of her ex-in-laws, raised her son, and even provided financial support to Xiaoman Lu during Lu's *toughest* years. Most impressively, she eventually became a highly respected CEO of a fashion company and a successful banker. It is a fascinating story still being discussed widely and continue to be "discovered" by the younger generations on the internet today. Pang-Mei Chang's book really helped people, especially those who grew up in China, to know the facts and understand better about the history and culture of China during the early 20th century.

We had a long discussion during the dinner around this topic. We guessed that the reason Richard gave us this book as a gift probably was not only because his daughter was the author, but it also revealed the history of his family. From the book, we also learned that Pang-Mei Natasha was requested by members of the old generation of the Chang family to show mercy for Zhi-Mo Xu, while she was preparing to write this book, despite the fact that Zhi-Mo Xu had brought an enormous amount of pain to Chang Yu-Yi, his own son, and the Chang family. This speaks volumes of the character of extraordinary decency of the Chang family. Richard told us how his daughters got interested in Chinese culture and how Richard and Sunny educated their children. We were very impressed with the fact that all their three children graduated from Harvard University.

Around spring of 2003, we took our oldest son to visit several university campuses to prepare for his college applications, including Yale University. Richard was very generous and arranged our visit. When we checked into hotel, we were surprised that the room had been paid for. When we asked him, he said "I have some extra funds to cover visits from collaborators." We were not quite sure what he meant. Many years later, as we became endowed chair professors after being recruited by Texas Tech University, we realized that he used his endowment fund to cover the hotel room expenses for us. This was another example of Richard's generosity and thoughtfulness. Not only did we get a chance to visit the campus, but we also visited Richard's lab and spent many hours with him. After we visited the Yale campus, we were surprised and delighted that Richard invited us to visit his vacation house or retirement-to-be house in Brattleboro, Vermont, which is about a 2 hour's drive from Yale University. We talked about children's education, and he also talked about his retirement plan. His vacation house is beautiful and sits right on the bank of a river. He told us that he loved diving and swimming. We also met for the first time Richard's wife, Sunny, in their vacation house. She was also so generous and prepared a lovely dinner for all of us. Richard and Sunny kindly invited us to visit anytime if we happen to be in the east coast of US. It was such a pleasant visit and one which we will never forget.

Looking back, we regretted that we did not keep up personal contact with Richard after around 2008 because of two events that occurred. Richard was retired in 2008 and naturally our professional interactions reduced. At the same time, we were recruited by TTU. With changes in email addresses and all the hectic days involved in lab moving and settling down in a new place, we lost contact with Richard. However, we did hear about Richard from our common colleagues. Prof. Mark Reed, Richard's colleague at Yale University, visited us at TTU in 2012. Prof. Mark Reed and we shared the same PhD advisor, Prof. Arnold Honig, at Syracuse University. He told us that Richard had an accident, while swimming in a pool near their retirement house. After that, we also met with Prof. Jun Han, another one of Richard's colleagues at Yale University, during a nitride semiconductor meeting and we were relieved to hear from Jun that Richard had recovered from the previous near drowning episode.

Prof. Richard Chang's contributions to science, his wisdom and generosity will always be remembered. His legacy will live on in a *new generation of researchers* who will continue to benefit from his work and seminal contributions. We miss him dearly.

Editors' note:
Pang-Mei Chang, *Bound Feet and Western Dress: A Memoir*, 215 pages, Anchor (1997).

Richard K. Chang: A Truly Wise, Humorous, and Kind Gentleman

Yong-Le Pan, Ph.D.
Senior Research Scientist, Army Research Laboratory
Adelphi, MD 20783 USA

Although a long commemorative paper "Richard K. Chang: A Memoriam" was recently published [*Journal of Quantitative Spectroscopy and Radiative Transfer* **255**, 107273 (2020)] in which we honored both the life and scientific legacy of Prof. Richard K. Chang, I still have more to write on the close interaction with him during my 12 year stay at Yale. He was one of the most important mentors in my life, one of the key individuals who selflessly helped my family, and a truly wise, humorous, and kind gentleman.

I first learned the name Richard K. Chang when I was a graduate student at East China Normal University in 1987. He was introduced to us by my advisor as one of the world's best known and highly respected scientists in the laser spectroscopic field. Later, I saw his name on the wall with Arthur L. Schawlow (1981 Nobel Prize in Physics), John L. Hall (2005 Nobel Prize in Physics), Y. R . Shen (Professor Emeritus of UC Berkeley), as one of the four honored professors overseas at the National Key Laboratory in Laser Physics and Optics, Fudan University. In the 1980s, this lab was considered by the Chinese Ministry and Education to be the best laboratory for laser physics and optics for the education of graduate students. Professor Zhi-Ming Zhang, the director of this key laboratory in the 1980s, made a connection with Richard in an international conference. Later on, four professors and two postdoctoral researchers from Fudan were trained and studied in Richard's lab, and two undergraduate students earned their Ph.D. under Richard at Yale.

The handsome man: from Yale Weekly Bulletin & Calender, Dec 16, 1991. Chang appointed to endowed post, the Henry II professor of Applied Physics

I started my academic career at the Department of Physics, Fudan University and performed research at Fudan's key laboratory right after I got my Ph.D. in 1993. Amazingly, I managed to get promoted to tenured associate professor in less than 3 years. In 1996, three young professors in the laboratory were recommended to apply for an open position in Richard's lab, and luckily, I was selected by Richard to join his research group under a J-1 visa as an exchange scholar.

When I arrived at New Haven on April 19 (Saturday), 1997, I was pleasantly surprised that Richard had reserved a room in the Duncan Hotel (Graduate New Haven now) and paid for two nights. He had even faxed from France a warm welcome letter to the front desk and appointed Dr. Paul Chen, a postdoctoral researcher in Richard's group who had also come from Fudan, to assist me in settling down. One week later, when he returned from France, I eventually met the professor I could hardly wait to see.

Before I had ever met Richard, I felt a little nervous as Professor Qian at Fudan, who had worked in Richard's group three times for about a year each time, told me that Richard was a very strict and hardworking person. Prof. Qian was right somehow: Richard taught and trained the students to very high standards, and he was very strict in all details regarding the research we were performing. In this way, generations of undergraduate and graduate students, young colleagues, and collaborators around the world — many of them now leaders at their universities, institutes, companies — learned from his serious scientific attitude and work ethic. Meanwhile, Richard was never excessively strict to his students and colleagues. He was always demanding but never unfairly critical, constantly patient but not harsh, not angry but smiling. I never heard him raise his voice to any students.

In 2007–8, we generally had a group meeting at 11 a.m. on Thursdays, but one of the Ph.D. students always arrived late. To accommodate him, Richard started the meeting at 12 p.m., but he was still late frequently. Eventually, Richard changed the start time to 1 p.m. with pizza to attract him to the meeting on time. From such a small example, you could see how lovely Richard was. He was a model of generosity, integrity, an admired and beloved true gentleman. We will never forget his calm, friendly, patient and polite demeanor.

I had a lot of fun in Richard's lab. Initially, I was assigned to work on a project to find the burning conditions to produce the highest combustion efficiency of gasoline for Texaco. About a year later, I was switched to work on bioaerosol detection and characterization based on fluorescence spectra and elastic scattering patterns from individual single aerosol particles in the 1–10 micron diameter size range. By detecting and synchronizing the arrival of flowing particles with a conditionally fired UV laser pulse, optimizing the detection efficiency etc., we were the first group who was able to measure single particle fluorescence spectra on-the-fly. One night, Richard brought his dog to the lab to test our newly developed single particle fluorescence spectrometer (SPFS), and we all got very excited that the system was able to detect the presence of the dog and human in the lab vs. an empty laboratory.

In 2003, Prof. Jean-Pierre Wolf from the University of Lyon and I were studying the time-resolved trajectory of ultrashort optical wave packets circulating near the surface of a dielectric microcavity using a femtosecond laser co-owned by Profs. Chang and Schmuttenmaer located in the Chemistry Building at Yale. We struggled to find the signal for weeks, and finally observed the expected phenomenon around 11 p.m. on a Friday night. Richard and Sunny immediately drove to Yale with a bottle of red wine from their Hamden home to celebrate the success on campus after hearing the news by phone. The

Richard and Sunny visited our first US house purchased at Cheshire, CT in 2003. From left to right are Richard, Sunny, Suping, and Yongle.

research life in Richard's lab was never boring, there were always surprises and amusements at any moment.

I cannot count how many parties Richard held for his group or guests, which I attended by myself or with my family, in the lab, at famous restaurants in/around New Haven, at his home in Hamden, CT, the country home in Brattleboro, VT, or the Mory's Club at Yale. Certain traditions were established: every New Year, he invited the whole group to have dinner; every graduation of his Ph.D. student, he asked the whole group to celebrate together. Work in Richard's group was serious science, but with a relaxed and enjoyable atmosphere. All these experiences remain unforgettable for our entire life.

Richard always thought about our futures and worried for us earlier and more completely than we did ourselves. When my family decided to stay in US at the end of 1999, we had to get my J-1 visa status waived, file an immigrant petition via an I-140 form, switch my status to the working H-1 visa, and finally apply for a green card using an I-485 form and update to permanent residence status. One key step among these processes was to get enough supporting letters from distinguished people to convince the immigration officers that the applicant is really an outstanding researcher/professor, who has extraordinary ability and will benefit the US interest if he/she remains in the US. Generally, it is customary for the advisor/boss to supply one letter. However, Richard did much more than anyone else. He chose all the important supporters for me, wrote a sample letter, then mailed and emailed to all with a formal cover letter to persuade them to write a letter to support my application. Meanwhile, he called everyone to make sure that they received his request package and had agreed to write a crucial letter for me. He followed up with them until the immigration attorney received all the required letters. With such strong support, which I had never heard of from other employees or students, my family received the green card at the end of 2001, much faster than most of the applicants at the time.

As we knew, Richard already had Parkinson's since the mid-1990s, and had the drowning incident in a swimming pool just after retiring in the summer of 2008. He had been in a coma for two days, stayed in the hospital for weeks, then in the physical therapy center for months. With Sunny's help, he recovered little by little through his tenacious willpower and arduous therapy and exercises. I drove from New Haven, CT to Brattleboro, VT to see him roughly every month during 2008–2009 before I started working in Maryland.

During such a very challenging recovery period, his mind remained in the 1960s–1970s for months. For example, he kept asking for his mother, who had passed away long before, to have lunch or dinner with us, or insisting on calling his late secretary to discuss work, but we never heard any complaint from him about anything. He was always smiling, patient, positive, happy, and generous even in such an extremely difficult time with some uncontrollable emotions. It showed that Richard's gentle manner was innate and deeply part of him, how he behaved at all times in his life.

We all learned from Richard K. Chang in our daily interactions with him, benefited from studying and working in his group, matured gradually and polished our behaviors steadily. We carried Richard's spirits from his lab to the world, as the old Chinese saying goes: "One must achieve proper behavior first, before one can achieve success in one's career." Such a philosophy not only benefits my career, but also is valuable in my family life.

I inherited some collaborative friendships from Richard that benefited my research career for a long time. Examples include ongoing teamwork with scientists Mr. Jerold Bottiger and Dr. Steve Christensen from Edgewood Chemical and Biological Center (ECBC), Dr. Jay Eversole from Navy Research Laboratory (NRL), Prof. Richard Flagan from California Institute of Technology, Prof. Christopher Sorensen from Kansas State University, Profs. Jean-Pierre Wolf and Giovanni Crosta from Europe. The bioaerosol study project also opened an opportunity for me to collaborate with researchers, Drs. Ronald Pinnick, Steven Hill, and Gorden Videen, at the US Army Research Laboratory (ARL).

In September 1998, Richard, Stephan Holler, and I drove to Adelphi to setup the newly developed SPFS system at ARL, where scientists from NRL, ECBC, Air Force Research Laboratory (AFRL), and the National Institute for Occupational Safety and Health (NIOSH) brought their interesting bioaerosol simulants, interference, and background aerosols to ARL to test the SPFS system. Disturbed by more than 10 people talking and walking around the lab, I could not find the signal by optimizing the alignment across time and space. Sweat was streaming down my face, it could be extremely embarrassing if I could not make the SPFS work, because the scientists came to ARL to learn how to obtain the dispersed fluorescence spectra from individual single flowing aerosol particles. Richard noticed my awkward situation and quickly, he invited all people to lunch and left me alone in the lab. Shortly thereafter, I got the signal and the wonderful spectra from single flowing particles as demonstrated on the computer screen in real-time, which they observed when they came back from lunch about two hours later.

It was this collaboration that built my connection, and finally got me hired by ARL in 2009 as a research physicist. Now I have a research group with several million dollars of equipment and parts distributed in a 4-laboratory room space. There are 4 sabbatical professors, 7 postdoctoral researchers, quite a few graduate students, undergraduate students, and high school students who have worked in my group since 2009. I try to carry forward Richard's good tradition to treat everyone nicely, and supply the best working conditions for them. Generally, the postdoctoral researchers in my group earned \$5K–\$10K/year more than others. If they cannot get an ideal job, I try to find additional funds to extend their stay for another half year. The research work runs very well for me at ARL. In 2013, I was honored to be elected an ARL fellow, a position usually given to the more eminent scientists and engineers in ARL.

It was Richard K. Chang, the truly wise, humorous, and very nice nobleman, who gave me the opportunity to work in the US, helped my family stay in the US, taught me to be a gentleman, mentored me to be successful in my career, and showed me how to be happy in my family life. He will live in our hearts forever.

Left to right: Yong-Le Pan, Richard K. Chang, Sung-Wen Chang, Su-Ping Chen, Hong-Yi Pan, and Giovanni F. Crosta.

Richard Chang's Influence on My Life

Seongsik Chang, Ph.D. Yale '98
Research Scientist, Apple Inc.
Cupertino, CA 95014, USA

November, 2020

Richard had a profound influence in my life, both scientifically and personally. His influence began when I was his graduate student (Ph.D. '98) and continued throughout my career and life. He was there to guide me through several critical stages of my life. Among all the values that he taught me, I can probably summarize his teaching into three: attitude toward scientific observations, compassion, and trust.

First, about attitude toward scientific observations. After one of my earlier thesis experiments, I didn't know what to do with my data. It was about laser emission from deformed microdroplets. My laser emission pattern was dramatically different from what everybody else expected. Light is supposed to emit at the highest curvature point, but it did not. I was so worried and blamed myself for not performing the experiment well. I was prepared to do the experiment again when I showed my data to Richard. Instead of him scolding me, he said "Seongsik, very interesting. You may just have discovered something new." The results of this experiment were later explained by Prof. Douglas Stone's group using chaos theory, so-called "dynamical eclipsing" and became a major part of my thesis. Since then, when I look at data, I accept it as it is without imposing a preconception. If existing theory doesn't explain the data, I challenge the theory. This early establishment of my attitude toward experimental observation in Richard's lab helped me do good scientific work. As an example, one of my first experiments at Hewlett Packard (HP) Labs came out to be completely opposite to what people expected, and eventually led to a device that made upcoming double-speed HP digital press possible.

Second, about compassion. It was during my graduate studies that I came out of the closet. I would eventually come out of the closet to almost everyone I knew, but in the beginning, I came out only to a select number of people. Richard was very important to me then (and still is very important to me now). I decided to come out to him. I said "Richard, I am sexually attracted to men, not women. Do you have a problem with that?" My tone was somewhat aggressive out of fear that he might not accept who I was. However, Richard said calmly, "I don't. I think it's in the gene. I know a gay man who takes care of my mother and he is a fine young man." As soon as he said that, I was very happily relieved and all my worries were gone. I felt Richard's compassion and acceptance. As I look back now, I was enormously lucky that I was surrounded by people who were open-minded and accepting during my coming out period at Yale. And Richard's acceptance meant a lot to me.

Third, about trust. When I started graduate studies with Richard, I was new to the field and confused about the complexity of the subject, whispering gallery modes. After meeting with Richard and Prof. Paul Fleury, Richard assigned an experiment to me. However, I had different ideas and I thought his assignment was not important at all. So, either I didn't do that experiment or I did something else. After Richard found out I didn't do what we agreed to do, he said, "Justin always does what he says he will do." At that time, I was kind of envious of Justin because Richard and Justin seemed to get along very well. I thought "Hmm, yes! No matter what, I should do what I promised to do." Later, I confessed to Richard that when Justin had wanted to borrow a motorized positioning system from me, I had lied and told him that I did not have it. That was because I wanted to keep it to myself so that I could have use of it anytime I wanted. Later, Justin confronted me about that. I recognize that he probably lost his trust in me. When Richard found out, he said, "Trust is very important among colleagues." Since then, whenever I interact with colleagues, I try my best to keep my promises to them. And I demand the same from my colleagues. And I truly believe that trust is the most important above all the work ethics, and trust takes the collaborated work further forward.

A few words about Richard's influence on me after I graduated with my Ph.D. in 1998, and visits with Richard and Sunny after I left Yale.

My first job was at a startup company founded by Seng-Tiong Ho. The company, Nanovision Technologies, worked on whispering gallery mode in semiconductor microdisk lasers. If there was any practical application of my thesis "Laser Emission from Deformed Microcavities," a semiconductor laser was the one. Although risky, this was one of the best options that I had out of several that Richard suggested. While I was very busy working at the start-ups Nanovation and Phosistor for the first four years after graduation, I continued to discuss with Richard about my career goals. In particular, I was still interested in a possible career in academia. When Phosistor didn't get further funding, I had a choice of going back to Northwestern University or doing something else. I called Richard with urgency, and Richard recommended that I stay in the Bay Area and look for a job in the Bay Area since Silicon Valley had lots of opportunities. Now when I look back, Richard's advice was right. After about four months, I landed a job at HP Labs, which was a dream job for me.

Living in the Bay Area, I had a chance to see Richard and Sunny from time to time. Richard's parents had lived in downtown San Francisco, near the Levi Strauss headquarters. Richard's sister passed away in November 2003, so it must have been 2004 when Richard and Sunny came to the Bay Area to sprinkle some of her cremains in the Pacific Ocean. I drove Richard and Sunny to a cemetery in Oakland Heights where Richard's parents were buried. Richard took me to Il Fornaio, an Italian restaurant in San Francisco near Levi's headquarter. I believe it was one of Richard's favorite restaurants in San Francisco. I felt special because he was sharing his personal stories about his mother, dad, and his sister.

A few years later, Richard retired while I was still working at HP Labs. During his

SeongSik Chang, David Leach, Roger Farrow, and Alfred Kwok in Vermont 2008.

retirement party at Yale, I told him that I was working on a device that increased the speed of the digital press by two times. And he continued to be very supportive and proud of me. A few days after, I heard news that Richard had a near-drowning episode in a pool near his vacation home in Vermont and he had lost some of his memories. That was a terribly devastating news for me. Not only was it very unfortunate for him and his family, but also it was a loss of my great mentor. I asked myself, who will I depend upon if I get into trouble?

Occasionally, I received news from Sunny about Richard's progress in his recovery. It sounded like through a modern rehabilitation program and acupuncture, he was able to recover much of his memories. Sunny told me that she occasionally reminded him about

me. And when I talked to Sunny over the phone, Sunny would put us on speaker mode so that Richard could participate in our conversation. I felt a little sad that Sunny had to ask Richard regularly if he remembered who I was. With time Richard made good progress on his recovery.

My first visit to see Richard after his accident was in Hamden. I visited with my friend Miguel who lived in Greenwich. Richard appeared to know me and Miguel. He had a little balance problem, but overall, he looked great.

I saw Richard and Sunny in 2013, when they visited Palo Alto to attend a memorial concert for Professor William Bennett's son, William Bennett Jr., who had passed away unexpectedly at a young age. He had suffered a massive stroke on stage while performing as the principal oboist for the San Francisco Symphony. I drove Richard and Sunny around and took pleasure in their company. They saw my lab and visited Mohiuddin's family in San Jose.

In February 2020, shortly before Richard passed away, I last saw Richard and Sunny in Singapore. I cherish that time. Thank you, Richard for everything that you did for me. I wish I was a better student and better achiever, but I did my best. I will miss you and the memory of you will be in my heart forever.

Richard K. Chang: The Teacher and the Knight of Science

Vladimir M. Shalaev
Distinguished Professor of Electrical and Computer Engineering
Purdue University, West Lafayette, IN 47907 USA
shalaev@ecn.purdue.edu

Richard was a world-changing scientist. He also was a true Teacher who inspired and supported many young researchers around him, including myself at the time. Even though I was not his student or postdoc, Richard would discuss my research, invite me to various conferences, write strong supporting letters for my promotion; he would always go an extra mile to support, encourage and help young scientists in the beginning of their careers.

Richard Chang's legacy in science is truly multifaceted. One of many areas where he did ground-breaking contributions is the nonlinear optics of microcavities. In the end of 90s, together with my colleagues, I showed that microcavities, with the so-called plasmonic (metallic) fractals inside, enable giant multiplicative enhancement due to two major contributions — the nanometer-scale localized plasmonic modes of fractals and the optical modes of cavities. Such unique combination enabled exciting phenomena, including for example lasing. Richard appreciated our work and invited me to Yale to give a seminar. I have to admit that at that time, I was perhaps too ambitious and perhaps a bit arrogant young scientist, that I guess did show itself in my arguments with theorists there. Richard was the very opposite — nice, polite, even somewhat humbled and still supportive during those discussions. With that, he gave me an example of a classy style on how to conduct scientific discussions. His fine personality was something special so that other people around him would always try to be better and do their best. I certainly had that feeling every time I met and talked to Richard.

His dedication to science was exemplary, if not extreme. Richard at some point mentioned that he worked so hard that he had to schedule time to meet with one of his own children. He was loved and admired as a scientist and a very fine person. I recall how our numerous discussions with Richard's and my common friend Joseph Zyss would every time end up with talking about Richard and how great person he was.

Richard was a true knight of science, a great teacher and one of finest persons, I ever met in my life. His scientific legacy will stay in the history of science.

Richard Chang: Scientist, Gentleman, Friend, and Mentor

Joseph Zyss
Professeur de Physique
Laboratoire de Photonique Quantique et Moléculaire (LPQM)
École normale supérieure de Cachan
94235 Cachan Cedex, France
zyss@ida.ens-cachan.fr

If I had to condense in a short and single statement what I hold to be one of the ultimate qualities for both a scientist and a human being, I would refer to his or her ability to sincerely and fully rejoice of someone else's achievements as though it were his or her own, albeit without the least appropriation. This often goes with a tendency to increase the other person's merit — somewhat undervaluing his or her own.

To me, no one exemplifies better than Richard this remarkable and seldom met moral and ethical trait as well as its day-to-day implementation in both professional and personal lives. Let me cite in this context the rather unique Hebrew verb "לפרגן", (pronounce "lefargen") of which I know no equivalent in any other language, that designates the ability to derive authentic satisfaction and genuinely rejoice from the successes of third parties. This verb is most likely deriving from the German "Vergnügen" for pleasure, which is being here enlarged from its eventually self-centered connotations to encompass a generous and outgoing feeling where the ego has little say. This seldom met psychological disposition nevertheless entails an indirect, however significant benefit by way of amply rewarded return in terms of mutually shared satisfaction. This is to say that I see here no inhuman sacrifice of the self here or even less some dubious sanctity that Richard who enjoyed life in all its small and great aspects, would have denied, but the key to a truly rewarding and happy life, one which is bound in turn to inspire other people's behavior and enlarge their own vision and horizon. Indeed, Richard was surrounded by an aura of respectful admiration and warm feelings leaving little room for wrong and mediocre behavior but triggering a quest for a higher level of moral and scientific standards in those who had the privilege to approach him. This being said, Richard, charismatic as he was in his gentle and soft manner, radiating authority in a higher form, was anything but a guru and I prefer to stop at that for fear of betraying his genuine modesty and the true nature of his imprint among us.

Richard's character and radiant personality had a long lasting influence not only on my own life, and on that of members of my family, be it my real family or call it my scientific family. With time, not only did we get to know each other better, but so did our respective families, from Paris to New-Haven and Brattleboro.

In the wake of these considerations, it so happened that, almost coincidentally, upon receiving some days ago due notice from the editor of PRL of the acceptance for publication of our most recent manuscript concerned with microlaser cavities of a new kind, my very first reaction was to think of Richard with gratitude for his inspiration but also, more selfishly I have to admit, how happy and proud I would have been to share with him the good news and further discuss with him the science therein and its possible implications. This revived in a way the loss, if only here in a somewhat childish but nonetheless a heartfelt manifestations. I miss the pleasure that he would have felt and its rebounding on my own that would have been translated into encouragements to progress and keep the momentum. Not to mention his precious and inspiring scientific views and advices that would have unveiled directions and advances that I might be missing based on my own limited resources. I will therefore content myself here by putting as a first reference [1] to this homage to Richard this work through a structure that he would have loved: a polymer based Möbius microlaser with radius 50 m, width 15 μm and thickness 3 μm. I feel both happy and proud to dedicate it to him as a tribute to the ongoing relevance and dynamics of a domain of research that he has initiated and developed much beyond his own group and his own days. In a higher sense, it may be said that Richard is somehow still with us through the lively pursuit of the directions of research that our photonic community has inherited from him.

No doubt that Richard would have loved to hear of developments where mathematics, physics and technology get together, including here a conceptual and practical toolbox imported from the formalism and math of general relativity and differential geometry, pushing forward his cherished microcavities into the new realm of a third dimension, including folding and chiral twists not to mention powerful esthetic and symbolic undertones exemplified by Möbius rings.

Let me now try and recall the scientific and personal circumstances of my first meetings with Richard in the late nineties. Whereas chance surely played a role, I was not fully unprepared for such a meeting in view of my earlier and then current interests in the physics of optically resonant cavities, albeit in another context than that of lasing. Our very first meeting took place in Eilat, a spectacular resort on the Red Sea at the junction of Israel, Jordan, Egypt and Saudi Arabia, making it a fascinating place from all points of view, from natural beauty to historical remembrances all the way from the antiquity to our times.

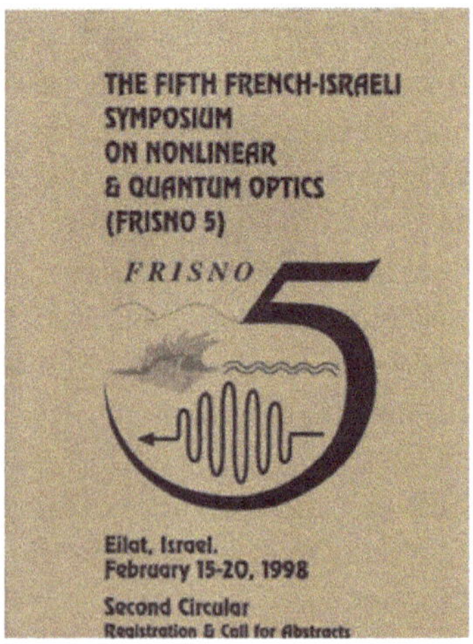

The occasion was provided by a biennial venue of the France Israel Symposia on Quantum and Nonlinear Optics in 1998, which was attended by Richard together with Sunny as well as by my wife Helene and myself. The stage was perfectly set for an enduring scientific and family friendship to this day. This event was made all the more memorable by the presence of many a great pioneer of our community, among whom Nicolaas

PARTIAL LIST OF INVITED TALKES:

Nicolaas Bloembergen:	*A Brief History of Breakdown by Laser Pulses*
Claude Boccara:	*Near-Field Optics: an Aertureless Approach*
Robert Boyd:	*Nonlinear Optical Processes in Optical Fibers and other Structured Materials*
Richard Chang:	*Optical Interactions Within Dielectric Micro-cavities: Cylinc Spheres, and Deformed Structures.*
Claude Cohen-Tannoudji:	*Ultra Cold Atoms: Recent Developments*
Richard Friend:	*TBA*
Theodor Hansch:	*Microscopic Traps for Neutral Atoms*
Aaron Lewis	*The Quest for Three Dimensional Nanometer Functional Ima*
Yaron Silberberg:	*Vector Solitons*
John Rarity:	*Advances in single photon sources and detectors (for qan information applications)*
Claude Weisbuch:	*TBA*
Urs Wild:	*Single Molecule Spectroscopy*

Bloembergen stood out. He had been Richard's PhD mentor and also played a significant and positive role in my own more modest story that needs not be detailed here. Suffice it to say that my relation to Richard is also part of a mutigenerational story of remarkable continuity and persistence over time.

I will get back later to the immediate sympathy and friendship developed in the wake of this meeting between us, but let me first stress the two main scientific factors that ensured the quasi instantaneous reactivity and subsequent by-product of this "collision", at least as from my own perspective. A fertile ground was somehow prepared on my side by my earlier interest in high quality optically resonant cavities in the context of nonlinear optical parametric processes and on the other hand by the development of a by-then well proven polymer-based technology mainly oriented at the time towards the development of electro-optic modulators for high speed optical communications.

Indeed, I had been active from the mid-eighties and on towards the realization of molecular crystal based optical parametric oscillator, in the wake of earlier molecular engineering research towards enhanced quadratic nonlinear efficiency. This goal was eventually reached in the mid-eighties leading to efficient tunable coherent infrared emission following optical pumping in the visible. Among other enabling features, it rested on the enhancement of optically nonlinear interactions within the active crystal by recycling of the generated infrared emission due to multipath interactions. This is achieved by the rebounding of light at two mirrors face to face with the required transparency and reflectivity at the appropriate wavelengths, very much like in a laser cavity. Although not akin to lasing as far as the physics goes, the engineering of parametric sources holds common features with laser sources.

Moreover, the eighties and nineties had witnessed the overwhelming development of integrated optics where the waveguiding of light through micron size optical microchips had become the standard for any component of practical and industrial interest. Semiconductors, in particular those of the III-V family such as Gallium Arsenide, Indium Phosphide and their alloys played a dominant role, due to their natural propensity to form epitaxially matched multilayer systems allowing for the piling up of layers with the required index and gain properties enabling miniature lasers, detectors, and modulators, among other required functions for telecom systems. I need not recall here that Richard as one the

International waters on Red Sea: site of first secret encounters between the Chang and Zyss clans for exploratory discussions as to the establishment of a French-US (micro-) billiard club, smiling faces and empty glasses indicating agreement on sight, with dear friends at the forefront: Late Daniel Chemla and wife Berit.

earlier graduate students of Nicolaas Bloembergen, the first one in nonlinear optics to my knowledge, had pioneered the field of semiconductor based nonlinear materials in early papers [2,3] and subsequent ones. Incidentally, he had also demonstrated and analyzed surface emission from nonlinear media, which predated in many aspects by almost one decade the domain that became widely known as surface nonlinear optics.

Very much in that same spirit, albeit limited to the less developed domain of organic nonlinear optics, functional polymers had somehow taken the lead over molecular crystals in view of their superior properties and ensuing potential towards integrated optics and miniature devices, in particular electro-optic modulators. The relative fragility of organics as compared to their more robust crystalline semiconductor counterparts would have discouraged efforts in that direction, had it not been for the flexibility and adaptability provided by precisely that same fragility. Indeed, epitaxial matching between semiconductor layers, with all its potential and accomplishment, entails rigid constraints that sets limits to the possibilities while ruling out more advanced 3D architectures as illustrated on Fig. 1. Such 3D geometries may not be of much practical interest today, but they open-up fascinating new physics explorations that semiconductor based multilayered systems could not provide, an advantage that Richard fully recognized and supported. This turned out to be the core and early cement of our forthcoming collaboration. Whereas my coworkers and I had developed a technology that enabled unlimited geometries, many more to be achieved on the future, Richard had developed the domain of semiconductor based micro-billiard lasers that he took to a level of sophistication that remains still unmatched, even with present day's tools [4]. This being said, the ability to electrically pump semiconductor lasers in elaborate electroded multilayer stacks remains a practical advantage that organic materials find it difficult to match to this day.

An additional feature which qualifies "soft" materials as opposed to their more compact inorganic counterparts is the ability to host molecular systems or quantum dots, up to a high concentration, endowing them with a range of properties from fluorescence and lasing to nonlinearities and no less relevant applications in chemical and biosensing. Lastly, the lower index range of functional polymers ranging from 1.5 to 1.7 depending on the nature and concentration of the embedded species, contrasts with that of inorganic semiconductors that range mostly in the 3 to 4 interval. Such lower values relax the out-coupling conditions from the cavity, thus providing access to a wealth of otherwise hidden information regarding the intracavity modes that sustain the emission, or their chaotic behavior versus the existence of a deterministic backbone of closed geometric orbits.

Richard had been aware for some time of the potential of soft materials as shown by his early demonstration of and analysis of the fluorescence from dye impregnated polystyrene microsphere back in 1980 [5]. This remarkable early achievement was the solid state facet of his work on exquisitely calibrated liquid droplets, that Richard had taken to a level of rigor and refinement which is evocative to me as much of art as of science [6].

After our first discussions at Eilat, it took us little time in Cachan to extend our already mature electro-optic thin films and waveguide technology onto billiard microcavities that Richard had pioneered in the realm of semiconductor structures. It was just a matter of designing new masks so as to vary the billiard contours. We could therefore produce in little time a wealth of cavity geometries of different sizes and molecular content, all based on in-house facility whereby a reduced fabrication cycle and inherent shape and material content flexibility appeared as a enticing assets for polymers.

Microcopy pictures of different planar cavity shapes with real colors featuring the respective spectra of inserted dye and dimensions in the 100 μm to 200 μm range. They illustrate the structural and spectral engineering potential of polymer based microbilliard lasers, the development of which Richard had contributed to and strongly encouraged.

This approach resulted in our first research that explored a variety of tailor made cavities [7], starting from the ellipse further deformed into quadrupoles and hexadecapoles. Large differences in the far-field emission patterns were accounted for by differing ray dynamics for each shape. Analyzing the differences in the appropriate phase space for ray motion, it was shown that the differing geometries of the unstable manifolds of periodic orbits are the decisive factors in determining the far-field pattern. This work benefited from an exemplary collaboration between Richard's experimental group and Doug Stone's theory group, the latter bringing in the full power of nonlinear dynamics and more specifically that of the Poincaré surface of section analysis in the appropriate ray optics phase space. This illustrates the collaborative spirit cultivated by Richard and his ability to develop mutually beneficial ties. With the help of Eugene Bogomolny from Orsay University, our theoretical support was subsequently extended onto the realm of semi-classical physics with the celebrated trace theorem playing a central role so as to quantitatively sort-out the contributions from co-existing modes and assign them to their matching orbits [8]. Among other collaborative outputs, I would like to cite our work with Richard on squares and spiral microcavities, the latter in an original dynamic mode together with the group of JeanPierre Wolf then at the University of Lyon [9,10], himself also a much appreciated member of Richard's extended family.

Let me now move to the more personal recollections that closely intertwines science and friendship, much I believe to their mutual benefit.

This photograph taken in 2007 embodies the essence of the unique blend of science

Sunny, Richard and myself on the foot-steps of the d'Alembert Institute's new building still in its infancy.

Joseph Zyss and Nicolaas Bloembergen at ENS Cachan.

and friendship that was characteristic of Richard's temper and approach in life. An outstanding mind must not necessarily go with a generous and outgoing character, but Richard's way illustrated a perfect combination of both. When I created the multi-disciplinary d'Alembert Institute in Cachan, now on the Saclay plateau, I benefited from the encouragements of Richard and it was like some sort of blessing for the launching of the institute to welcome him and Sunny a few days before the official inauguration of the dedicated building.

Likewise and not surprisingly, it was Nicolaas Bloembergen, himself Richard's mentor who had installed my laboratory in Cachan by paying a special visit and giving a seminar, when I joined the Ecole Normale Supérieure in 1999 on its then Cachan campus. How not to feel elated when both scientifically and morally supported by such a lineage, but also humbled and challenged to deliver which I very much doubt I lived-up to. My only claim is to have passed on the torch to a more gifted generation of researchers and professors among which Isabelle Ledoux and Mélanie Lebental stand out respectively for physical chemistry and for microlaser physics, both highly appreciated by Richard and the latter in his direct lineage. They are now in turn nurturing a new generation of even younger researchers that one may want to call Richard's scientific great-little children.

At that same level, albeit in different realms, I cannot not cite here alongside Richard, Jacques Friedel, Daniel Chemla, Bob Silbey and Chuck Shank, to whom I can never be grateful enough for their mentoring, trust, friendship and help in many aspects of my personal and scientific life. I like to fancy them discussing science in life in some imaginary Olympus. Who said that scientists must be fully rational thinkers and abstain from supposedly unproductive daydreaming wanderings?

Finally, I like to remember my visits to Richard and Sunny's home in Brattleboro, sometimes with my wife Hélène, and the many pleasurable times spent together, from science discussed on their deck overlooking the breathtaking view of a river flowing through a forest and hills that were suggestive to me of an earlier era on the continent, biking with Richard uphill to buy honey from a local producer, or enjoying a lobster dinner where I am afraid that my poor experimental skills were publicly exposed, albeit with no trace of malevolence from my dear hosts who must have pretended to ignore it.

Life after death is a matter of personal belief but not one that is very popular in our scientific community. I like to believe that the Hebrew concept of “םלוע אבה” (pronounce “Olam haba”), meaning literally the world to come, simply refers to the future, that of our children and their own descendants, and whether one will have contributed to it, be remembered in the future and leave a durable imprint. In that respect, Richard’s heritage appears as a gold mine, to be further exploited and developed by future generations be they aware of his role or not. On that basis, Richard is living on and has indeed his share of the world to come.

Joseph Zyss
Paris-Saclay, Tel Aviv.

References

[1] *Möbius Strip Microlasers: A Testbed for Non-Euclidean Photonics.* Y. Song, Y. Monceaux, S. Bittner, K. Chao, H. M. Reynoso de la Cruz, C. Lafargue *et al.*, *Phys. Rev. Lett.* **127** (20), 203901 (2021).

[2] *Dispersion of the optical nonlinearity in semiconductors*, R. K. Chang, J. Ducuing and N. Bloembergen, *Phys. Rev. Lett.* **15**(9), 415 (1965).

[3] *Experimental Verification of the Laws for the Reflected Intensity of Second-Harmonic Light*, R. K. Chang and N. Bloembergen, *Phys. Rev.* **144**(2), 775 (1966).

[4] *Stimulated emission and lasing in whispering-gallery modes of GaN microdisk cavities*, Seongsik Chang, Nathan B. Rex, and Richard Chang, Gabel Chong and Louis J. Guido, *Appl. Phys. Lett.* **75**(2), 166 (1999).

[5] *Observation of Structure Resonances in the Fluorescence Spectra from Microspheres*, R. Benner, P. Barber, J. Owen, R. Chang, *Phys. Rev. Lett.* **44**, 475 (1980).

[6] *Ray Chaos and Q Spoiling in Lasing Droplets*, A. Mekis, J.U. Nöckel, G. Chen, A. D. Stone, and R. K. Chang, *Phys. Rev. Lett.* **75**(14), 2682 (1995).

[7] *Dramatic shape sensitivity of directional emission patterns from similarly deformed cylindrical polymer lasers*, H. Schwefel, N. Rex, H. Türeci, R. Chang, A. Stone, T. Benmessaoud and J. Zyss, *J. Opt. Soc. Am. B* **21** (5), 923 (2004).

[8] *Trace formula for dielectric cavities. II. Regular, pseudo-integrable, and chaotic examples*, E. Bogomolny, N. Djellali, R. Dubertrand, I. Gozhyk, M. Lebental, C. Schmit, C. Ulysse, J. Zyss, *Phys. Rev. E* **83**, 036208 (2011).

[9] *Direct evidence of open ray orbits in a square two-dimensional resonator of dye-doped polymers*, G. D. Chern, A. W. Poon, R. K. Chang, T. Ben Messaoud, O. Alloschery, E. Toussaere, J. Zyss, S. Y. Kuo, *Opt. Lett.* (29), 1674 (2004).

[10] *Deciphering output coupling mechanisms in spiral microcavities with femtosecond light bullets*, F. Courvoisier, V. Boutou, J. P. Wolf, R. K. Chang, J. Zyss, *Optics Letters* **30**(7), 738 (2005).

Letter to Richard on the Occasion of His Retirement from Yale in June 2008

Pam Becker
Researcher, Institute for Neurodegenerative Disorders
New Haven, CT 06510 USA
pbecker@indd.org

Dear Dr. Chang,

It is amazing that while you were doing ground breaking research in physics, you were also participating in research with us at IND. You faithfully completed the diaries, put up with the numerous blood draws and ECG's and tolerated our intrusion with grace and good humor. And you created legends. The best story ever of why the pills were not returned is yours: "They were inadvertently dropped off the boat in the Mekong River".

I have looked forward to your visits. You have provided wise counsel about grandchildren learning to talk in more than one language at a time. Your stories of travels have opened new vistas for me; I have enjoyed them immensely. I hope that you will continue to enjoy traveling, enjoy your lovely wife Sunny and watch your grandchildren grow in love and peace.

With great affection and admiration
Pam

Letter to PY Nicole on the Passing of Richard

Laura Zung
CEO, Lancaster Health Group
Chicago, IL 60630 USA
laura@zung.org

Dear PY,

Here's my favorite memory of your father. We came to visit you in Vermont when Ming-Wei and Ming-An were in a Circus Summer camp, and we drove over from the Adirondacks to see their final performance. We stayed up late that night visiting, and I remember that the house followed the river and was long and skinny. The house was filled with fabulous Chinese art and western furniture. It was the merging of cultures. I think we even went swimming that night in a pool. We played cards and chatted late. Anyway, that doesn't matter.

The next morning, we celebrate with breakfast in your parent's dining room. Denis mans the Nespresso Machine and waxes poetically about the merits of the purple capsules. All the kids are eating cereal and talking at once. The adults are eating toast, fruit, dumplings, and leftover Chinese food from the night before. Your sister is there with your niece, and the newspaper is being passed around. You are talking about acupuncture and Chinese medicine.

I am sitting next to your father and he's telling me with grace and humor about his research and his graduate students. They had recently had a party for him, and he talks animatedly and smiles from the heart. His eyes shine. In my other ear, your mom whispers to me that your dad can't remember which grandchild is which, but he can remember his graduate students. I feel the love of your family all around.

And I bask in the moment of a big Chinese family. These are the moments that I love in my own family. The hodgepodge of people and conversations and the mix of western and Chinese food. Everyone is talking at once and enjoying the hubbub around the table. Pass the Toast. Pass a section of the paper. Who wants a fried egg? Acupuncture can help with a lot of things. Ming-Wei was great as the ring master. Ordinary moments are the most beautiful.

I came home from that trip and bought a Nespresso machine, and I use the purple capsules. If Denis says they are the best, why try anything else? I think of that moment often when I am brewing my coffee in the morning.

Godspeed Dr. Chang. On to your next adventure.

Love, Lz

Optics at Yale with Richard Chang

Stephen Holler, Ph.D. Yale '99
Associate Professor of Physics and Engineering Physics
Fordham University, Bronx, NY 10458 USA
sholler@fordham.edu

The May 1994 issue of *Physics Today* was a special issue that focused on nonlinear optics. When I came across this issue while a junior at Polytechnic University, I knew then where I wanted to go to graduate school and with whom I wanted to work. It had to be Yale, and it had to be with Richard Chang. I was already doing research on aerosols and whispering gallery modes, otherwise known by the more accurate, but less sexy, appellation: morphology dependent resonances. The pioneering work in these areas being performed in Richard's lab, the Center for Laser Diagnostics, brought together my existing work with my newfound interest in nonlinear optics.

I clearly remember the day I met Richard for the first time. It was when I traveled to Yale for an interview for grad school. I remember taking a cab from Union Station over to Becton Center. It was really a short trip that I would come to walk many times during my time at Yale, but this was a different trip loaded with nerves, anxiety, and the promise of the next step in my education. Arriving at Becton Center, I eagerly walked through the glass doors to the lobby. This was it. I pushed the elevator button, and when the doors opened, I pressed the 4. The doors closed, and I was on my way up. As I rode up those four floors, I chuckled as I thought of a comment a friend had made to me. The lab I was working in, the Microparticle Photophysics Laboratory, occupied the basement of Nichols Hall at Poly, and my friend had said to me in the days before I departed: "All physics labs are in the basement. You're going to go to Yale and get locked in the basement. Just wait, you'll see." So, I rode to the fourth floor thinking that the next step of my journey would get me out of the basement, out of the dungeon lab.

The door opened to the fourth floor, and I turned out of the elevator to Pat Brodka's office. I introduced myself, and she promptly brought me to Richard's office next door. That was my first encounter with the great man. We talked briefly in his office before he asked if I would like to see the lab. "Of course," I replied. With that we headed out of his office, and I naively expected to walk down hall to a lab. Instead, we walked over to the elevator. Bing! The doors opened, we entered, and Richard proceeded to push the button…B…we were headed down to the basement, just as my friend had predicted. All was good. I saw the labs, met the group, and knew that this is where I needed to be.

When I finally did get to Yale to begin my studies, I remember meeting Richard again. We sat in his office discussing research projects. We talked about this new work that was being planned to correspond with the theoretical work that Doug Stone was doing on asymmetric resonant cavities (ARCs). It was going to be challenging experimental work that involved coupling fibers and microcavities, like research being performed at Poly that I was aware of but not working on. However, this work would involve investigating chaotic trajectories of light within the cavity. It was interesting work that had implications in the growing area of optical communications. As I left Richard's office that day, I had a challenge laid out before me. My foundation was set, just not in stone.

I was part of the group that first semester but not doing any research yet. I began working in the lab in the spring, but rather than the ARC project, I was shifted towards aerosol work: the characterization of bioaerosol particles using light scattering and fluorescence. While the ARC project was exciting and novel, the aerosol work had important implications particularly as it was Army funded and there was much concern about biological weapons during the late 90s. The shift in my assigned research project did not matter to me. I was just happy to be at Yale, to be working with Richard, and I was eager to get started. What I did work on wound up shaping me and propelling me through my career in ways that, I am certain, would not have been possible had I stayed on the ARC project. This project made all the difference in who I am, and I have only Richard to thank for it.

Except for a couple of side projects, I never did get to incorporate nonlinear optics into my research, but I still would not change anything about my research or experience at Yale. The diversity of experimental techniques I was exposed to, and the wisdom imparted by Richard were invaluable. To this day my work deals with many of the same aspects that were first explored in Richard's lab. In fact, lately, I find myself gravitating to many of the same tools we employed in those projects more than two decades ago.

When thinking about the current projects that I'm engaged in, I often think about their origins. This is particularly true for the work I do in two-dimensional light scattering. This work started early in my time in Richard's lab and was initially focused on multi-spectral scattering to elucidate information about an object. It all started with an imaging fiber bundle and a Raman shifter to convert some of the second harmonic light of the Nd:YAG to a red beam so that we could obtain multi-spectral one-dimensional scattering patterns. With limitations in both spectral and angular range, we transitioned to a dye laser to obtain broader spectral coverage, but the resolution was too limited, and the signal washed out. We then pivoted to a configuration that dropped the multi-spectral character and allowed us to study scattering in two angular dimensions with a CCD array.

Since we were interested in studying aerosol clusters (*e.g.*, aggregates of bacterial spores), the complex speckle patterns created by interference and multiple scattering necessitated the inclusion of a second angular dimension to reveal insights into the character of the scattering. I remember standing in Richard's office looking at the image data Yongle Pan and I had acquired. We were writing the first of the papers on this topic and we were considering what to call this. The importance of names was not lost on Richard. As we kicked around ideas, he finally asked "have you ever been to Taos [New Mexico]? These remind me of the mountains there." Thus, was born two-dimensional angular optical scattering (TAOS). The speckle pattern in our images evoked an image in Richard's mind of a mountainous region in the southwest United States. Although I have yet to travel to Taos, NM, I have been skiing in nearby Santa Fe in the Sangre de Cristo mountains, which were Richard's inspiration for the name of our approach.

The work on TAOS continued after I left Yale. Kevin Aptowicz introduced some innovative optical designs to expand the angular coverage to obtain nearly full forward and backward scattering simultaneously. Kevin also performed multi-spectral imaging with multiple cameras, and I obtained simultaneous red and green TAOS imaging of flowing aerosols with a single-color camera in the early 2000s. The complexity of the patterns means classification remains a challenge, but we remain focused on achieving this goal with reasonable accuracy. Some early work proved promising, and Kevin and I are currently collaborating on this issue using calculated scattering patterns from a variety of cluster morphologies. In addition, Matt Berg (Kansas State University) and I introduced simultaneous TAOS and digital holographic imaging thus obtaining information about particle shape and orientation along with the exact scattering TAOS pattern.

Over two decades have passed since we first introduced TAOS the work continues to grow and evolve. While the sun may have set on Richard, his legacy endures and shines as bright as the brilliant glow that emanates from the Sangre de Cristo mountains at sunset.

Letter to Richard on the Occasion of His Retirement from Yale

April 18, 2008

Dear Richard,

Congratulations on your long and distinguished career. I know that you will continue to experience life and wonder with the same vigor and enthusiasm that marked your career. Reflecting on this momentous time, I am reminded of how deeply you influenced my nascent career. It is safe to say I would not be where I am were it not for you (and that's a good thing).

I was fortunate enough as an undergraduate to experience and develop an appreciation for experimental science. This background provided me the guidance and discipline to know what I wanted to do after college. Well, I suppose I am still not sure exactly what I want to do as there are many interesting things I would love to explore. However, I did know that the research being performed in your laboratory was interesting and that I wanted to work with you (on whatever) at Yale. I found myself with even greater fortune when that became a reality.

Now the profound impact you had on my life and career can be traced back to a time shortly after I arrived at Yale. I remember talking with you in your office about this new project that was to be starting up, and about a system which I would study. This, of course, was the asymmetric resonant cavity and the waveguide coupled micro-resonator. As it was my first semester at Yale and I was deeply mired in coursework, my work on this project would begin months later.

In the intervening months, Paul decided to leave for Corning and Thomas was heading off to MIT leaving a vacuum on the bioaerosol project. Andrew took up the micro-resonator work and I was moved (and I use the term loosely as I never worked on the micro-resonator project) over to the fluorescence and scattering project. It was this reassignment that altered the course of my career and set the stage for my journeys across the country, and for that I am truly grateful.

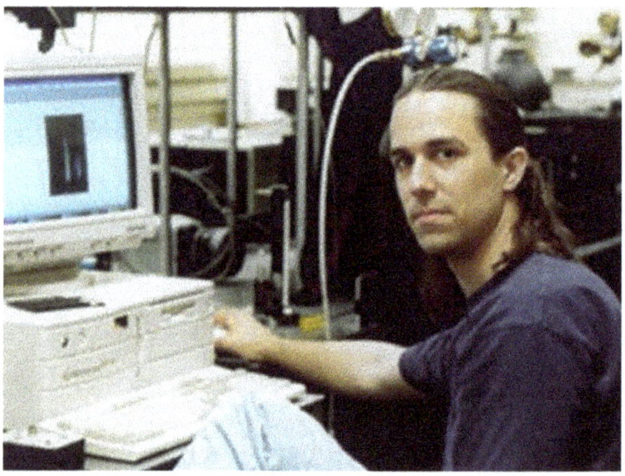

A photo I found that Yongle took early on in the light scattering work (pre-TAOS). This was from our attempt to use the coherent fiber bundle to look at simultaneous, multi-color light scattering from a single particle using a Nd:YAG second harmonic for the green and Raman shifter to obtain the red.

The bioaerosol project provided me with a superior group of people to work with from all over the world. My good fortune continued by being able to work closely with people like Yongle, Steve, Ron, Jerry, and Jean-Pierre, providing me the opportunity to learn from some of the best. It expanded my horizons and brought me into contact with those people I now call colleagues and partners. Your insight into an important problem provided me with the necessary tools, not only for the research, but also for the networking and collaboration.

The move to the bioaerosol project truly did have a deep and lasting effect on the course my career has taken. Were I not in Boston presenting the results of our work, I would not have met a Yale graduate from JPL (Steve Fuerstenau) who passed my contact info to a small contract R&D firm in the Bay Area (Los Gatos Research). Were it not for the bioaerosol work, I would not have gone to Sandia, where I was able to experience first-hand the mountains near Taos, the name-sake of the two-dimensional light scattering work. And finally, had I not gone to LGR I would not have met the people I call partners in my latest endeavor called NovaWave Technologies.

Yes, Richard, your guidance and mentoring has had a deep influence on my career. It is still early, but I will always remember where I have come from and where I have been. Where I am going is anybody's guess, but wherever it is, I am certain that your influence will always be felt, and for that I will always be grateful.

Best wishes and thanks for everything,

Steve

Letter to Sunny on Richard's Passing

Harald Schwefel, Ph.D. Yale '04
Associate Professor of Physics
University of Otago, Wellington 6021, New Zealand
harald.schwefel@otago.ac.nz

Dear Sunny,

I heard through Grace and Hakan of the sad news about Richard's passing. Now thinking of him, I am flooded with the warm memories that I have of the time I could work with him at Yale (1998–2005). I remember fondly the visit of Hakan, Esin, and me to your house in Brattleboro, VT in 2003 and our hike in the Black Mountains. I attached some photos of that time.

Left to right: Sunny, Richard, Esin, and Hakan Türeci.

Following Yale, I went back to Germany, where we met again in 2006 and 2008. Since 2015 my family (my wife Nicole and children Carlotta 9, Greta 7, and Paul 3) and I moved to New Zealand, and we love it here.

I cherish nice memories of Richard and I am very grateful that I got to know Richard a little bit.

Left to right: Harald, Richard, Esin Türeci, and Sunny.

I wish you lots of strength for the coming time and remain,

with warm wishes and sympathy,

Harald

Left to right: Greta, Carlotta, and Paul Schwefel.

Memory in Les Houches

Yoshihisa Yamamoto
Director of Physics and Informatics Laboratories NTT Research, Inc. and Professor, Stanford University, Palo Alto, CA 94301, USA
yoshihisa.yamamoto@ntt-research.com

One of my favorite conference locations around the world is Les Houches in France. My first participation in Les Houches Physics School was from May 23 to June 2 in 1995, when I met late Professor Richard Chang for the first time. The conference scope was "quantum optics in confined systems" and both of us were invited to give a series of lectures. I gave four lectures on amplitude squeezed state generation in semiconductor lasers and controlled spontaneous emission of excitons in semiconductor microcavities.

At Les Houches Physics School, we normally have a long afternoon break. Richard and I had a pleasant walk along hiking courses in French Alps and discussed various ideas related to our own presentations. In 1995, the famous Shor's algorithm was already discovered, and people started to search for various micro/nano-scale coherent devices to observe artificially created quantum effects. My scientific interest is slightly different from those emerging and later main-stream quantum information science and technology. I said, "Understanding how classical robustness surfaces from quantum substrates is a fascinating open question than observing various quantum phenomena in artificial and often macroscopic systems, that are carefully isolated from environments. I would like to explore the power of dissipation and quantum-to-classical transition in an open system rather than the power of coherence in an isolated system in the context of future quantum computing." Richard agreed with me and said: "Your idea is very unique and certainly a more natural approach than the schemes that majority of our community is taking." I was pleased to hear his positive response.

There was no lecture in the weekend at Les Houches Physics School. On that Sunday, I went to the Vallée Blanche in Chamonix Mont Blanc for 20 km downhill ski with a group of students. I invited Richard to come with us, but he wisely declined. A late May is already a summer season in Mont Blanc. After getting off the famous rope way, we climbed up about an hour to see the Italian border. From there we began a down-hill ski tour toward Chamonix, but soon after we started, one student slipped, fell, and broke his leg. On top of the mountain, there was no hope the rescue will come to bring him down to Chamonix. We spent one hour to fix his broken leg by wooden plates and make him ski by himself. By the time we started the tour again, the sun was already very high, and the snow started melting. It was a very hard condition for ordinary skiers like me. I normally fall only once or twice in an entire season, but I fell down thirty-three times on this single day. There was a dangerous crevasse, where we had to ski across a narrow snow bridge one by one. One student told me two people were killed yesterday by the avalanche here. We were fortunate to come back safely to the School with only one injured student. When I returned to the School in late Sunday afternoon, Richard told me, “You are the most courageous professor at this School.” I replied to him, “No, I was just ignorant how dangerous it is to go to the Vallée Blanche at this time of the year.”

Fudan University, Shanghai, China, 1997

Chang Family Album

Ron Shen, Theodore Hänsch, Richard and Sunny at Fudan University in June 1997.

Ron Shen, Theodore Hänsch, and Richard visiting Fudan University in June 1997.

Richard, Theodore Hänsch, and Ron Shen visiting Fudan University in June 1997.

Remembrances of Professor Richard Chang

Christopher Sorensen
Cortelyou-Rust University Distinguished Professor
University Distinguished Teaching Scholar
Department of Physics, Kansas State University
Manhattan, KS 66503 USA
sor@phys.ksu.edu

Long before I met Richard, I had great admiration for his work. I cannot remember exactly when or where, but I first met him at a conference dedicated to scattering optics in the late nineties. There he was the great Professor Chang! By chance, I eventually found myself sitting at a table with him and Otto Glatter talking about light scattering. Apparently, he did not know how great he was because he was kindly interested in my work! And indeed, as I grew to learn with every subsequent encounter, he was humble and straightforward with no pretenses.

Richard's scientific work inspires me to recall Blake who wrote "to see a world in a grain of sand...". Richard saw a world in an aerosol droplet, and he opened that world to the rest of us. Richard pioneered the study of whispering gallery modes (WGMs) in laser excited droplets. He showed how these resonant modes, due to an integer number of wavelengths around the interior circumference of the droplet, are sensitive to particle size and shape. He then took great advantage of these modes for a very wide variety of phenomena including enhanced Raman scattering, nonlinear optical phenomena, and microlasers. This work is both opportunistic, with this novel phenomenon, and very clever. Moreover, since many others have followed his lead, it is certainly seminal.

Richard also confronted the problem of aerosol detection and identification of aerosols, especially biohazards. Richard used optical methods to solve this very difficult but important problem. Again, one finds path-breaking contributions of two-dimensional (2D) scattered light detection and a bioaerosol sorter.

Richard as the person so impressed me that, I read the book his daughter, Pang-Mei Natasha Chang, wrote "*Bound Feet & Western Dress*". This memoir about her grand-aunt's persistent rise to eminence provided me with respect for the Chang family.

Inventive, seminal, and very often beautiful are proper adjectives to describe his work. Grace, humanity, and intelligence are proper descriptors of the man, Richard Chang.

Letter to Richard on His Retirement from Yale in June 2008

Charles Ahn
Professor of Applied Physics
Yale University, New Haven, CT 06520 USA
charles.ahn@yale.edu

Yale University

Department of Applied Physics
Becton Center
P.O. Box 208284
New Haven, Connecticut 06520-8284

May 2, 2008

Dear Richard,

I am delighted to write this letter as we celebrate your long, distinguished career at Yale. While I am happy for you as you approach retirement, I will miss your presence in the department. You represent the best of what a Yale professor should be.

Everybody is familiar with your scientific and departmental accomplishments; what I would like to recall here is the charismatic, charming way in which you have led. I remember when I first arrived at Yale, we had a discussion about Harvard, which at the time was our common ground. We talked about the flair and distinction of that place and how Yale strives for those same qualities. You have carried yourself with a charisma and distinction that would make both institutions proud. Our department is a better place for it.

In addition, I appreciate very much your sense of propriety, integrity, and dignity, and how you have imbued the department with those sensibilities through your quick and subtle sense of humor. I have talked with Hui Cao on occasion since she arrived here this semester, and we have shared stories about your helpful mentoring. She and I both appreciate your humor and advice that have helped us get acclimated to the place. We have shared some laughs about how sage and subtle your advice has been, advice that has made my life much easier here. I imagine the same is true for Hui.

In closing, I thank you so much for your support over the years, and I wish you all the best.

Sincerely,

Charles Ahn

Letter to Richard Chang on the Occasion of His Retirement in June, 2008

Janet Pan
Associate Professor of Electrical Engineering,
Yale University, New Haven, CT 06520 USA

Dear Richard,

I'll never forget your enthusiasm in discussing my work last autumn. Thank you so very much for taking an interest in my work and in my welfare. Thanks for all your advice about writing papers, giving talks, and future research directions. Thanks also for the many dinners at your house. You are a wonderful mentor!

I hope you and Sunny enjoy a well deserved retirement.

Best wishes,

Janet

Memories of Richard Chang in 2021

I was an Assistant Professor and Associate Professor at Yale University between 1999 and 2009. During this time, Richard Chang was my mentor. More importantly, Richard was my role model as a teacher, a research scientist, and as a friend.

Although Richard and I attended MIT more than 30 years apart, we still knew some of the same people. One of the first questions that Richard asked about my MIT days was whether I knew Prof. Campbell Searle. Prof. Searle was Richard's undergraduate academic advisor. Richard even showed me the wedding present that he had received from Prof. Searle. Like Prof. Searle, who was the Chairman of the Undergraduate Educational Policy Committee in MIT EECS during my college years, Richard understood the importance of undergraduate teaching. One of my most memorable experiences occurred when Richard asked me to help give oral exams to the undergraduates in his Electromagnetic Wave Theory class.

Oral exams were not part of my MIT undergraduate experience, because MIT classes were too large. However, the undergraduate enrollment in Yale Applied Physics was small enough for (almost) individual instruction. As a teacher, I learned that oral exams and written exams provided complementary information about a student. We could adjust the difficulty of an oral exam (in real time) to match a student's grasp of the intellectual material. Moreover, an oral exam would allow us, as teachers, to talk with, and teach, students on an individual basis. Since I found Richard's oral exams to be such a positive teaching tool, Peter Schultheiss and I adopted oral exams as part of our Circuits and Systems Laboratory class. I am ever grateful, and awed, to have had such conscientious teachers as my role models: Richard, Peter Schultheiss, and Campbell Searle.

In research, Richard and I were never "officially" collaborators. However, Richard was still a wonderful mentor. I remember that Richard helped me to rehearse for an Invited Talk at an APS March Meeting. Richard also proofread at least two of my journal articles. During my last years at Yale, Richard wanted me to arrive at some strong research results. We spent countless hours in his office, and over lunch, discussing my work (sometimes with Hui Cao). Richard even volunteered to have his Senior Research Scientist, Dr. Yongle Pan, help me out. Richard also gave advice about how to choose collaborators (answer: there are very few good collaborators).

Richard always emphasized the importance of being honorable and keeping one's word.

Finally, Richard was a very special friend. I spent many happy afternoons with Richard and Sunny at their houses in both Hamden and Vermont. I remember getting a tour of the Vermont house after recent renovations. I think Paul Fleury, Bob Wheeler, and Peter Kindlmann were there on that day too. Sometimes, Richard's children would join us for dinner at a nice restaurant in Vermont. I cannot thank them enough for their generosity. I wish all the best to Sunny, Nicole, Natasha, Kevin, and their families.

Letter to Sunny on Richard's Passing

Yves and Daniele Fasquelle
Paris, France

Dear Sunny,

Denis just called to tell us the sad news and Daniele and I want to express our feelings in such a difficult moment for you and all those who love Richard.

How many wonderful moments we had with Richard and you! We remember the first time we met and, as Daniele has often said, after less than one hour with you, it was just like if we had known Richard and you for years. We remember the wedding in Sainte Colombe, France and the marvelous trip we had through your beautiful country a few weeks after. We remember the numerous times we went in Brattleboro, biking with Richard, so proud to show us he was able to climb the hills. We were admiring his courage and his personal strength to fight his illness: never complaining, always ready to help people. We remember the visit that Richard organized to show Daniele the Yale musical instruments museum; Daniele was able to play the piano on the instrument which had been made for Liszt. We remember all the discussions we had with Richard and you, showing how close we were. As an engineer I was so impressed by Richard's works on laser. We were also admiring his kindness with Ming-Wei and Ming-An, our shared grandchildren.

After so many years of suffering, he has gone peacefully. And we know that, during all these years, you devoted all your time and energy to stay with him and help him. Richard was surrounded with love. I believe that now the love of our God gives him peace and joy with his parents, his sister Maejeane, and all those he loves.

The departure of Richard creates an aching void in your heart and the hearts of your children and grandchildren. For us too, because a friend, a member of our family is gone.

Sunny, your pain is ours.
Je vous embrasse de tout mon cœur,
Yves and Daniele

Letter to Richard on the Occasion of His Retirement in June 2008

Jean-Pierre Wolf
Professor of Applied Physics
Université de Genève, GAP-Biophotonics
CH-1211 Genève 4, Switzerland
jean-pierre.wolf@unige.ch

Dear Richard, I would love to describe here all the wonderful memories we have together, but this would probably take the whole book. Instead, let me just put down some places (but not all, because of available space again ;-) around the world, as keywords, where we shared great and unique moments together: Becton, Hamden, Lyon and Brattleboro (with its famous Chang Castle, its gym, hikes and movie theatres) of course, but also, the Taft apartments, Willoughby (the physics crazy projects generator), Panda, the Connecticut River and its canoes, Providence (with or without high ranking restaurants), Newport, Adelphi, Baltimore, San Francisco, Brighton (not so sunny but with Sunny), Bruxelles (and his famous chocolate and ... museum!), Rouen (ultrafast visit), Paris (from Lidar at The Eiffel Tower to the Institut de France), Nice (and its famous car rentals), St Tropez (amazing sanitary logistics), St Valery en Caux, the Auvergne's Volcanoes and Castles, Bocuse, Geneva, Lugano, ...and many more. Our passion for Science, lasers and droplets brought us together but very soon our deep friendship took over, which is one of the most beautiful things in life.

With all my friendship, Jean-Pierre

Memories with the Chang Family in 2022

Richard and Jean-Pierre at Villa Villemanzy in Lyon in 1997.

In 1995, I was a physics Professor in Lyon but I was carrying out experiments at the Freie Universitat Berlin with my friend Ludger Woeste on the generation of harmonics in microdroplets, using femtosecond lasers. A young graduate student, Jerome Kasparian, was doing his PhD on this subject and spent most of his time in the laboratory, so that he did not get the chance to see anything from this splendid city that is Berlin. Jerome and I found in the literature some exciting results from Richard and Steve (Hill) on the very same topic, so that when Jerome planned his PhD examination in 1997, we asked Richard if he would agree to come to Lyon as member of the Jury. That was the beginning of our friendship. When Richard came to Lyon, we immediately found a lot of common passions in life, besides micro-cavities and lasers: friends, family, art, culture, gastronomy and jokes. Since then, we never lost contact.

About half a year later, I broke with my girlfriend (more precisely, she broke with me ;-) and I decided that I needed a change of mind. What could be a better idea than a sabbatical leave at Yale in Richard's group? Richard enthusiastically encouraged me to do so and I submitted a NATO proposal to support my stay in New Haven. It was accepted in summer 1999. But before heading to Yale, I had the great pleasure and privilege to be invited to Nicole's wedding with Denis in Normandy, where I could meet for the first time the whole Chang and Fasquelle families. At this beautiful wedding and fantastic party, I decided to take pictures like a "free lance paparazzo" (;-)), one of my favorite hobby, and to prepare a wedding album as a present for the fresh and happy married.

In September 1999, I finally got my J1 visa and flew to JFK, where a limo was waiting for me, booked by Richard: What a welcome treat! I felt like the Wolf of Wall Street. I arrived 3 hours later at the Duncan hotel in New Haven, up Chapel Street; and it felt a little less so ;-). But who cared? A new life was ahead.

Then came the wonderful sabbatical time at Yale, with Richard and Sunny. I got a

small health issue also, and they were there, as they always are: "Don't worry, we'll take care of you, you will come to our place, we prepared a room for you". And they did take care of me marvelously, as if I was family. And when the small health issue was gone, I still felt part of the family. Richard and Sunny invited me often to Brattleboro over the weekends. They were planning an addition for the house, which required quite a bit of attention. I met their friends and Sunny's family from Vermont, and we often had this super delicious Chinese food at Panda North. I met Sunny's mother too, who was so sweet, strong and positive as her daughter. Sometimes, we were heading to the movie theater in Brattleboro and watched some movies there, if possible thrillers, or going for a hike, or to the gym, or canoeing, or just having a nice glass of wine together and talking. Or writing scientific papers. I was invited to Brattleboro also for Christmas. Vermont was buried in snow. It felt so good.

Lab life was great fun too, working mostly with Steve Holler and Yongle Pan, but also chatting with other collaborators of the group like Andrew Poon. We were chasing bio-aerosols by analyzing their scattering properties and fluorescence emission. It was a time where people felt really concerned about bio-terrorism, and major efforts were dedicated nationwide to efficiently detect pathogen bacteria in air. Part of our work was also basic research oriented. Together with Steve Hill from ARL, we were investigating some weird properties of droplets, while emitting multi-photon excited fluorescence. Was really an exciting time with plenty of results, which we published in our favorite journals. Together with Richard, we then thought about a really cool experiment: watching the collision of femtosecond pulses in a micrometric droplet as a function of their traveling time. Yes, I know, sounds crazy. For the experiment, we joined Charlie Schmuttenmaer at the Chemistry department, who had such a laser. After weeks of hard and harder work, Yongle finally managed to find the signal. It was late at night, and he was still in the lab. On our side, we were having dinner at Richard's place with Doug Stone and his

wife. The atmosphere became electric. We were calling Yongle every half hour to know if he observed additional peaks in the plots. Because we were observing collisions between optical pulses, we baptized this experiment the "click-click" experiment.

During my joyful stay, I also made a lot of new friends: Steve Hill, Yongle Pan, Steve Holler, of course, but also Steve Arnold and Michel Devoret. Michel was at Yale on sabbatical at the same time as me, and we were not only "the two frenchies", as Dan Prober was saying with a friendly and gentle smile, we became close friends. We kept in touch since then and Marie-Helene and Michel surprised me 10 years later, for instance, when they came for my 50th birthday party in Geneva. Actually there was a third Frenchie there, Francois Courvoisier. Francois was a student from the Ecole Normale Superieure de Cachan in the group of Joseph Zyss and was doing an internship in Richard's lab. He was so good that I tried to attract him to Lyon for a PhD, which he did some years later. He is now a CNRS researcher at the University of Franche Comte and has his own very successful research group.

Steve Hill, Richard, and Jean-Pierre in Vermont.

As a side note, besides good food, good wine, hiking, going to the gym, going to the movie theater and playing with lasers, Richard and I had the same passion for coffee. I remember soon after my arrival at Yale, we met with colleagues in the faculty room in the morning and had a coffee there. Was very nice, but after the meeting, I asked Richard: "sorry Richard, but, isn't there a Starbuck in New Haven to get an espresso?". And Richard answered: "ah, you like "real" coffee?" And he brought me to this lovely and excellent coffee shop round the corner, called Willoughby! From this time, we often found difficult physics problems that could only be solved around a "real coffee" there! But then, Sunny and Richard introduced me to "real tea". I never drank green tea that had this delicate, refined, yet vigorous flavor before: a revelation in my life!

During my stay, we also visited friends, went to conferences, and even did some vacation in San Francisco, where I visited for the first time the famous Chinatown. Some memorable conferences were for me the OSA conference in Providence, where Mae Jean

Jean-Pierre Wolf and Richard in San Francisco.

joined us and spent her days finding the best restaurants in town, and of course the CLEO-Europe in Nice. During our stay in Nice, we decided to take one day off and visit a bit the surroundings, which I know pretty well. We then rented a car and drove to Saint Tropez on the Corniche, and its spectacular view on the Calanques. In Saint Tropez, after a small walk in the city and the castle, we found a cute small restaurant where we had lunch. It turned out that this restaurant had the smallest bathroom we ever experienced, so that Richard named it the "nano-bathroom restaurant"! It was a marvelous journey, lots of fun, and I must admit, I never regretted having missed some scientific talks for it!

After my return to my regular job at the University Lyon, I very often turned back to Yale for visiting Richard, finishing experiments, writing papers, etc. I will always remember the words from Richard when I was arriving in New Haven: "Hi Jean-Pierre, welcome back home!". And it really felt like this. One year later, we had the great pleasure of welcoming Richard and Sunny in Lyon for a sabbatical leave. We did some nice experiments in the lab using femtosecond lasers, but also visited Maejeane, Nicole and Denis in Paris. We also made a nice tour around the volcanoes in Auvergne, and had a great time there. For this, to feel like "adapted" to the region, we spent the nights in a castle. Richard and Sunny had a VIP room in one of the round towers in the picture.

Some time later, I was awarded the Grand

Prix de Physique de l'Academie des Sciences, and I invited Maejeane and Richard to come with Noemi and me to the ceremony at the Institut de France in Paris. No need to say that we had a nice dinner and a top notch bottle of wine together afterwards to celebrate!

In September 2003, it was my turn to get married, with Noemi. The ceremony and the party took place in Lugano, in the Italian speaking part of Switzerland, where Noemi grew up. Sunny, Mae Jean, Richard and Kevin made us the extraordinary pleasure to join for this "once in a lifetime" event. It was fantastic to have them there with us, sharing these unique emotions. We then met several times (but not enough) for special occasions, in Paris, Brussels, Antibes, Monte Carlo (with my kids, born 2009) and it felt each time as if we had had dinner together the night before. The privilege of really close friends. One of these special occasions was Richard's retirement party in June 2008. So many people, of all generations, and all looking at Richard with these smiling and shiny eyes, full of admiration and wonderful memories. Among them, Nicolaas Bloembergen, 88 at that time, Nobel laureate and supervisor of Richard at Harvard, who presented a brilliant talk from the very beginning of non-linear optics to the latest attosecond spectroscopy techniques! We all then drove up to Vermont in a joyful atmosphere. Nico even came with us in the afternoon for a long and friendly walk in the hills around Brattleboro.

The last time I could visit Richard and Sunny was in summer 2013. I took the chance of being on the East Coast for a Gordon Conference to rent a car and drove to Hamden.

Was really nice to talk, have a walk in the surrounding forest and spend again some time together. Not enough time though. As usual. After the move to Singapore, we mainly kept contact via e-mail. Exchanging the last news with Sunny and Richard, but also with Nicole, Natasha, and Kevin, was and always is a very special moment for us. And we will keep it so, and Richard will be with us.

Note to Richard on the Occasion of His Retirement in June 2008

Véronique Boutou, Ph.D.
Chargée de recherche, Institut Néel
CNRS UPR2940
38042 Grenoble Cedex 9, France
veronique.boutou@neel.cnrs.fr

Véronique Boutou with her sons Félix and Dario in January 2008.

Véronique Boutou's habilitation defense in Lyon in October 2006.

Letter to Richard on the Occasion of His Retirement in June 2008

François Courvoisier, Ph.D.
CNRS research director, Institut FEMTO-ST
Département d'Optique, Université de Franche-Comté,
F-25030 Besançon, France
francois.courvoisier@femto-st.fr

Dear Richard,

There are many many things to tell you, even if, all in all, we didn't meet that many times. The thing is that each of our meetings left me so many *souvenirs*, that I really want to thank you for all you gave me. You are really one of those "big guys" that you meet once and remember all your life.

Your simplicity, your welcome, your kindness associated to a so brilliant career makes you really admirable.

I do not know if you remember the first time I met you in your office, but I will always. In a few minutes, you explained to me — at this time even not graduated — your research themes, the state of the art, the applications ... and you made it so concise and clear that it really made a strong impression on me. I was realizing that I was one of the hundreds of students that you had during your career, but you gave me as much care as if I was the first.

Look at the blackboard! Remember those squares? May 2000.

What I loved in working with you is this communicative excitation and enthusiasm that you always have. Each time we met, I really enjoyed in advance to find again this specific excitement. For all this, it is not exaggerated to say that you are one of my 2–3 models in life and science. When I was this student in the plane going from France to these unknown US, I really would not have imagined coming back with so much.

It is also thanks to you that I met Jean-Pierre: "François, you should have lunch with Jean-Pierre today." It was a very nice idea, Richard!

And for this and all that precedes, you are my "grandfather in research". That is why, I see you retire with a bit of emotion.

Adeline, Hugo, Louise, and I wish you a happy retirement and wish you will spend lots of time with your family and especially your grandchildren!

François Courvoisier

Best wishes from all of us!
Louise, Adeline, Hugo, and François.

For Richard

Dianne Terrace
Yoga instructor and Movement specialist
North Haven, CT 06473 USA

Throughout my career as a Yoga instructor and a movement specialist, I have come into contact with many people who have left lasting impressions, both positive and negative. But none has had a greater impact upon me than Dr. Richard Chang, whom I met in 2002. I think of the person I was "before" Dr. Chang and the person I am "after" Dr. Chang.

We have all had teachers that have in some way contributed to our lives, but working with such a brilliant professor who was more than willing to become a student of mine changed the way I looked at the world and myself as an instructor. Before I met Dr. Chang, I looked at my students as people that would learn and perhaps benefit from the disciplines I have studied for years. After working with Dr. Chang over the years, I realized that if a professor as distinguished as Dr. Chang was willing to become a student and learn from me, then surely those people I regarded as students also had something to teach me.

This renowned professor taught me that everyone has the capacity to teach as well as learn from others. He was always observant, hard-working and dedicated to slowing the advance of the Parkinson's disease he faced so valiantly. His humility, combined with his courage and utter enjoyment of every facet of his life has made me a better instructor and a better human being. I can still hear his voice as I left every training session, "Thank you so much for coming to work with me." I will always miss you, dear friend.

Meeting Richard Chang

Cynthia Griffin
Licensed Massage Therapist
Natural Health Therapies
Plantsville, CT 06479 USA

When I think of Richard, I smile and feel truly blessed to have crossed paths with such a wonderful human being. He was an unexpected teacher and dear friend.

I met Richard Chang in 2005 in New Haven CT while working at a Wellness Center. He requested a Traditional Thai Massage session, and I was the only licensed massage therapist at that facility who specialized in that modality.

Our first meeting was etched in my mind. When I walked into the reception area to greet my new client, I saw a gracious man who was a little unstable on his feet but was determined to receive what I had to offer. Richard revealed that he already had a Traditional Thai Massage practitioner named Bobbi but was looking for an additional experienced practitioner to add to his toolbox of natural therapies to help him with his Parkinson's disease.

At that time, I was newly divorced, and trying to figure out how to be a single mother of two children. I lived with my daughters in Southington CT, which had little ethnic diversity. I grew up in Southington. My blond hair and Irish/Norwegian/European background allowed me to blend in with the majority of others who lived there.

My longing to be out of my small box as much as possible made it an easy decision to pursue an open position in New Haven. The facility offered many alternative healing modalities and had a yoga studio right next door. This was my chance to work in a city with lots of diversity in people, restaurants and Yale University.

I continued to have Thai massage sessions with Richard at the wellness center. I also began seeing him at his home in Hamden where I met his wife Sunny. She was a strong intelligent woman with an inviting personality. Their home reminded me of the time I spent in China. Its Chinese style architecture had a sense of simplicity and flow. It also included cherished family heirlooms. I always felt very welcomed at their home, like one of the family.

Over the years, Richard told me stories of his life with humility and humor. I noticed a common theme of turning big and small challenges into gifts of knowledge for future reference. He talked about his childhood in China, his experiences as a young man arriving in America, his adult life as a scientist and teacher of physics at Yale. When he spoke of his wife and family there was such a sense of pride and love in his words. All these conversations fueled my desire to learn more about a world outside of my realm and personally embody the belief that anything could be possible.

I got the sense Richard was raised by a strong courageous mother. He told me that he was born into a family that did well in China. When the Chinese revolution started, his whole world changed. His mother and father decided to flee to safety with the children and a nanny. During this time, his mother carried several pieces of jewelry that were later sold to finance their survival during the revolution period.

Richard continued his story with a grin on his face. He said, "My mother decided that I needed to go to school in America. Otherwise, I would turn into a derelict like the young men without direction or a proper education." I wondered what the Chinese word for derelict was and how that was translated. He continued, "I arrived in the US with not much but the clothes on my back, one suitcase and a goal, which was education." Again, I was in awe at the strength and courage it must have taken to move forward all alone, at a young age, in a new country with an unfamiliar culture. As we know today, Richard excelled with a passion in Physics and a desire to teach and influence others.

I never noticed any ego or sense of grandeur in Richard. He often would talk to me about Physics, which was beyond my comprehension. Even though I could not keep up with his thought process, I thoroughly enjoyed the excitement and passion that showed through his words. I often told him I did not quite get what he was saying. Then we both would have a good laugh.

One day we had a conversation about Eastern Philosophy and Buddhism. I questioned why there was a hierarchy in prayer or honoring religious spiritual beings and human beings of the past and present. I understood why you would honor your Spiritual Being first but, why honor your teachers before your parents? Is not that, disrespectful to the family structure?

This is my interpretation of his reply. Our parents form our bodies. Our teachers' lessons can direct our journey in life, and what happens after we leave our bodies in death. During our life on Earth, everyone and everything we come into contact with, can be one of our teachers, if we look a little deeper. Some teachers have a structured intellectual curriculum that is followed and learned from. Other teachers exude knowledge by just being who they are, what they do, and how they treat others. Teachers can come in forms other than humans, like other sentient beings or experiences, we have. We just have to notice what is being taught. This cycle of learning will continue on-and-on.

I could tell how important family was to the Changs. They all enjoyed world traveling and they felt that vacation time together was very important, especially because his children were spread out in different countries. He told me how his family arranged a hot air balloon safari ride during a family vacation so they could all enjoy the adventure together without Richard over-exerting himself. Again, no boundaries got in the way of what was most important.

Despite the challenges of Parkinson's disease, he continued to work and travel to present scientific lectures. He asked for help if needed from people he trusted without holding back because of false ego or pride. I was in awe of his perseverance, his support from his wife and the wonderful group of people that helped him along the way.

I was honored to be asked to go to his retirement party alongside his loved ones, colleagues, coworkers and friends. What a joyous occasion to honor such a wonderful accomplished man. When I looked around the room, I saw a group of diverse people from all walks of life that had a common bond of a teacher, role model and friend.

At that time, I was dating someone, and asked him to be my "plus-one" to the party. As he did with everyone else, Richard graciously greeted us with inviting conversation and gratitude for sharing his special celebration.

In conversation later, Ricard revealed to me that my date reminded him of a "Mongolian Warlord". I still laugh when I think of that comment. Richard was right. He was not the right guy for me. I took that comment like one that would come from a dear, observant friend who could use humor to gently plant a seed of truism to push me into reality.

Soon after the retirement party, Richard and Sunny asked me if I would like to go to their home in Brattleboro for a visit. It was the 4th of July. I accepted and insisted on giving Richard a Thai massage session while I was there. My sister accompanied me to the Changs' second home in Vermont — another house renovated with an Eastern flare and decorated with spectacular family artwork. The four of us had a nice walk downtown and great conversation while dining at a local restaurant.

Later in the afternoon, Richard received a Thai massage from me. Then it was time to say our goodbyes. During this conversation he said he was not sure when he would see me again because he might not be in Hamden much. I told him how happy I was that he was retired now and could be anywhere he wished and, if it could be worked out, I would love to visit him and Sunny in Vermont again.

On the drive home, I had a very heavy feeling and somehow knew that something was going to happen. This feeling did not go away. I carried it around for a long time and was curious to know what it was.

Sometime later, I received an email from the group at Yale that set up the retirement party. This email stated that Richard had a speech he wanted to read during the party but was too exhausted to present it. It continued by saying, after his drowning accident, Yale thought it was a good time to present the speech to everyone by email.

I was confused by the words "drowning accident" and was fearful that Richard had passed. I called every contact number I had for the Changs including his other massage therapist and Yale. I left messages asking what had happened. Bobbi, Richard's other Thai Massage Therapist, told me that he was alive and in recovery after a near death drowning and a coma. One of many complications he was now suffering from was loss of memory. He awoke from his coma thinking it was sometime in the 1980s. He did not recall anything after that.

After months of intense rehabilitation and continued daily cognitive exercises at home, Sunny asked Bobbi and me to come up to Vermont to try to spark some memories in Richard. We both jumped at the invite and the chance to see if we could be of help. We spent the day having light conversation and took a drive and a scenic walk. Unfortunately, Richard did not remember either of us, but he was still his gracious self, like the man I experienced on the first day we met in New Haven. As always, he was genuine, authentic and humble. Richard was so heartfelt when he said to me, "I apologize. I know, I know you, I'm sorry but I can't recall from where." I felt honored that he had some deep sense of knowing me even if it was not cognitive.

I met with the Changs one more time in Hamden before they decided to move to Singapore, to be closer to family and the availability for therapies not easily accessed in the US. After all, family is the best medicine. Sunny continued to keep the lines of communication open through yearly emails. These emails were a glimpse into the happenings of the whole Chang family and the joyous occasions and achievements of the budding grandchildren.

When I received the email that Richard had passed, I knew his energy would continue on. I may not understand the dynamics of physics, but I do understand the cause and effect of how people can connect and affect each other's lives intellectually, mentally, and energetically. Richard's teachings will continue in the cycle of scientific learning. His essence will still be honored and felt by family and friends.

When one reads this collaborative book, it will show how Richard's life will continue to be a positive testament of what a true teacher is. The stories he shared with me will be in my mind forever, and the gratitude we shared for each other will remain in my heart. I was honored to have him as a client, friend and teacher of life's lessons.

Thank you, Richard Chang!

Note to Richard on the Occasion of His Retirement in June 2008

Dear Richard,

Sometimes you meet people quite by accident or so you think. Some say these meetings are all predestined and according to life's plan. Maybe I was meant to learn and practice Thai Yoga Massage so I could meet the most amazing person I know. You continually tell me how grateful you are for our sessions but I am truly the blessed one. You have taught me so much just by your presence, words and kind actions.

Happy Retirement!

I wish you love, happiness and good health,
Cynthia Griffin

Note to Richard on the Occasion of His Retirement in June 2008

Gina N. LaRossa, Yale B.S. '01
Assistant Professor of Medicine
Washington University School of Medicine
St. Louis, MO 63110, USA
glarossa@dom.wustl.edu

Thank you for being such a great professor and mentor. My best college memories came from people like you. I wish you an interesting, busy and happy retirement, and I hope to see you if I am ever near New Haven. Let me know if you are ever going to be near St. Louis.

Fondly,
Gina LaRossa

My daughter, Gina, had great respect and affection for Richard when she was an undergrad at Yale. He was her professor, mentor and research advisor, and they have kept in touch over the years. He was the kind of teacher who enriches the college experience and makes it memorable for students.

She is currently a first-year resident at Barnes Jewish Hospital in St. Lewis, and I know how bad she feels that she cannot get time off to attend the retirement party. She had asked me if I could find the photos that we took with her, Dr. Chang and another student when Dr. and Mrs. Chang had us all over for champagne after graduation in 2001.

Denise LaRossa

Alex Potter and Gina LaRossa celebrate graduation in May 2001 with Dr. Chang.

My Reflections of Richard

Andrew Wing on Poon, Ph.D. Yale '01
Professor and Head, Department of Electronic and Computer Engineering
The Hong Kong University of Science and Technology, Hong Kong, China
eeawpoon@ust.hk

My stories with Richard began when I was a graduate student at Yale. I was most privileged being one of Richard's later graduate students from 1996 to 2001. Little did I know then that Richard's deep influence on me and his friendship would have continued for the next two decades. Richard had a profound impact on my growth, academic career, development, and life. He has always been, and will continue to be, my role model as a scholar, a mentor, and a gentleman. He was a rare human being who touched and inspired so many and left a legacy by his four decades long of great scientific work yet remained so humble. He put his scientific work into real use and made a lasting technological impact to the detection of biological aerosols after the 9/11 attack. Richard taught us by example through his own long fight with Parkinson for 25 years how to bring out the very best from one's life and work, even when the situation may not be the most favorable. He faced life's challenges with a smile, a crystal-clear mind, and an always positive attitude believing that tomorrow would be a better day. He lighted up the paths for many, and surely for me.

Looking back, I see many fond memories of Richard. Knowingly and unknowingly, those memories keep guiding me in face of new challenges and giving me confidence in my sense of judgment. In difficult times, they cheer me up and get me going. Richard set very high bars for his students and had all the expectations that we will reach them in time. He made us "leaders, not troopers," and lifted us to become a better scientist, a better person.

I joined the RKC lab at the late phase of Richard's droplet-microcavity research. Instead of continuing with droplet nonlinear optics or lasing work, Richard assigned me to study the optical resonances in asymmetric resonant cavities. At the time, the topic of asymmetric resonant cavities in a stadium or quadrupolar shape was of major research interests for directional-emission microcavity lasers (first observed from Richard's pioneering lasing droplet work in the mid 80's). The new shape of microcavities also attracted interests for studying the chaotic ray-wave nonlinear dynamics (analogous to quantum chaos). The global tidal wave of asymmetric resonant cavities spanning the late 90's to 2000's began at Yale Applied Physics. The RKC lab drove the experimental efforts while Professor Doug Stone's group worked on the ray chaos theory. I became one of Richard's graduate students along with Seongsik Chang, and later Grace Chern Metcalf, who were well exposed at the epicenter of the latest asymmetric resonant cavity optics and theoretical concepts.

My experiment was to measure the optical resonances with a spectral resolution superior to the conventional method using a spectrometer. We wanted to examine the input- and output-coupling dependence of the asymmetric resonant cavities as a function of the cavity orientational angle and the scattering angle. All these should be studied in a passive cavity situated in free space, so that we could ensure that there are no complications from the nonlinearity or gain of the cavity medium or perturbations from a coupling structure.

Richard already had in mind then such a thesis work would have shed light on understanding the missing phase-matching requirement for exciting optical resonances in chaotic or non-regular open round-trip paths. The work would have paved the way for its technological use as optical channel add-drop filters, a building-block component for the then rapidly emerging wavelength-division-multiplexed optical-fiber communications behind the Internet.

To study the sharp optical resonances in an experiment, Richard suggested me to measure the elastic light scattering spectra at an angle by illuminating a glass fiber (custom-made with a quadrupolar-shaped cross section) from the side using a laser beam. That sounded like a perfectly doable experiment. I thought I could finish my thesis work quick enough.

Early lessons from Richard: One of the first lessons I learned from Richard was how to do an experiment with elegance. I was setting up the light-scattering experiment for the first time. I meant to mount a glass fiber for illumination by a laser beam from the side. Knowing how brittle the hair-thin glass fiber is, my initial attempt was to stick the fiber on a stage using a double-sticky tape. I thought naively that was a smart method. Richard saw it. With a dismay, he commented that one should be smart enough to mount a fiber with a more elegant method than using a double-sticky tape. Since then, I have never used a double-sticky tape in my experiments. Now, I will give my students the same look as Richard once did whenever I see my students using double-sticky tapes on their experimental setups or not applying themselves to solve a problem with elegance.

Richard was keen on experiments done with the best available and cost-effective tool. He equipped the laboratory with state-of-art gadgets if they were needed in our experiments and as far as the research budget allowed. The RKC lab had a mix of new and old "toys." For example, I was the lucky graduate student who unpacked the "present" and used in my experiments the New Focus external-cavity tunable diode laser, which was just available in the market in the mid 90's. With this new laser, I could measure the elastic scattering spectra from the fiber by fine-scanning the laser wavelength in seconds and minutes and detecting the scattered light intensity by a photodetector. This method enabled by the latest technology then allows the spectral resolution to far exceed the conventional method (using a broadband light source and a spectrometer), which Richard used in the early 80's for his pioneering work studying optical resonances in circular glass fibers by elastic light scattering.

The laser-wavelength-scanning method now constitutes the standard technique for laboratories measuring sharp optical resonances from microcavities. When I started my laboratory in 2001, the same New Focus laser was the first piece of equipment I purchased using almost my entire startup fund. I learned from Richard then solving problems in an elegant way often requires leveraging the best available "toys." Over the past two decades as a professor, I never shy from investing in my laboratory with equipment if they are justified.

There was a side story to my early setting up of the experiment. As the scattered light is weak, the experiment required a lock-in amplifier to enhance the signal-to-noise ratio. The amplifier we used then was the state-of-art Stanford SRS system (I was again the lucky graduate student who unpacked the "present"). I thought then it would be even better if we could do the experiment with two of these, perhaps to use an extra one for measuring the scattered light signal at a different angle simultaneously. Obviously, I was not thinking through my experiment carefully. I proposed to Richard to purchase an additional lock-in amplifier for the experiment. His response was a stern look. He basically said I had not thought it through. I realized I must have let him down. Throughout the rest of my thesis experiments, I never found the need of having a second lock-in amplifier. In my lab now, I have so far only one SRS lock-in amplifier for my students' experiments, and it works fine.

"Have you heard of Fano resonances?": Working on experiments is never easy. Working with a giant like Richard did not make an experiment easier. Back in spring/summer 1997, when I first observed odd-looking sharp asymmetric resonance line shapes in my scattering spectra from circular glass fibers, I thought initially (and was told by fellow graduate student Steve Holler) that they might be experimental artifacts. I was still fooling around with taking data by the lock-in amplifier then. The resonance line shapes, unlike the typical symmetric "Lorentzian" peaks, were peaks accompanied with sharp dips on either left or right side. Some even appeared as sharp dips instead of peaks. I had no ideas of what my data were telling me. I even doubted whether they were properly measured.

I showed my data to Richard one day during his daily visit to the lab (usually around 11 a.m.). He did not really say too much then, or perhaps he just asked me to continue with the experiments. A few days later, again during his daily visit to the lab, he asked me in a casual manner whether I had heard of "Fano resonances." I had no ideas. However, I did know of and even had a brief encounter with a "Professor Fano" from the University of Chicago, where I studied for my bachelor's degree in physics. Without elaborating further, Richard just asked me to look up "Fano resonances" from the library. Those were the pre-Google-search days.

I went upstairs to the Becton Engineering Library. After randomly hitting a few textbooks, I found "Fano resonances" from perhaps an atomic physics book or an advanced quantum mechanics book. The reference directed me to look up the original Physical Review 1961 paper by Ugo Fano on coherent quantum interference in atomic resonance transitions. I quickly looked up the paper and indeed found after pages of long equations the asymmetric resonance line shapes like what I observed. Richard must have known this!

I realized then that although Fano resonances were a known atomic resonance phenomenon, the essence of the asymmetric resonance line shape is not a quantum property. It is due to a coherent interference between a resonance and a background. This makes Fano resonances applicable to my light scattering experiments, which measure the coherent interference between the glass microcavity resonances and the non-resonant background scattering. I spent the following days or weeks to work out a simple theoretical model based on Fano's insights on coherent wave interference between a resonance and a background. To my delight, my "toy" model could perfectly account for all the asymmetric resonance line shape features I observed from experiments. I proudly showed my work to Richard. He was delighted.

So, I thought we should quickly publish our findings on Fano resonances. That would have been my first publication. To my knowledge, that would have been the first literature explaining the simple essential physics of coherent interference for Fano resonances, in the contest of elastic light scattering from microcavities. Richard, with his high standard to doing science, however, was not just satisfied. He asked me to press further and to come up with an application of Fano resonances before we publish the work. Richard was the true scientist who was not easily satisfied by just understanding a phenomenon but would have gone on to look for and to demonstrate applications of the scientific principles. He taught his students by many of his own work the art and beauty of applied science.

As a junior graduate student then, I fell short of realizing the potential applications of Fano resonances. Today, Fano resonances in photonics has been widely used for sensitive optical resonance-based sensing. Then, distracted by many other experiments, we put aside writing up the possible Fano resonance paper. Fano resonances remained an integral feature in my thesis work, and in my later work with my students.

As a follow-up action, Richard asked me to send an email to Professor Fano to share with him our experimental finding and modeling work. Professor Fano replied my email after a few days. He told me how delighted he was to see his resonances measured from elastic light scattering spectra. What a pity that I lost Professor Fano's email! I also wrote up a section on Fano resonances in my thesis.

My first publication on Fano resonances appeared in 2003, after my student and I revisited the work using a prism-coupled glass fiber-microcavity. Later in 2007, my other student and I demonstrated the use of Fano resonances in an integrated silicon photonic switch. We eventually found applications of Fano resonances, which Richard first showed me the way a decade earlier.

Setbacks from experiments: Richard then wanted me to figure out the missing phase-matching mechanism for optical resonances in asymmetric resonant cavities from my experiments using the deformed-shaped glass fibers custom-made by Corning. Richard always had a high confidence in his students, and he expected us to strive up to his high standard. The experiment turned out to be far from trivial. There were also substantial practical limitations, including the lack of different samples. Corning only made for us one single batch of glass fiber samples, with one quadrupolar shape, one size and one specialty glass. There were plenty of number of samples though, so I could keep repeating my experiments. Added to this was the complexity of the multiple optical resonances observed from the scattering spectra. The resonances are sensitively dependent on the quadrupolar shape orientational angle, the fiber tilting angle, the scattering light angle, and the laser beam focusing condition and incident location. It took me the next three years day in and day out repeating all kinds of experiments that came to mind. I accumulated stacks of data and laboratory logbooks detailing my observations and data analysis. Yet, I was not able to conclude on anything substantial.

Richard was very supportive and extremely patient. Despite the lack of obvious progress, he and I regularly discussed about the latest data and my plan for the next experiment. In the meantime, the Stone group published a front-cover Science paper with the Federico Capasso group then at the Bell Lab. They reported directional emissions from the so-called "bowtie mode" of a stadium-shaped semiconductor quantum-cascade laser. Richard and I tried to relate their results to our resonant scattering spectra from the glass fiber, but in vain. The bowtie mode cannot be confined in my quadrupolar-shaped glass microcavities. Neither did the chaotic ray dynamic model from the Stone group provide us with clear insights into the phase-matching condition for optical resonances. Without a wave-optics theory or a numerical modeling tool such as the finite-difference time-domain method (which became commercially available in the early 2000s), Richard and I could only resort to our limited experiments. We persevered despite repeated failures while others were miles ahead.

Exploring new concepts: Richard always walked with us and always trusted our scientific judgements. He was always willing to listen to my half-baked ideas and always encouraged me to further develop them. Richard's openness and confidence in his students' work led me to explore other related concepts and ideas while continuing the experiments on asymmetric resonant cavities.

For example, we studied the optical resonances from a tilted circular glass fiber. We observed from the light scattering spectra resonance blueshifts that we attribute to open round-trip trajectories that spiral along the circular cylindrical fiber. We termed such resonances as "spiral resonances." They are like the well-known resonance blueshifts in open "zig-zag" round-trip trajectories in an angle-tilted Fabry-Perot cavity or etalon. The insight we learnt was that an open round-trip trajectory can sustain resonances if the wavefront matches and overlaps upon a round trip. This is essentially the phase-matching condition Richard must have envisioned for the asymmetric resonant cavities and hoped that I could demonstrate from experiments.

Later, I proposed to Richard to explore optical resonances in square-shaped micropillar cavities using a square-shaped glass fiber. The idea was stimulated by my fellow colleague graduate student Nathan Rex's work. Nathan was measuring lasing emissions from square-shaped GaN micropillars. This time, we found commercially available square-shaped glass fibers. Based on the wavefront-matching concept learnt from "spiral resonances," the RKC lab demonstrated and explained multiple optical modes in square-shaped micropillar cavities, with multiple wavefront-matched open four-bounce round-trip trajectories. Whereas the conventional phase-matching condition would have predicted single optical modes in square-shaped micropillar cavities, with closed four-bounce round-trip trajectories. This work became the bulk of my thesis.

An idea on microspiral cavities: We did not give up the asymmetric resonant cavities. One main objective of the work was to come up with a microcavity design that enables unidirectional emission. With the quadrupolar or stadium shape, the four-fold rotational symmetry would still give four or multiple directional beams. By late 1999, Richard and I began to contemplate what if one further reduces the symmetry of the asymmetric resonant cavities to the absence of rotational symmetry while preserving the light cavity confinement? Could such an asymmetric microcavity emit light in a unidirectional manner?

Our idea then, intuitive, and half-baked, was a spiral-shaped microcavity. The Archimedean spiral has a linearly changing radius over the azimuthal angle. There is a notch on one side. Without any experimental or theoretical support on our idea, Richard and I believed that the cavity light travelling along the spiral circumference in the direction that sees the notch could preferentially emit unidirectionally from the notch. Whereas the cavity light travelling in the other direction tends to remain inside the spiral cavity. Doug Stone initially thought the microspiral cavity would not work at all for a laser as the notch would cause too much loss. Richard thought we should do an experiment.

To do the experiment, Richard found a new collaborator, Professor Joseph Zyss in France. Professor Zyss was an expert on organic light emitters. The Zyss group was able to fabricate organic microstructures on a chip. So, by early 2000, I started drawing spirals on a layout using a computer-aided design tool. I designed a photolithography mask with various spiral-shaped microcavities. We had the mask made at the Cornell Nanofabrication Facility and shipped it to Professor Zyss' laboratory to fabricate organic microspiral laser structures on a glass substrate. That was my first encounter with integrated photonics and nanofabrication. It turns out that for the next two decades my students and I would be striving on integrated photonics and nanofabrication, fabricating microcavities on silicon wafers.

After waiting for many long months, the first polymer microspiral sample arrived at the RKC lab in late 2000. I began some preliminary experiments using the photonic chip while busy wrapping up my thesis work. I passed my thesis defense in May 2001 and returned to Hong Kong in June 2001. The microspiral work was continued by Grace Chern Metcalf. In 2003–2004, Grace and Richard demonstrated beautifully AlGaN microspiral lasers. They observed unidirectional lasing emission from the microspiral notch.

Thesis defense, and other fond memories of Richard: My thesis defense was a bit more dramatic than I would have thought. Perhaps I had not consulted with Professor Doug Stone sufficiently enough on various theoretical interpretations of my thesis work, he initially was not so impressed by the work. In a two-hour-long session with my thesis examination committee, Doug Stone and Professor Robert Grober drilled me to defend my thesis. I ended up filling out six white boards (three sliding on the front and three at the back) with equations and schematics. Richard, being completely confident in his students, just asked the committee members "to take out a baseball bat and hit hard." He kept silent the whole time except nodding his head occasionally. I believe I passed the defense with flying colors. Richard told me later that Doug Stone wrote on the thesis examination report a very high compliment on my thesis work.

There are many other lessons and fond memories of Richard. Richard once shared with us the story of Doug Stone asking him why he chose to work on liquid droplet optics but not other hot topics. Richard's answer was he worked on whatever he was interested in and let others follow his direction later. Indeed, Richard's liquid droplet research is phenomenally fruitful, both academically and technologically. On the one hand, his droplet-microcavity work helped lay down the foundation for optical microcavities, which now find applications as photonic components for communications, sensing and computing. On the other hand, Richard, and his associates, notably research associate Dr. Yongle Pan, studying microparticles with complicated morphologies and biological contents have found numerous creative ways to laser-based diagnosis of particles and bioaerosols on the fly, single particles at a time. This resulted in a laser-based bioaerosol-detection system currently installed in the San Francisco airport after the 9/11 attack. Richard was a master applied scientist who demonstrated the deepest scientific insights and found practical use of them to solve real-world challenges. I cannot help but wonder what would Richard have done to detect COVID-19 virus in aerosols?

I remember well Richard's lucid and inspiring lectures on quantum electronics and nonlinear optics. Richard always made the most subtle concepts crystal clear. I am still constantly referring to Richard's lecture notes for my teaching and research, and still find useful insights from them. Before each lecture, Richard would ensure each of us received a xerox-copy of his hand-written transparency slides. One can see Richard must have spent lots of time and great effort preparing his lectures, making his lectures seemed effortless. I always strive to give lectures like Richard did.

Richard was always humble and was a gentleman. I never saw him talking about any of his many distinguished achievements or profound impact on so diverse scientific and technology fields. He was straight to his students, but a stern look was usually sufficient to make us realize our stupidity.

I remember having many light sandwich lunches with Richard while sitting on a bench by the street discussing science (Richard often treated us sandwiches from a small store nearby). I also cherish our occasional "fancy" lunches at Mory's, usually as a celebration after a thesis defense.

Richard's tap on his chair to indicate time is up for the meeting (or we had not thought through what we were saying and to try again next time) is memorable. I would never forget his hand-written corrections on our many manuscripts. On one of our preprints from *Optics Letters*, Richard's last name was spelled wrong as "Change." He asked me to tell the editor to change it.

Visiting Andrew and Lina's home, Hong Kong, 17 May 2004.

Richard's friendship and influence after Yale: My story with Richard continued after I returned to Hong Kong as an assistant professor in the Hong Kong University of Science and Technology (HKUST). I continued experimental research on optical microcavities. I came up with new twists on fabricating and integrating microcavities on silicon wafers using the university's cleanroom facility. "Be a leader, not a trooper" and "shed light on the paths of others" have been my guiding principles over the years following Richard's teaching. Picture shows Richard visiting HKUST with Andrew and his students in the laboratory, Hong Kong, 17 May 2004.

I was immensely grateful and honored to have hosted a few visits by Richard and Sunny between 2004 and 2013. Richard and Sunny had five visits to Hong Kong on 16–18 May 2004, 14 Feb 2007, Jan 2011, 27–28 May 2012, and Apr 2013. For each visit we indulged ourselves at The Peninsula Hotel, where Richard had fond childhood memories. Richard interacted with my students during his visits, commented on their work and listened to their presentations.

The Peninsula Hotel, Hong Kong, 17 May 2004.

In 2008, Richard invited me to coauthor an article on microspiral lasers for *OSA Optics and Photonics News*. This turned out to be a memorable article. During the final drafting of the article Richard was nearly drowned in a swimming pool, just about a week after his retirement from Yale. The first thing Richard asked after he awoke from coma was whether Andrew had received his revisions on the manuscript.

During Richard's last two trips to Hong Kong, we were able to spend a few quality days working together on writing a book, which Richard wished to be a monograph on the principles and applications of microcavities and microparticles. To Richard,

Visiting HKUST at Andrew's laboratory, Hong Kong, 17 May 2004.

microcavities and microparticles are like two sides of a coin yet they evolve into so diverse scientific and technological communities. I hope I could help finish his book vision one day, in a form that Richard would have approved.

Visiting HKUST and listening to Andrew's students' presentations, Hong Kong, 11 Jan 2011.

Interactions at OSA Annual Meeting, Tucson, Arizona, USA, 7 Oct 2003. I was holding a drawing of the Yale campus, a gift from Richard and Sunny for my wedding in 2002. Left: Professor Alfred Kwok, Richard's former graduate student.

Dinner at the Poons, Hong Kong, 27 May 2012.

The Hong Kong Exhibition Center, Hong Kong, 27 May 2012.

Hosting a 4-day zoom workshop on microcavities in 2022. This has been postponed from December 2019.

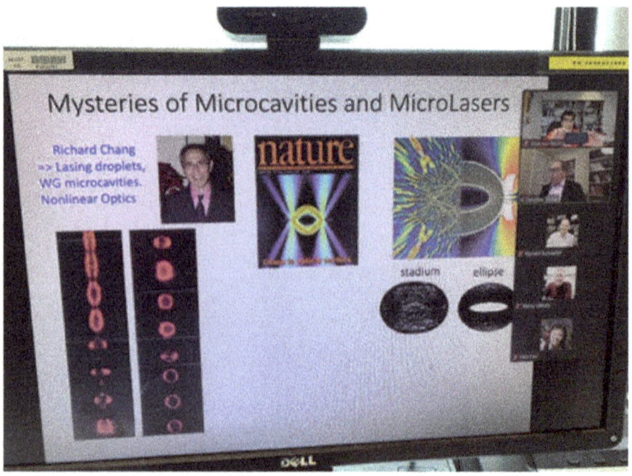

Richard's work on lasing droplets, whispering gallery microcavities, and nonlinear optics in Zoom, Andrew, Doug, Harald, Kerry, and Hui participating.

Letter to Richard Chang on the Occasion of His Retirement in June 2008

April 11, 2008

Dear Richard,

Even from a young career that is barely 1/6 of yours in time, I see how almost impossible it is to keep running a research group in the frontier, writing good papers and grant proposals, juggling administrative tasks, maintaining good relationships with colleagues all around and inspiring the next generation. Richard, you have achieved all these exceptionally over your very distinguished career. As your former student, I have the greatest fortune to learn from you and see how a great mind in science works and contributes. Now as a young professor, and for many years to come, I will strive to deliver my best, knowing that it will be insignificant against your high standard. Let me tell you once again how deeply grateful I am for everything that you have done for me over the years, and I am very proud and honored to celebrate your retirement with you as your student and friend.

Looking back, I see many fond memories that we both cherish. May be I won't just list them here as they are an integral part of me. But knowingly and unknowingly, these memories keep guiding me in face of new challenges and giving me confidence in my sense of judgment. In difficult times, they cheer me up and get me going. Richard, you have always been there for me, since the early years when we worked on the ARC and square micropillars and the spiral cavity ideas, and later my career development and most of all the tenure. At each and every step, you always find the right moment to inspire me and make me see a bigger and better picture.

When I learnt that you finally made up your mind to retire from the front line, I understand that you must have mixed feelings. But now you may finally free up yourself from endless grant proposals and review meetings, and only give lectures when you feel like to give one and only for the best students. I am sure the lab that you keep running will keep you busy with science that you most enjoy. Of course, I will be most happy and proud to see our creative effort in spiral-shaped microcavities will continue to make great impact on the microcavity community, and now the silicon photonics community as well!

Richard and Sunny visiting Andrew, Matthew, and Lina, dinner at The Peninsula Hong Kong, 14 February 2007.

"Be the leader not the troopers" and "make

good students better" are two important attitudes in research and teaching I learnt from you. In my own ways, I will try hard to continue to contribute to science and inspire the next generation. Hopefully, in time, I could also experience in many ways some of your joy and excitement in academia.

Lina and I wish you and Sunny all the happiness and longevity. I am sure you will enjoy traveling around the world and visiting many exotic places. Be sure to come see us in Clear Water Bay whenever you want to indulge yourself in The Peninsula. I will certainly find my way to New Haven and Vermont in my CLEO trips. For now, I am just looking forward to your retirement party and writing our joint manuscript!

With All My Best Wishes
Andrew

Letter to Sunny

Michel Devoret
F.W. Beinecke Professor of Applied Physics
Yale University, New Haven, CT 06520 USA
michel.devoret@yale.edu

September 13, 2021

Dear Sunny,

Thank you for your kind letter. We imagine how difficult it must be for you, missing Richard. I will be very happy to participate in the memorial book and I hope this letter to you will serve in this role. I vividly remember the rare and wonderful humanity with which Richard was fulfilling his role as professor, mentor and researcher. I not only admire what Richard has accomplished in his career, but I would like to also stress that he was among US scientists a role model for me and others. When I travel to visit universities in the US and the rest of the world, I often receive testimonies of how much Richard was known and appreciated in the Physics community. At a more personal level, my family and I were very much touched by the way Richard and you welcomed us in your houses in Connecticut and in Vermont. We have such very fond memories of the visits at your house in Brattleboro, a town we now go to at least every year in the Fall. We also treasure the remembrances of the dinners we had at your charming house in Hamden! On our visits at Brattleboro, we had the great pleasure, thanks to you, of meeting Jean-Pierre Wolf. Jean-Pierre and his wife Noémie are now treasured friends of ours!

To come back to Richard, I can now appreciate he was always — and most crucially, he gave the inspiring example that it was important to be — well above the too common mediocrity of university life. He was embodying fundamental human qualities that made our Applied Physics Department at Yale a welcoming and stimulating community. Reflecting upon these years where we met you and Richard, I realize the situation has recently evolved downhill, unfortunately. Maybe this impression is simply due to the fact that two decades ago we had chosen to join a very unique and singular place, which Richard had largely contributed to create, and we now face the normal, average academic life in American universities. I am almost sure Richard would have whimsically laughed off my current pessimistic remarks about the evolution of our department and invoked the Second Law of Thermodynamics! His generosity, humor and optimism live on inside us.

With my warmest regards,
Michel

Note: Michel Devoret is the F.W. Beinecke Professor of Applied Physics at Yale University. He was born in France, and he now holds American citizenship. In 1999, he moved from Paris to New Haven with his family for a one-year sabbatical at Yale. It was on this occasion that he first met Richard and Sunny Chang. Michel was so warmly welcomed by members of the Applied Physics Department like Richard that he decided to accept Yale's offer of creating a new lab. Michel joined Richard's department in January of 2002 and has remained a professor at Yale since then.

Letter to Richard Chang on the Occasion of His Retirement in June 2008

April 19, 2008

Dear Richard,

We are very sorry not to be able to participate in the celebration of your career organized by the department and your colleagues. I have to be in Paris for this College de France teaching and I hope I will be able to convey in writing what I would have preferred to say in person on the occasion of the party given in your honor. Please forgive my any awkward prose. I would like to tell you how much I admire what you have accomplished at Yale. US scientists such as you have been very important to me in my own career as role models. As a matter of fact, when I travel to visit universities in the US or in the rest of the world, I often get testimonies of how much you are known and appreciated in the physics community. At a more personal level, my family and I have been very touched by the way you and Sunny have welcomed us in Connecticut and Vermont. Your friendship is very dear to us. It has also been wonderful, thanks to you, to meet new friends like Jean-Pierre and Noemie. Recently, we have found great pleasure in participating in the welcoming of the Cachan delegation you have so well organized. Your hospitality, generosity and open-mindedness are fantastic.

We all wish you and Sunny a lot of happiness. Your probability of presence in Becton may be changing, but not the place you occupy in our hearts.

With our warmest thoughts,
Michel

Letter to Sunny

Hakan Türeci, Ph.D. Yale '03
Professor of Electrical and Computer Engineering
Princeton University, NJ 08544 USA
tureci@princeton.edu

February 19, 2020

Dear Sunny,

Esin and I are utterly heartbroken and saddened by the news of Richard's passing. I really loved him and feel very fortunate to have known him. Though I am not very good with faces, his beautiful smiling face is surprisingly crisp in my memory.

I was in Doug Stone's research group 1999–2003 and had the fortune to collaborate with Richard on several projects. Those collaborations and my interactions with Richard as a scientist and mentor during that period set a defining example for me. But perhaps as importantly, Richard's human side has left with us a profound and lasting impression of how to live this life. We cherish the memory of when you and Richard hosted us so graciously at your home in Vermont and vividly remember feeling the strong values that tied your family together.

Please keep us informed if there is an event that is planned in the US for Richard. We would very much want to be part of it.

We hope you are surrounded by people you love as you navigate through this profound loss. Our sincere condolences to you and your family,

Hakan and Esin

Letter to Richard Chang on the Occasion of His Retirement in June 2008

Looking back, the period of my life at Yale was an extraordinary period. And knowing you was one of the best things that befell me in this period. The details are not important, but I would like you to know that I have a very simple, basic model of a great scientist in my mind: an extraordinarily strong intellectual who not only does great science, but also shines with his humanity upon those around him. I am just not so sure whether I had this model in my mind all along, or whether it materialized after knowing you.

Richard, it was a pleasure working with you! I just wish I had come to Yale about 20 years earlier and had a chance to work with you longer.

Now that you will be traveling full-time, we would love to see you and Sunny wherever we will be.

Hakan

Richard Chang and Bugs — A Random Reminiscence

Arto V. Nurmikko
L. Herbert Ballou University Professor of Engineering
School of Engineering, Brown University
Providence, RI 02912 USA
arto_nurmikko@brown.edu

When people spoke about the 21st century becoming the Century of Biology, they probably did not have pandemic-generating viruses in mind. Rather, developments such as gene editing were predicted to compete in impact with the Internet and artificial intelligence, the two other contemporary 'revolutionary and transformative' technologies. But as the past two years have shown, propagation of humble viruses, particularly through airborne means, gives 'bugs' the power and efficacy of indiscriminate spread, which leaves Social Media in a trail of dust. Of course, it is not entirely accurate to label viruses as 'bugs' as they are sort of the biologist's Schrödinger's cat: dead as well as alive at the same time. But that appellation surely applies to bacteria.

What connection, you ask, could there be between bacteria and one Richard Chang — one of the very finest scientists I have known? A personal answer lies in the way that Richard and I got to know each other some twenty years ago.

You may recall that shortly after 9/11 (2001) there was the "Anthrax scare" (2001), packages of 'white powder' sent to the White House and other targets across the country. Multiple Federal agencies were duly tasked to develop countermeasures, DARPA being one such agency. And so it came to pass that a proposal by a bunch of university types and a handful of corporate research scientists managed to get a multiyear, well-endowed contract, to develop a portable optical sensor system. The goal was to demonstrate a scalable sensor, which would be able to 'sniff' the presence of airborne Anthrax, in the event that the pathogen might be released by bad actors as an aerosol cloud, say over a major metropolitan area. While I was the paper-pusher for the particular DARPA-funded team, Richard was its brains.

I certainly knew of Richard long before actually getting to collaborate with him on the 'Anthrax detector'. Here was a pioneer in the technique of stimulated Raman scattering (SRS), having collaborated with his PhD mentor, the Nobel Prize in Physics 1981 winner, at Harvard, Nico Bloembergen. But what I only got to know when putting together the team for the DARPA proposal was that, he had actually figured out how to apply sophisticated laser-based optical techniques to spectroscopically fingerprint 'bad bugs' in his laboratory — including Anthrax, the toxin tucked away within the molecular capsid of an enveloping spore until inhaled. Now, bringing actual Anthrax spores to Yale campus even for the noble science of high-resolution spectroscopy on those sleeping albeit lethal microparticles was unwelcome, to put it mildly. But Richard and his team had found a benevolent close proxy in Bacillus Subtilis, a sort of Mr. Hyde to the Dr. Jekyll of the actual Bacillus Anthracis. Which then became the material system which our team deployed as the target, to detect minute quantities of the stuff floating in small volumes of laboratory air while engineering a compact system based on ultraviolet (UV)-induced fluorescence spectroscopy. Samples of the 'Mr. Hyde bug' would be irradiated by arrays of microscale UV light emitting diodes (LEDs), a new generation of semiconductor light source developed by several pioneers in the team. It was challenging engineering work, nose-to-the grindstone kind of work. The project outcome after four years was a technical tour-de-force success, in so many ways thanks to Richard being the team's guiding light — our wise man.

I made many a pilgrimage to gain scientific insights from the wise teacher. We met in his office at Yale or mine at Brown, splitting the difference at least once in favor of a roadside motel in Mystic, CT. I remember having usually rambled on about this or that technical challenge, looking for a key to a solution. Richard would listen, politely and patiently, often without much visible reaction. And then, after I had run out of gas in my soliloquy, he would deliver a characteristically short and incisive comment which cut through the issue like a hot knife through butter (I know, I should use the 'focusing like a laser beam' metaphor here, since he knew a thing or two about lasers).

And then there were those semi-annual DARPA program reviews. The program manager was a U.S. Army Lt. Col. who had earlier served as a tank commander overseas. When he turned toward you, one could almost feel the turret of an Abrams rotating, the cannon taking direct aim at the target. He was brusque to a fault and had little patience, especially when we tried to explain why this or that milestone had not quite been reached. Except when Richard spoke, quietly, yet so clearly and so slowly patiently, and it suddenly felt as if the Abrams had just reverted to being a children's toy. In general, Richard had that aura of scientific authority about him when interacting with people, the power of his scientific knowledge like a halo.

One arrived at those DARPA program reviews always with a bit of a queasy feeling in the stomach. Typically, there would be several teams funded by the same program, all competing against each other, since it usually transpired that one or more would get the axe after the review.

Over the years, I have been to more such uncomfortable meetings than I care to count. Except for the one on Daufuskie island off the coast of Savannah, a resort of sorts in Georgia. For this particular DARPA review, we were housed in small bungalows, each participant to his or her own. Unusually for the occasion, spouses were allowed to come along, Richard had his Sunny, and I had my wife Pirkko.

Sitting there in the open-air restaurant near our adjacent bungalows in the relative warmth of a late southern autumn, the four of us had a most enjoyable and leisurely lunch. A momentary tonic and a counterpoint to a high-tension morning. Richard told us about his ancestral roots in China and for a blessed couple of hours the conversation steered light years away from why he and I had to be there in the first place. And that is one lasting image I have of Richard, sitting there on that veranda, peering through those wire-rimmed glasses with that friendly yet slightly enigmatic smile on his face, calm as the Buddha.

The Chinese Belief in *Yuan**

Edmund Yeh
Professor of Electrical and Computer Engineering
Northeastern University, Boston, MA 02115 USA
eyeh@ece.neu.edu

It has been one of my life's great blessings to have met and known Richard and his family. The Chinese believe in *yuan*, and it is hard to argue in retrospect that this did not have a role to play in my becoming a friend of the Chang family.

While still a high school student in Houston, Texas, I had read, in the leather-bound *China Weekly Review* back issues at Rice University, about the politics of Zhang Junmai, or Carsun Chang, the leading constitutionalist and pioneer of the "Third Force" in Republican China. As an undergraduate at Stanford, my favorite reading space was at the Hoover Institution, where Zhang Jia Ao, or Chang Kia Ngau, the leading banker of Republican China, was a well-remembered presence who spent his last years in sunny Palo Alto. While studying abroad at Cambridge and cycling over the delicate bridges of the River Cam, I inevitably fell under the spell of Xu Zhimo's immortal poetry. And as a graduate student in the other Cambridge, I came upon the marvelous *Bound Feet and Western Dress* by Pang-Mei Natasha Chang, while browsing the China section of the former Wordsworth Bookstore in Harvard Square.

I had little idea that these strands of experience were soon to converge. Before arriving at Yale as a fresh faculty member, I had come across Richard's impressive photo and wondered what a remarkable personality stood behind it. I was soon introduced to Richard by a colleague, who noted that one cannot expect a better combination of the gentleman and scholar in the same person. Meeting Richard was to experience sharpness, urbanity and generosity at the same time. His achievements as a researcher and teacher at Yale were legendary, but his kindness and equanimity were equally legendary. I was a young academic just starting out, and Richard generously offered his time and advice, introduced me to his research team and invited me to develop joint research proposals. We shared many interests in science and engineering, but invariably our discussions also veered off to China, Shanghai (where we had common roots), and the Chang family.

Richard loved to share his favorite Shanghai memories: riding his bike as a child on the balcony of the family flat at the Picardie Apartments, which stood at the triangular tip of the former French Concession. There, the Changs were neighbors with Du Yuesheng**, who needed a place of quick refuge in case things got complicated in the Chinese sector, and were visited by Zhou Enlai, who worked hard to seek the support of third parties. Richard shared vignettes of his family history: how fond his father was of Xu Zhimo, in spite of the travails of Xu's marriage to Richard's aunt Zhang Youyi (Chang Yu-i), the heroine of *Bound Feet*; and how Youyi found domestic happiness later in life with Dr. Su in Hong Kong. I told Richard about the uncanny connections between our families: of how my grandfather had studied, in his youth, under Zhang Junmai, and had been a member of the Constituent National Assembly, which in 1946 ratified the Republic of China Constitution principally drafted by Zhang Junmai, and later, while heading the Shanghai branch of the Ministry of Economic Affairs, had interacted extensively and even traded offices with Zhang Jia Ao.

Not long after my arrival in New Haven, Richard introduced me to Sunny and his wonderfully loving family. Each gathering with the Changs is a joyful memory. At the Chang vacation home in Brattleboro, I enjoyed a fine day with Richard's colleagues and friends amid the ravishing fall foliage and sparkling waters of the Connecticut River. At an event in Trumbull College, where I was Resident Fellow, we listened to Pang-Mei Natasha's poignant reading from *Bound Feet*. On a visit to Beijing, I met PangYuan Nicole and her husband Denis at his restaurant Brasserie Flo. At the Shangri-Lee Restaurant in Cheshire, CT where our families had a lively gathering, the owners quietly told me of how Richard had helped them through difficult times. There were many other such memorable occasions, but time passed, quickly. I last saw Richard in 2009, shortly after his recovery from a swimming accident. I had called Sunny to express my well wishes, and Richard kindly asked me and my wife to dinner that very night. I am grateful he did. It was wonderful to see, and it is wonderful to remember Richard smiling as he left the restaurant, in the loving arms of Sunny.

My wife and I moved to Boston in 2011, and over the years we happily learned of Sunny and Richard's many travels and their prodigious grandchildren. We are very happy Richard was able to enjoy in his golden years the balmy weather and fine food in Singapore, surrounded by his loving family.

Editors' note:
*Yuan describes the immediate chemical connection, being on the same wavelength among people.
**Du Yuesheng, also known by Dou Yu-Seng, or Tu Yueh-sheng, or Du Yueh-sheng, nicknamed "Big-Eared Du", was a Chinese mob boss who spent much of his life in Shanghai. https://en.wikipedia.org/wiki/Du_Yuesheng.

Letter to Sunny

Maria P. Rao
Administrative Assistant in the Department of Applied Physics
Yale University, New Haven, CT 06520 USA
maria.rao@yale.edu

May 6, 2021

Hi Sunny,

I am so glad you contacted me to share my feelings about Richard in the book. I am truly honored.

I will always remember Richard with great memories. He was the kindest and most considerate person I've ever met.

A true gentleman. He would invite me to share his healthy snacks in his office and at one point he gave me a small ceramic bowl made in Sorrento, Italy. (See photo on the right.) I will cherish this bowl with fond memories. "Richard, you are missed by so many staff, students and your peers. You were a mentor to so many without knowing it. Your legacy lives on forever. Thank you for memories and positive example you provided each and every day."

Thank you again dear Sunny and hopefully we will be able to see each other again,

Love, Maria

"I LIVE FOR THE MOMENT AND THE FUTURE" by Maria P. Rao

Photo for Richard Chang on the Occasion of His Retirement in June 2008

Maria Rao and Theresa Evangeliste in Becton Center.

Welcome to the Madness

Kevin Aptowicz, Ph.D. Yale '05
Professor of Physics, West Chester University, West Chester, Pennsylvania 19383 USA
kaptowicz@wcupa.edu

Kevin with wife Katie Eyer and son Cian.

I had the pleasure of being Richard's graduate student from 2002 until 2005. During my first semester, my coursework was totally dominating my time. I had already done some research with Richard, so I was integrated into the group, but I struggled to contribute while classes were in session. About mid-semester, Richard asked if I could meet with him. We chatted in his office about a fellowship opportunity through Sandia National Labs that would contribute significant to the costs of my education. My current project fit well with the mission of the fellowship and we outlined what I should submit in my application. The only hitch was that the application for the fellowship was due in two days, right in the midst of midterms! That night I performed a quick literature search until I was kicked out of the library and then I stayed the rest of the night and wrote a draft of the research plan for the fellowship. At about 5:00 a.m., I emailed it to Richard and went to bed. Upon arriving on campus in the afternoon, I found the paper waiting for me on my desk, edited by Richard. On the top of the paper Richard had written, "The proposal looks good. Welcome to the madness!" I immediately identified with that phrase and I knew that Richard understood and appreciated all the effort that I put into the application. It was as if I had been welcomed into the world of graduate physics — that I can now stand proudly as a physics graduate student knowing that I was carrying my weight.

Throughout my time as a PhD student, I had many more experiences with Richard which showed his ability to empathize with you while encouraging you to reach even higher. He was truly an outstanding mentor, always pushing you to achieve more, but also leaving enough room to grow in your own direction. He also was exceptionally generous and gracious, with everyone from visiting scholars to struggling graduate students.

I'm now a faculty member in the physics department at West Chester University and taped to the wall in my office is that quote. It is a constant reminder to me that life at times can be quite chaotic, but that the madness is all par for the course. I've also been known to write the quote on the works of my students when the semester becomes particularly hectic for them. This is just one small example of how Richard's actions continue to affect me every day.

- Kevin Aptowicz

Building Community

June 26, 2021

I am now in my fifteenth year as a professor and academia can be brutal. Grant proposals, paper deadlines, complex collaborations, broken equipment, tight budgets ... and I haven't gotten to the teaching! It is easy to be overwhelmed or lose sight of the bigger picture. But what is the bigger picture? What is the signal from the noise? It is in these moments that I think about Richard. What would Richard do? How would Richard navigate these turbulent waters?

I spent four years working with Richard and watching him navigate academic life. I watched Richard as he interacted with new grad students and old collaborators. I watched as Richard astutely navigated a snafu with a collaborator or invited the parents of a new grad student to his home for dinner. I watched Richard as he learned about a paper being rejected or a grant proposal getting funded. One particular element stood out to me from this time. Richard never lost track that science is fundamentally a social endeavor. It's about people as well as ideas. Although scientific ideas were important to Richard's success (and indeed he was a brilliant scientific mind) it was how he treated people that made him special. He was present when he was with you. You felt seen. He supported you as a person just as much as he did a scientist.

As the physics community wrestles with issues of diversity and inclusion, I think of Richard's ability to empower his students and foster within them a sense of belonging. As a community, we need more of that. We need to follow in Richard's footsteps and make sure our students are seen and supported. I will forever be grateful to the fortuitous chain of events that lead me into Richard's lab. Although I miss him dearly, the primary emotion I feel is gratitude. Thank you, Richard, for all you have done for me. I will try my hardest to uphold your legacy as I navigate my academic life and build community.

From left to right: Hermes Huang, Sunny and Richard Chang, Grace Chern Metcalfe, Kevin Aptowicz, and Hakan Türeci.

Letter to Richard Chang on the Occasion of His Retirement in June 2008

Grace Metcalfe, Ph.D. Yale '05

Dear Richard,

Grace Chern Metcalfe and Kevin Aptowicz displaying their graduation gift from Richard, May 2005.

When I was looking for a graduate research advisor, I remember emailing you asking if we could meet and discuss your research. Although you were out of the country at the time, you emailed me back immediately. (It always impressed me how well-connected you stayed to the lab and your students, regardless of where you were in the world!) You suggested I first talk to your then graduate students Andrew Poon and Nathan Rex. The very first thing Andrew said to me was, "Richard is THE BEST advisor you'll ever find." I was struck by how passionately Andrew praised you. At that time, I could not understand how a person could inspire such respect and admiration.

But during my four years working in the RKC lab, I realized what Andrew was trying to tell me. Your enthusiasm for science is absolutely infectious. When my research wasn't going well, I found myself saying, "I'm going upstairs to find Richard." Because after every meeting with you, I would leave the room re-energized and enthused again about our projects. You inspire me by example to be not just good at science, but a good scientist: someone who upholds their ethics, integrity, and humility. But your influence extends further than science. I would watch you treat a high school student (with Chemical room

Kevin Aptowicz and his wife Katie Eyer, Richard and Sunny, Grace (Chern) Metcalfe and her husband Michael Metcalfe, May 2005.

Richard accepting the Memory Book from Grace at the retirement dinner in New Haven, June 2008. It was a very nice surprise.

duty!) with as much respect as you would a Yale faculty member. When you found out I loved to teach, you allowed me to co-teach the "Optical Properties of Semiconductors" with you. (This experience helped me immensely when I taught a graduate-level special topics course on terahertz at the University of Maryland in the spring of '06.) I especially remember a time in graduate school that was very difficult for me. And I appreciate and thank you for responding with such understanding and compassion. I thank you for your mentorship, advice, encouragement, kindness, and much, much more…

I will also always fondly remember the wonderful dinners at your Hamden home, the fun trips to Brattleboro (with biking and kayaking! Michael and I are hooked now!), your witty anecdotes, your family's kindness and openheartedness, and of course you and your family's participation at our wedding in Ireland! Michael and I look forward to many more memories with you and your family.

With warmth and gratitude,

Grace

Grace presenting the Memory Book at the retirement dinner in New Haven, June 2008. There were over 80 contributions to the Memory Book.

Preface to the Memory Book for Richard Chang on the Occasion of His Retirement in June 2008

June 12, 2008

Dear Richard,

It has been a distinct honor and pleasure to prepare this "memory" book for you on the occasion of your retirement. The response to our request for contributions was overwhelming with over 80 submissions. But that is not the least bit surprising as it is a reflection of your brilliance as a scientist and compassion as a friend. To help commemorate your official tenure at Yale, your family, friends and colleagues have contributed some of the most touching letters, stories, and photos which show how you have positively touched each of our lives. I hope you enjoy reading this book in your "free" time (free of grant proposals, committee obligations, and meetings!), as much as I have enjoyed putting it together with Nicole and with the invaluable help of Giselle DeVito, Maria Gubitosi, Marshall Long, as well as all the numerous contributors to this book.

With my warmest regards,

Grace

Grace (Chern) Metcalfe

The Email Requesting Contributions to the Memory Book

Dear Friends,

As a retirement gift for Richard, we (Grace (Chern) Metcalfe, PhD '05 and Richard's daughter Nicole) are putting together a book of photos and letters written by Richard's family and friends. We ask for your contribution to help create this memorable keepsake.

Here are a few ideas for contributions:

- your favorite Richard-related story/memory
- a letter to Richard (perhaps about how he has influenced your life/career)
- your favorite story that Richard told you
- photos (of a Richard-related event, of yourself and your family, or even your favorite picture of yourself)

Of course, you are not limited to the suggestions above, and more than one contribution is fine. The letters can be as long or short as you want. Any type of contribution would be greatly appreciated by us, but most importantly, by Richard.

We ask you to contribute by ***April 18th*** so we can have the book ready in time for Richard's retirement celebration on June 12th.

We look forward to your photos and letters to make this book complete.

Best wishes,

Grace Metcalfe and Nicole Chang

My Interactions with Richard Chang

Jean-Claude Auger, Ph.D.
Kyolaris Research,
Hong Kong, 999077 China

I got to know Richard, in 1999, thanks to Gorden Videen. It was during the last year of my doctoral research work in Paris. At that time, I was so desperate to comprehend light scattering issues that I had decided to send questions to all the scientists whose emails I had found in publications. Gorden was one of the researchers who did take the time to nicely answer them. After a while he introduced me to Richard with whom he was working on two-dimensional angular optical scattering (TAOS) technique. That was the beginning of fruitful scientific collaboration and numerous rich human interactions with Richard.

It was nearly four years after our first email exchange, in May 2003, that I finally met Richard for the first time in his office at the Yale Applied Physics Department. I was very impressed by his personality. I met Richard for the second time in Salobreña in the south of Spain during the 8th Conference on Electromagnetic and Light Scattering by Nonspherical Particles, in May 2005. I think it is also the first time I met Sunny. We continued to discuss on TAOS identification and decided to work even more closely. A few months after the congress, Richard sent me an invitation letter to work with his team at the Applied Physics Department. At that time, I was working in a R&D center for a Latin American chemical industrial company in Mexico City. The job was very interesting, but I did not hesitate too long before accepting Richard's invitation. I could not miss this opportunity to directly interact with him and learn from him and his team members.

I have great memories from the months I spent in New-Haven during 2006. Richard and Sunny were always very kind. They helped me solve every problem I encountered. There is one singular moment I shared with Richard that I wish to mention here. We were spending the weekend in Richard and Sunny's second home, in a city

Richard, Sunny, and me in Vermont, May 2006.

whose name (Editors' note: Brattleboro, Vermont) I totally forgot about two or three hours driving from New-Haven.

Richard wanted to show me around the area. We left with his car. I was driving. We had a very long and rich discussion, on many diverse topics, from philosophy and expectation of life to the study of sciences and world history. Richard started telling me about his youth, his departure from China and the different people he had met when he was a student. At one moment he mentioned the tragic destiny of one of his best friends. The tone of his voice changed. The emotion was so intense that I had to stop the car on the side of the road. Fortunately, we were alone on a small country road. This moment overwhelmed me. After a few minutes of silence, I restarted the car. We continued our discussion by talking about other things. The beauty and may be also the oddity of life, is that the same day, during the evening, we went to see *Mission Impossible III* in theaters. Not so much of an intellectual movie after a so deep and intense discussion.

The last time I met Richard was at the Progress in Electromagnetics Research Symposium (PIERS) that occurred in 2007 in Beijing where I also had the opportunity to spend time with him and his family.

I am very grateful that I have been able to know Richard. In addition to being a great scientist, he was a very cultured and humble man of great kindness. He also had a lot of charisma and a great sense of humor. I am also very glad that through Richard I have been able to know Sunny.

Letter to Kevin and Sunny

Vanessa Wood, B.S. Yale '05
Vice President for Knowledge Transfer and Corporate Relations
Professor of Information Technology and Electrical Engineering
ETH Zürich, 8092 Switzerland
vwood@ethz.ch

January 31, 2021

Dear Kevin and Sunny,

I am so very sad to hear about Richard. I would be absolutely honored to contribute to the memorial book. Please let me know what you need.

Richard was such a mentor to me and an important figure in my career. I have thought about him often and, in fact, when I was asked to give a talk about role models that shaped me in May last year, I spoke about him. Richard introduced me to what it meant to be in the scientific community, he encouraged me to stay in science when I was considering other career paths, he was the one to told me apply to and go to MIT EECS for the PhD, and somehow back in 2004 predicted that I would be a professor at major European university.

I also remember with fondness how you Sunny and Richard invited me to the Chinese restaurant in Hamden and your home for delicious meals with the PhD students in Richard's group. I was so saddened to learn of Richard's accident right after his retirement celebrations, but I am so pleased to hear that you and Richard were able to move to Singapore and spend a number of years with family.

Perhaps you would be charmed to know that the guy I met at MIT and brought to Richard's retirement parties (you may remember Keith as the mechanical engineer who brought the carbon fiber boat to the party at your home in Brattleboro) and I are now married and have a 6 month old daughter, Claire Amelia. I remember Richard giving him a bit of a quiz to check him out. Keith was a founder at Markforged (www.markforged.com) so we did long distance Boston-Zurich for a while, but the pandemic now has us "stuck" in Switzerland. We are looking forward to traveling to the US to introduce Claire to our families. Below are some pictures.

My thoughts and love,
Vanessa

Letter to Richard Chang on the Occasion of His Retirement in June 2008

Dear Richard,

When I applied for the NSF Graduate Fellowship, I wrote in my personal statement that I hoped one day to be as inspirational a mentor to young scientists as you were for me. Of course, I am a long way from being a mentor, but I am happy to reflect on specific ways you helped me to gain confidence and to make the most of my studies and opportunities. Without your support I may not have had the confidence to continue my studies of science and engineering.

In 2002 you recommended me for a summer research experience at the Laboratoire de Photonique Quantique et Moléculaire (LPQM) in Cachan, France. The following spring you allowed me to take your optoelectronics class as a sophomore. During the summer of 2003 you let me join a research project in your lab, working with your wonderful graduate student, Grace, among others. You offered me exposure to a professional society at the Conference on Lasers and Electro-Optics (CLEO) in 2003. In the summer 2004, you arranged for a summer of research abroad at Laboratoire de spectrométrie ionique et moléculaire (LASIM) in Lyon, France. These opportunities deepened my desire and confidence to pursue a career in science. I was only an undergraduate, but you welcomed me to the community of scientists of which I continue to feel privileged to be part. Now, as a graduate student, I think back on your guidance and continue to strive to live up to the expectations you set for me and all of your students.

I was a hesitant young woman beginning my science studies, and you bolstered my confidence. Perhaps it was your intuition that I needed some encouragement, but your help made me more comfortable through many periods where I would have otherwise felt quite isolated and unsure of myself.

In addition to academic guidance, your wit and caring marked my undergraduate experience. When I was applying to graduate school, you and Sunny, perhaps sensing how nervous I was about the process, told me the story of your application to graduate school. As I recall, you applied to Harvard on somewhat of a whim, filling out the application form after the deadline and using a complimentary Gulf Oil pencil you found in your car. You assuaged my fears with laughter. I am also grateful that you took the time to consider my interests and personality and to recommend that I apply to the Electrical Engineering and

Computer Science department at MIT. It was not a program I was considering, but I am pleased to be a part of it.

It has been a distinct honor and a great pleasure to be your student. Thank you for the inspiration you continue to provide.

With warmest regards,
Vanessa Wood

Grace Chern Metcalfe, Kevin Aptowicz, Vanessa Wood, and Richard, May 2005.

Interactions with Richard

Giselle Maillet
Assistant to the Chairman
Department of Applied Physics
Yale University, New Haven, CT 06520 USA
giselle.maillet@yale.edu

I first met Richard in June 2007, when I took on the role of the Chair's Assistant after my predecessor, Jayne Miller, retired. Richard's office was directly next to mine, and he was one of the very first to welcome me to the Department. As was Richard's style, he immediately made me feel a part of the department "family", sharing stories of how the department was formed and how he came to be at Yale. Not once did I feel a disparity in my relationship with Richard, as an administrative assistant to a faculty member, he treated me as if I was a colleague. Just one year later Richard retired, and it was at his retirement party that I recognized, more deeply than I already had, what a remarkable human being Richard was! Person after person shared stories of Richard and his family welcoming them into their home, easing them into a new life in New Haven and making them feel like "family". Richard maintained lifelong relationships with so many of his students and postdocs, not just superficially but as if they were his own children. Such kindness, selflessness, and love for others are rare qualities in this world and Richard overflowed with them. I feel exceedingly blessed to have known Richard and will forever hold him dear to my heart.

Personal Memory of Richard K. Chang

Hui Cao
John C. Malone Professor of Applied Physics,
Professor of Physics, and Professor of Electrical Engineering
Yale University, New Haven, CT 06520 USA
hui.cao@yale.edu

Before I met Richard, I had read many of his papers and admired his research achievements. I finally met him in 2002 at the Physics of Quantum Electronics (PQE) conference that was held in Snowbird, Utah. At that time, I was a junior faculty at Northwestern University. My first impression of Richard was that he was modest, warm and friendly, despite the fact that he was a pioneer and world leader in optical microcavity lasers and applications.

During my interview at Yale in 2007, I learned that Richard had planned to retire soon, and I would inherit most of his lab space should I come to Yale. Richard showed me his lab and we had an extensive discussion. I hesitated about the move, as I was happy at Northwestern and not sure about Yale's expectation of me. In a follow-up visit, Richard and his wife Sunny took me to their beautiful summer house in Vermont, and we spent a weekend there. I felt at home, and had long, frank conversations with them about Yale, Applied Physics, and New Haven. Finally, I decided to move to Yale, but I knew I had some big shoes to fill.

Thanksgiving party at Richard and Sunny's home in Hamden, CT with their daughter Nicole and her children, and my mother and my research team, November 28, 2013.

Richard had helped me tremendously with the move of my group from Northwestern to Yale, and provided invaluable advice and suggestion of the lab space renovation at Yale. He treated my team members as his own and invited us to a wonderful Thanksgiving party at his house. We had so much fun with him, Sunny and their children and grandchildren. They were such gracious hosts that we felt like a big happy family (see the photo).

It was a great pity that I did not have a chance to collaborate with Richard after moving to Yale, because of his rapid declining health. I was amazed how strong and resilient he and Sunny had remained in front of all sorts of difficulties and challenges. Over the years, Richard and Sunny became dear friends to my parents and me. They visited Peking University and saw the campus where I grew up and studied. My mother and I visited Richard and Sunny after they moved to Singapore.

I was devastated when I heard that Richard passed away in 2020. He will always live in my heart. Next year I will organize a session in memory of Richard at the PQE conference where I first met Richard [1]. Many years ago, Bob Grober, a former professor of Applied Physics at Yale, told me that if there were to be a picture next to the definition of 'gentleman' in any dictionary, it must be the photo of Richard. I couldn't agree more: Richard was a true gentleman in every sense.

Program of the Memorial Session for Richard Chang [1]
51st Winter Colloquium on the Physics of Quantum Electronics
January 10–14, 2022
Snowbird, Utah, USA

Monday Morning Invited Session 2

Breakout Session 1: *In Memory of Richard Chang*
Location: Ballrooms 1 — Hui Cao, Chair

12:00 **Douglas Stone**, *Yale University*, "Chaotic Micro-Cavities and Micro-Lasers: An Overview After 25 Years" [210]

12:20 **Yong-Le Pan**, *CCDC-U.S. Army Research Laboratory*, "Optical-trap laser spectroscopic technologies for studying the properties of single airborne micron cavity and particles" [178]

12:40 **Andrew Poon**, *Hong Kong University of Science and Technology*, "Micro-Cavity Optics in Silicon PHotonics" [189]

[1] Abstract book of the 51st Winter Colloquium on the Physics of Quantum Electronics, Jan. 10–14, 2022, Snowbird, Utah, USA.

Letter to Richard Chang on the Occasion of His Retirement in June 2008

"Although I arrived at Yale only four months ago, I can tell how much Richard is loved and respected here by his colleagues and students." — Hui

Dear Richard, You are my role model! Hui

From right to left: Hui, Richard, Sunny, Sunny's mother, and Hui's father.

Dear Richard and Sunny, My parents were delighted to meet you!

From right to left: Hui's mother, Richard, Sunny, Sunny's mother, and Hui's father.

Remembrances

Gustavo E. Fernandes, Ph.D. Yale '08
Circadian Positioning Systems
Newport, RI 02840 USA

I first met Richard in 1999, my freshman year at Yale. I'd been paired with him probably because I'd mentioned I wanted to be a Physicist in my application. My memories of that time are punctuated by all the new things that I was experiencing, as well as the all too frequent feelings of inappropriateness and embarrassment, which I've come to understand are somewhat common to eighteen-year-olds. Everything was new to me. First time living away from my family, first time at college (and Yale at that), and first time being fully immersed into the American culture, a culture that was very foreign to me but that I always greatly admired.

I still remember that first meeting with Richard. I arrived at his office door after wandering about lost in Becton Center for a while. His door was open and as soon as I peeked inside his office Richard greeted with a big smile. He said "Bom dia!" with a flawless Portuguese accent and invited me in. We talked for maybe 20 minutes. He told me about his work and asked about my interests. I remember him asking where I was from and then telling me about the time he had lived in Brazil before coming to the US to attend high school at age fifteen and half. Most of all, I remember how he made my uncomfortable eighteen-year-old self feel at ease and capable of communicating with him on the same wavelength (no pun intended), something that was rare for me in those days. I left that first meeting with the feeling that I had just met a true gentleman.

I believe that we had a couple other orientational meetings that year, but I would have to wait until the second year of my PhD, 2005, to connect with Richard again. This time around I was in a crusade to find a research group that I liked and that I felt I could become a valuable member of. Time was running out. It was expected that students would find a research group earlier, during the first year of their PhD. I had been involved with the nuclear physics group, Wright Nuclear Structure Laboratory (WNSL) — that bunker on the side of a mound up on Science Hill, and felt that I might end up doing nuclear or high energy physics in some capacity, though those prospects didn't seem too exciting to me. I had, however, decided to take Richard's Optics class that semester and became fascinated with the topic. In the first days of that class, I had recalled my previous encounter with Richard, and after pondering about what a career in optics would be like I didn't skip a beat in asking him if he'd let me join his group. The rest is history.

Between work and the exquisite meals at the Shangri-Lee, I couldn't have asked for a better adviser than Richard. He was always patient, understanding, and supportive. He was demanding of excellence too. He treated us as part of his family, always included us in making decisions that affected us and the lab and was always prepared to stand up for us when others attempted to "throw their weight about" and make unreasonable demands of us. He was a true leader and inspiration.

I very much cherish my memories with Richard and my experiences as his student and mentee. I had a wonderful time as part of his lab. I am grateful for all the inspiration and support he gave us, his students and, though he is no longer physically with us, I am certain that his memory will continue to inspire us to give our best effort in whatever we do. May he rest in peace.

Letter to Richard Chang on the Occasion of His Retirement in June 2008

The three years that I spent as a Ph. D student with Richard were a truly remarkable learning experience. I will always remember fondly the discussions we had and how much they helped shape me as a scientist and as an individual.

I also remember fondly the first time I met Richard, during my first semester as an undergraduate at Yale. I went to knock on his office door in Becton to ask him to approve my course schedule, and he greeted me by saying "- Bom dia!" with a flawless Portuguese accent.

Thank you Richard, and best wishes in your retirement.

Gustavo

From Spiral Micropillars to Non-Hermitian Photonics

Qinghai Song
Professor, College of Science
Harbin Institute of Technology
Shenzhen, 518055 China
qinghai.song@hit.edu.cn

In the summer of 2004, I was still a PhD student supervised by Prof. Lei Xu in Fudan University. I was entrusted by my mentor to welcome Prof. Chang and Sunny at Hongqiao airport. That was the first time I met Prof. Chang. We talked a lot during the trip from airport to Fudan University. That was also the first time, I realized that a big scientist was also very humorous. In Fudan University, Prof. Chang had given a talk about the spiral microlaser, which closely related to his very famous paper "Unidirectional lasing from InGaN multiple-quantum-well spiral-shaped micropillars" [1]. From the talk, I learnt one of the important research directions in photonics, i.e., the impacts of topology on the angular momentum of light. Of course, I couldn't fully understand the importance of spiral resonator beyond the unidirectional laser emission at that time.

In 2005, Prof. Chang kindly offered me a chance to visit Yale University and study there for 7 months. At Yale University, I got the chance to study GaN based spiral microlaser for the first time. Compared with the conventional whispering gallery mode (WGM) microcavity, the spiral resonator mostly supports quasi-WGMs propagating in one direction. As a result, the GaN microlasers mainly emit unidirectionally at the notch area. Since the power of microlaser is typically very low, the outputs are sometimes amplified with an external amplifier. Based on the unique properties of spiral resonator, Prof. Chang proposed the configuration of directly integrating microlaser and amplifier (see Fig. 1). I was involved in this project and studied with Dr. Grace Chern at the beginning and with Dr. Gustavo Fernandes at the end. We found the emissions from the spiral microdisk could be mostly collected by the semicircle. By increasing the injection current on semicircle, the output power has been greatly enhanced by more than an order of magnitude [2]. By controlling the currents on spiral microcavity and the semicircle, both the intensity and wavelength of the outputs can be switched [3]. These results have been experimentally achieved within several months. In the summer, Prof. Chang invited me to visit his house in Vermont, where we sat together and prepared the drafts. He discussed every detail with me and showed me how to write and revise the manuscript. I really learnt a lot from him, including deep thinking in science and having patience toward students. These valuable lessons affected me and shaped my entire academic career.

After the 7 months' visit, I went back to Fudan University. I realized that spiral cavity

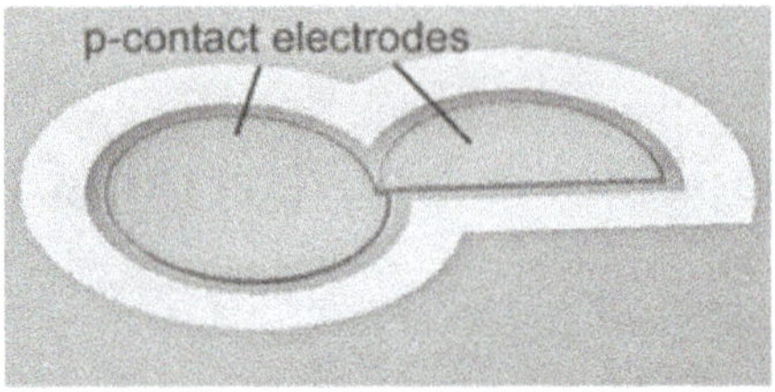

Figure 1: The scanning electron microscope (SEM) image of the spiral laser and amplifier [2] (used with permission).

was a good candidate of chaotic system, which matched my previous research on disordered systems very well. As a result, I brought the research on spiral microcavity back to Fudan and kept working on this topic. Based on Prof. Chang's configuration, we replaced the semicircle to a straight waveguide. The waveguide could efficiently collect the microlaser emission as well. More strikingly, it was also compatible to the on-chip integrated systems. The corresponding experimental results have been summarized as one important chapter in my PhD dissertation.

Actually, the concept of waveguide connected WGMs microdisk kept inspiring my research. In 2012, I became a full professor of Harbin Institute of Technology and started to run my own research group. When we were looking for a system to realize high quality factor (Q-factor) and the unidirectional emission simultaneously, I quickly remembered the research on spiral microcavity and proposed the following configuration. As depicted in Fig. 2, we replaced the spiral cavity and semicircle in Fig. 1 with a quadrupole microcavity and waveguide directly. As the phase space structures shown in Fig. 2(a), some structures such as period4 stable islands are independent to the connection waveguide and can maintain high Q-factor resonances confined by total internal reflection (TIR) within them. The tunneled light shall form a chaotic layer around (blue dots) the stable islands and then be collected by the waveguide efficiently. This extension from a spiral cavity has perfectly tackled the conventional trade-off between high Q-factor and highly directional emission [4].

Similar configuration has also been applied to couple light from waveguide back to the resonator. Based on the time-reversal process of a laser, we have successfully injected

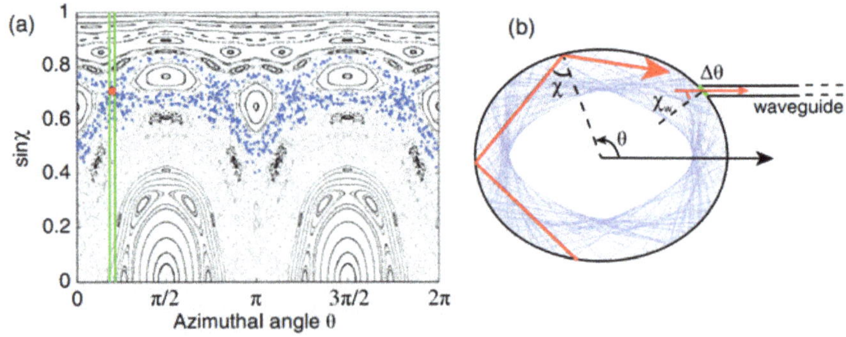

Figure 2: The waveguide connected microdisk. (a) and (b) show the corresponding phase space structures and the spatial configuration [4] (used with permission).

light to a microdisk without considering the conventional phase matching conditions [5]. Both the high Q-factors and high injection efficiency have been experimentally realized. Intriguingly, the break of phase matching condition makes our configuration more suitable for the ultra-wideband light coupling, which is almost impossible for the evanescent coupling scheme.

Besides the configuration, the physics in the spiral microcavity is also fruitful. Prof. Chang and Prof. Douglas Stone (also from Yale) revealed that the asymmetric scattering/reflection at the notch could impact the angular momentum distributions [1]. The standing waves in circular microdisk turns to propagating waves along either clockwise (CW) or counter-clockwise (CCW) direction. Soon after, people start to realize such propagating waves related to an important concept, i.e., the optical chirality [6]. With a simple 2×2 non-Hermitian Hamiltonian, the effort of asymmetric scattering on optical chirality has been built and explained. Tailoring the asymmetric scattering coefficient (η), the optical chirality can be increased from 0 in conventional WGM microcavity to 1. The case with $\eta = 1$ is actually the well-known exceptional point (EP), which inspired many important fundamental and practical breakthroughs in nanophotonics.

An important progress is about the optical sensing. Around the exceptional point (EP), the sensitivity is proportional to the square root of perturbation instead of the perturbation itself. Such a dependence is extremely essential for ultrasmall perturbation, where the square root can be much larger than its initial value. Based on this concept, we have further extended the research of spiral microlaser to optical sensors. Since the spiral cavity usually supports relatively low Q-factor modes, we couple a WGM cavity to the spiral cavity and reveal that the chirality can be perfectly transferred from the spiral one to the circular cavity without degrading the Q-factors. With the improvement, both the high sensitivity of high Q-factor modes and the enhancement at the exceptional point (EP) can be utilized simultaneously [7].

The study with Prof. Chang keeps inspiring my research. After understanding the physics of chirality in spiral cavity, we started to construct a different configuration beyond the conventional one composed of a tunable laser, an evanescent coupling system, a photodetector, and an oscilloscope. To reduce the system dimensions and simplify the detection, we have introduced the laser emission into the spiral cavity system [8, 9]. For the first time, we have built the one-to-one correspondence between farfield angular distribution and the nearfield light propagation. We found the chirality calculated from farfield emission (α_2) was almost identical to the nearfield one (α_1). As a result, we pointed out that the nanoparticle could be detected with the laser performances.

Figure 3(a) shows the configuration for the detection system. The spiral cavity has strong asymmetric scattering loss but only worked on the evanescent waves that exponentially decay away from the cavity boundary. The nanoparticle is extremely small and can be considered as a symmetric dipole. As the nanoparticle directly affects the resonant mode, the weighted scattering cross sections of spiral cavity and nanoparticle are comparable. With the increase of nanoparticle size, the components of symmetric scattering increase as well. As a result, the chirality that can be measured with farfield patterns decreases.

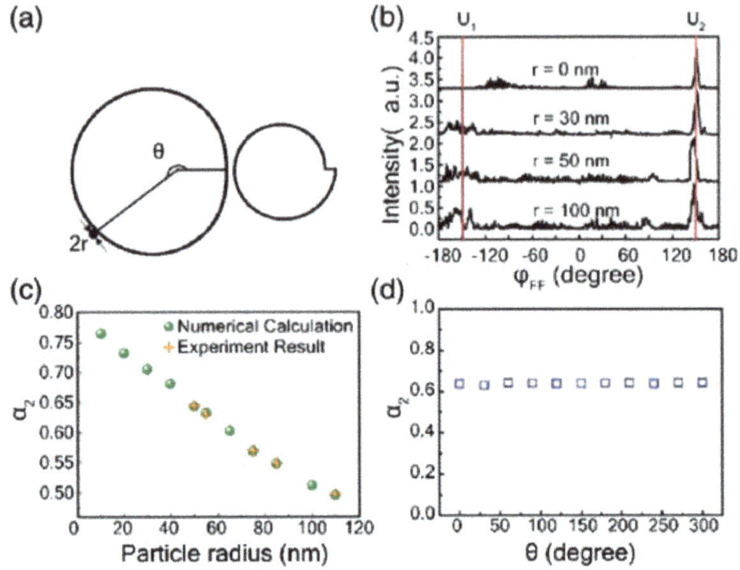

Figure 3: The coupled system based far-field nanoparticle sensor. (a) The schematic picture of the microcavity system, (b) the influence of nanoparticle on the far field emissions, (c) the change of optical chirality as a function of nanoparticle size, and (d) the position dependence of the chirality on the nanoparticle [9] (used with permission).

Compared with the previous report, the spiral cavity has the following inherent advantages. (1) The introduction of exceptional point (EP) can resolve nanoparticles with higher resolution. (2) The laser emission can bring near field information to far field, dramatically simplifying the detection system. (3) The far field laser emissions can be monitored by the back-focal plane imaging. As a result, the nanoparticle size can be measured by taking pictures without employing any spectral information [9].

The last advantage lies in the chaotic nature of spiral microcavity. In literature, two scatters on a WGM microcavity can interact with each other via the WGMs, introducing the unpreferred position dependence. As shown in Fig. 3(d), the spiral system can perfectly suppress such dependence and produce the same chirality wherever the nanoparticle is attached. This is attributed to the chaotic nature of the spiral cavity. In contrast to the WGM cavity, the scattering from the notch area experiences numerous and random orbits along the spiral cavity. Such orbits typically have different orbit lengths and introduce different phase shifts. While the orbits changes with the position of nanoparticle, the chaotic nature introduces an average of such phase changes and make the overall value to be negligible [9].

In summary, I have worked together with Prof. Chang for about 7 months at Yale University. After the visit, Prof. Chang and I kept up with scientific discussions for a long time. His spirit and his working style kept inspiring my research and my career development. I have learnt far more than the above works from Prof. Chang, and I believe the valuable lessons learnt from Prof. Chang will keep impacting my entire academic career.

References

1. G. D. Chern, H. E. Türeci, A. Douglas Stone, and R. K. Chang, Unidirectional lasing from InGaN multiple-quantum-well spiral-shaped micropillars. *Appl. Phys. Lett.* **83**, 1710 (2003).

2. G. D. Chern, G. E. Fernandes, R. K. Chang, Q. H. Song, L. Xu, M. Kneissl, and N. M. Johnson, High-Q-preserving coupling between a spiral and a semicircle μ-cavity. *Opt. Lett.* **32**, 1093–1095 (2007).

3. G. E. Fernandes, L. Guyot, G. D. Chern, M. Kneissl, N. M. Johnson, Q. Song, L. Xu, and R. K. Chang, Wavelength and intensity switching in directly coupled semiconductor microdisk lasers. *Opt. Lett.* **33**, 605–607 (2008).

4. Q. Song, L. Ge, B. Redding, and H. Cao, Channeling chaotic rays into waveguides for efficient collection of microcavity emission. *Phys. Rev. Lett.* **108**, 243902 (2012).

5. S. Liu, W. Sun, Y. Wang, X. Yu, K. Xu, Y. Huang, S. Xiao, and Q. Song, End-fire injection of light into high-Q silicon microdisks. *Optica* **5**, 6120616 (2018).

6. J. Wiersig, S. W. Kim, and M. Hentschel, Asymmetric scattering and nonorthogonal mode patterns in optical microspirals, *Phys. Rev.* **A 78**, 053809 (2008).

7. Q. Song, N. Zhang, H. Zhai, S. Liu, Z. Gu, K. Wang, S. Sun, Z. Chen, M. Li, and S. Xiao, The combination of high Q factor and chirality in twin cavities and microcavity chain. *Sci. Rep.* **4**, 6493 (2014).

8. N. Zhang, S. Liu, K. Wang, Z. Gu, M. Li, N. Yi, S. Xiao, Q. Song, Single Nanoparticle Detection Using Far-field Emission of Photonic Molecule around the Exceptional Point. *Sci. Rep.* **5**, 11912 (2015).

9. N. Zhang, Z. Gu, S. Liu, Y. Wang, S. Wang, Z. Duan, W. Sun, Y. F. Xiao, S. Xiao, and Q. H. Song, Far-field single nanoparticle detection and sizing. *Optica* **4**, 1151–1156 (2017).

Letter to Richard Chang on the Occasion of His Retirement in June 2008

Hermes Huang, Ph.D. Yale '09
Director of Science and Technology
Smiths Detection Inc.
Danbury, CT 06810 USA
hermes.huang@smithsdetection.com

The American Association for Aerosol Research (AAAR) award ceremony in Austin, Texas 2005.

Richard is probably more sympathetic and understanding as an advisor then I could possibly hope for. He's allowed me to continue with all of my outside activities in music and frisbee, even though it definitely cuts into time I could be spending on research. At the same time, he always has tons of good ideas, and is always pushing me to take some risks in my research and to explore new territory. Unfortunately, it would probably take three of me to execute all of his good ideas.

Richard receiving the American Association for Aerosol Research (AAAR) David Sinclair award in Austin, Texas 2005.

Last summer when we attended a conference in Graz, Austria, I applied for a Yale travel grant, and got a whopping $100 grant. After he found out about this, he was so taken aback that he went ahead and set up a travel fund of his own, so that in the future when graduate students need to travel for a conference which isn't covered by grant money, they may be able to do so without spending so much of their own money. This action speaks to both Richard's compassion towards his students, as well as his "get things done" attitude. If he sees a solution to a problem, he will act on it.

I'm honored to be Richard's last student, and I hope my research was able to end his tenure at Yale on a good note.

-Hermes

Resume

Richard K. Chang[†]
Henry Ford II Professor Emeritus of Applied Physics

Address: Yale University
Department of Applied Physics
P.O. Box 208284
New Haven, CT 06520, USA
Telephone: +1 (203) 432-4272
Fax: +1 (203) 432-4274
Email: richard.chang@yale.edu

Education:

1961	M.I.T.	B.S. in Electrical Engineering
1962	Harvard University	M.S. in Applied Physics
1965	Harvard University	Ph.D. in Applied Physics (Adviser: Nicholaas Bloembergen)

Appointments:

Harvard University

1965–1966	Postdoctoral Fellow, Division of Engineering and Applied Physics (with Peter Pershan)

Yale University

1966–1970	Assistant Professor of Engineering and Applied Science
1970–1976	Associate Professor of Engineering and Applied Science
1976–	Professor of Applied Physics
1981–1991	Director, Center for Laser diagnostics
1984–	Professor of Electrical Engineering
1989–	Professor of Physics
1991–2008	Henry Ford II Professor of Applied Physics
2008–2020	Henry Ford II Professor Emeritus of Applied Physics

Gordon Research Conference

1989	Chair, The Physics and Chemistry of Laser Diagnostics in Combustion

Fellowships

1967–1969	Alfred P. Sloan Foundation Research Fellowship
1969–1970	Yale University Faculty Fellowship

Professional Societies

Eta Kappa Nu, Sigma Xi, Tau Beta Pi, American Physical Society (Fellow, 1975), IEEE, OPTICA (former Optical Society of America) (Fellow, 1979), Director at Large, 1983–1986, Connecticut Academy of Science and Engineering (2004).

†Passed away

Books Edited

Surface Enhanced Raman Scattering (Plenum Press, 1982).
Optical Effects Associated with Small Particles (World Scientific Publishing, 1988).
Optical Processes in Microcavities (World Scientific Publishing, 1996).

Awards

Elected to Connecticut Academy of Science and Engineering (2004).
American Association for Aerosol Research (AAAR) David Sinclair Award (2005).

Patents

1. R. K. Chang, Y. L. Pan, S. C. Hill, "Method and System for Sorting Particles Sampled from Air," US 7,410,063 B1, awarded August 12, 2008.
2. S. C. Hill, R. K. Chang, H. Huang, and Y. L. Pan, "Classifier for Particles in a Liquid Using a Large-Orifice Droplet Ejector". Published patent application: WO2007012000, January 25, 2007.
3. R. K. Chang, J. P. Wolf, V. Boutou, and Y. L. Pan, "Systems and methods for sorting aerosols," US 7,190,450 B2, awarded March 13, 2007.
4. S.C. Hill, R.K. Chang, and J.P. Wolf, "Aerosol Particle Analyzer for Measuring the Amount of Analyte in Airborne Particles," US 7,153,475 B2, awarded December 26, 2006.
5. S. C. Hill, R. G. Pinnick, Y. L. Pan, K. B. Aptowicz, K. P. Gurton, and R. K. Chang, "Method and instrumentation for determining absorption and morphology of individual airborne particles," US 7,126,687 B2, awarded Oct. 24, 2006.
6. R. K. Chang, Y. L. Pan, R.G. Pinnick, and S. C. Hill, "Method and instrumentation for measuring fluorescence spectra of individual airborne particles sampled from ambient air," US 6,947,134 B2, awarded Sept. 20, 2005.
7. R. K. Chang, Y. L. Pan, R. G. Pinnick and S. C. Hill, "Aerosol fluorescence spectrum analyzer for rapid measurement of single airborne particles," US 6,532,067, awarded March 11, 2003.
8. A. D. Stone, R. K. Chang, and J. Nöckel, "Asymmetric Resonant Optical Cavity Apparatus," US 6,259,717, awarded July 10, 2001.
9. R. K. Chang and R. G. Stafford, "Raman Scattering System and Method for Aerosol Monitoring," US 4,127,329, awarded November 28, 1978.

Publications in Journals, Proceedings, Book Chapters, and Technical Reports

IBM publication on semiconductor optoelectronics

1. R.K. Chang, "Trapping Effects and Lifetime of CdSe Photoconductors by Modulation Techniques," (B.S. Thesis) IBM Research Report RC-930 (1963).

Harvard University publications on nonlinear optics

2. N. Bloembergen, R.K. Chang, J. Ducuing, and P. Lallemand, "Nonlineairites Optiques de Quelques Composes III-V," *Physics of Semiconductors*, M. Hulin, ed. (Dunod, Paris, 1964), p. 121.
3. R.K. Chang, J. Ducuing, and N. Bloembergen, "Relative Phase Measurement between Fundamental and Second-Harmonic Light," *Phys. Rev. Lett.* **15**, 6 (1965).
4. R.K. Chang, J. Ducuing, and N. Bloembergen, "Dispersion of the Optical Nonlinearity in Semiconductors," *Phys. Rev. Lett.* **15**, 415 (1965).
5. N. Bloembergen, R.K. Chang, and J. Ducuing, "Dispersion of the Optical Nonlinearity in Semiconductors," *Physics of Quantum Electronics*, P.L. Kelley, B. Lax, and P.E. Tannenwald, eds. (McGraw-Hill, New York, 1966), p. 67.
6. N. Bloembergen and R.K. Chang, "Second-Harmonic Generation of Light from Surface Layers of Media with Inversion Symmetry," *Physics of Quantum Electronics*, P.L. Kelley, B. Lax, and P.E. Tannenwald, eds. (McGraw-Hill, New York, 1966) p. 80.
7. R.K. Chang and N. Bloembergen, "Experimental Verification of the Laws for the Reflected Intensity of Second-Harmonic Light," *Phys. Rev.* **144**, 775 (1966).
8. R.K. Chang, "Dispersion of Nonlinear Susceptibility," Cruft Laboratory, Harvard University, Technical Report No. 491 (1965).
9. N. Bloembergen, R.K. Chang, and C.H. Lee, "Second-Harmonic Generation of Light in Reflection from Media with Inversion Symmetry," *Phys. Rev. Lett.* **16**, 986 (1966).
10. R.K. Chang, B. Lacina, and P.S. Pershan, "Raman Scattering from Mixed Crystals $(Ca_xSr_{1-x})F_2$ and $(Sr_xBa_{1-x})F_2$," *Phys. Rev. Lett.* **17**, 755 (1966).
11. C.H. Lee, R.K. Chang, and N. Bloembergen, "Nonlinear Electroreflectance in Silicon and Silver," *Phys. Rev. Lett.* **18**, 167 (1967).

Yale University publications on nonlinear optics

12. J. Schwartz, C.S. Naiman, and R.K. Chang, "Effects of Internal Induced Absorption on Laser Emission," *Appl. Phys. Lett.* **11**, 242 (1967).
13. R.K. Chang and L.K. Galbraith, "Optical Third-Harmonic Generation in Dye Solutions," *Phys. Rev.* **171**, 993 (1968).
14. N. Bloembergen, R.K. Chang, S.S. Jha, and C.H. Lee, "Optical Second-Harmonic Generation in Reflection from Media with Inversion Symmetry," *Phys. Rev.* **174**, 813 (1968).
15. R.K. Chang, J.M. Ralston, and D.E. Keating, "Temperature Dependence of Raman Linewidth and Intensity of Semiconductors," in *Light Scattering Spectra of Solids*, G.B. Wright, ed. (Springer-Verlag, New York, 1969), p. 369.
16. J.M. Ralston and R.K. Chang, "Optical Limiting in Semiconductors," *Appl. Phys. Lett.* **15**, 164 (1969).

17. J.M. Ralston and R.K. Chang, "Nd:Laser Induced Absorption in Semiconductors and Aqueous PrCl3 and NdCl3," *J. Opto-Electronics* **1**, 182 (1969).

18. R.K. Chang, "Tunable Lasers," *McGraw-Hill Yearbook of Science and Technology*, D.N. Lapedes, eds. (McGraw-Hill, New York, 1970), p. 232.

19. J. Schwartz, W. Weiler, and R.K. Chang, "Laser-Pulse Shaping Using Inducible Absorption with Controlled Q-Switch Time Behavior," *IEEE J. Quantum Electronics* QE-6, 442 (1970).

20. J.M. Ralston and R.K. Chang, "Spontaneous Raman Scattering Efficiency and Stimulated Scattering in Silicon," *Phys. Rev.* **B 2**, 1858 (1970).

21. J.M. Ralston, R.L. Wadsack, and R.K. Chang, "Resonant Cancellation of Raman Scattering from CdS and Si," *Phys. Rev. Lett.* **25**, 814 (1970).

22. B.E. Argyle, R.L. Wadsack, and R.K. Chang, "Raman Determination of 2F7/2 Crystal Field Levels in YbAlG and YbGaG," *J. Appl. Phys.* **42**, 1478 (1971).

23. R.L. Wadsack, J.L. Lewis, B.E. Argyle, and R.K. Chang, "Phonons and Crystal Field Levels in Dysprosium Garnets by Raman Scattering," *Phys. Rev.* **3**, 4342 (1971).

24. B.E. Argyle, J.L. Lewis, R.L. Wadsack, and R.K. Chang, "Electronic Raman Studies of Antiferromagnetic Dysprosium Aluminum Garnet below TN and with and without an Applied Magnetic Field," *Phys. Rev.* **B 4**, 3035 (1971).

25. F.G. Parsons and R.K. Chang, "Measurement of the Nonlinear Susceptibility Dispersion by Dye Lasers," *Opt. Commun.* **3**, 173 (1971).

26. D.G. Fouche and R.K. Chang, "Relative Raman Cross Section for N2, O2, CO, CO2, SO2, and H2S," *Appl. Phys. Lett.* **18**, 579 (1971).

27. R.K. Chang, "Q-Switched Pulses Stretched to Order," *Laser Focus* **7**, 32 (1971).

28. J.L. Lewis, R.L. Wadsack, and R.K. Chang, "Resonant Raman Cancellation in ZnS and ZnSe," in *Light Scattering in Solids*, M. Balkanski, ed. (Flammarion Sciences, Paris, 1971), p. 41.

29. F.G. Parsons, E. Yi Chen, and R.K. Chang, "Dispersion of Nonlinear Optical Susceptibilities in Hexagonal II-VI Semiconductors," *Phys. Rev. Lett.* **27**, 1436 (1971).

Yale University publications on Raman scattering

30. R.L. Wadsack and R.K. Chang, "Resonance Electronic Raman Effect in DyA1G," *Solid State Commun.* **10**, 45 (1972).

31. D.G. Fouche and R.K. Chang, "Relative Raman Cross Section for 03, CH4, C3H8, NO, N20, and H2," *Appl. Phys. Lett.* **20**, 256 (1972).

32. D.G. Fouche and R.K. Chang, "Observation of Resonance Raman Scattering below the Dissociation Limit in I2 Vapor," *Phys. Rev. Lett.* **29**, 536 (1972).

33. R.J. Kobliska, S.A. Solin, M. Selders, R.K. Chang, R. Alben, M.F. Thorpe, and D. Weaire, "Raman Scattering from Phonons in Polymorphs of Si and Ge," *Phys. Rev. Lett.* **29**, 725 (1972).

34. D.G. Fouche, A. Herzenberg, and R.K. Chang, "Inelastic Photon Scattering by a Polyatomic Molecule: NO2," *J. Appl. Phys.* **43**, 3846 (1972).

35. R.K. Chang and D.G. Fouche, "Gains in Detecting Pollution," *Laser Focus* **8**, 43 (1972).

36. R.H. Callender, S.S. Sussman, M. Selders, and R.K. Chang, "Dispersion of Raman Cross Section in CdS and ZnO over a Wide Energy Range," *Phys. Rev.* **B 7**, 3788 (1973).

37. J.-J. Song, P.B. Klein, R.L. Wadsack, M. Selders, S. Mroczkowski, and R.K. Chang, "Raman-Active Phonons in Aluminum, Gallium, and Iron Garnets," *J. Opt. Soc. Am.* **63**, 1135 (1973).

38. M. Selders, E. Yi Chen, and R.K. Chang, "Dispersion of Raman Scattering in CdTe and InP below the Absorption Edge," *Solid State Commun.* **12**, 1057 (1973).

39. S.S. Sussman, R. Alben, M. Selders, R.K. Chang, and R.H. Callender, "Wavelength and Concentration Dependence of Raman Scattering from CdS1xSex," *Solid State Commun.* **13**, 799 (1973).

40. P.B. Klein, H. Masui, J.-J. Song, and R.K. Chang, "Selective Resonant Enhancement in the Two-Phonon Raman Spectrum of Si and GaP," *Solid State Commun.* **14**, 1163 (1974).

41. R.K. Chang, E. Yi Chen, F.G. Parsons, and M. Selders, "Dispersion of the Second Harmonic and Displacive Coefficients in Semiconductors," in *Polaritons, Proceedings of the First Taormina Research Conference on the Structure of Matter, Taormina, Italy, 1972*, E. Burstein and F. DeMartini, eds. (Pergamon Press, New York, 1974), p. 319.

42. H. Masui, P.B. Klein, R.K. Chang, R.H. Callender, and R.J. Chicotka, "Second-Order Raman Effect in AlP, AlAs, and AlSb," in *Proceedings of the Twelfth International Conference on the Physics of Semiconductors*, M.H. Pilkuhn, ed. (B.G. Teubner, Stuttgart, 1974), p. 509.

43. P.B. Klein, J.-J. Song, R.K. Chang, and R.H. Callender, "Multiphonon Raman Scattering in Si and AlAs," in *Proceedings of the Third International Conference on Light Scattering in Solids*, M. Balkanski, R.C.C. Leite, and S.P.S. Porto, eds. (Flammarion Sciences, Paris, 1976), p. 93.

44. R.G. Stafford and R.K. Chang, "Absolute Raman Scattering Cross Sections of Sulfate and Bisulfate and Their Application to Aqueous Aerosol Monitoring," in *International Conference on Environmental Sensing and Assessment*, Vol. 2 (IEEE, New York, 1976), p. 23.

45. E. Kultsar, E. Yi Chen, and R.K. Chang, "Wavelength Deviation Sensing Detector," *Appl. Opt.* **15**, 862 (1976).

46. R.G. Stafford, H. Masui, R.L. Farrow, R.K. Chang, and L.G. van Uitert, "Coherent Emission from Ho3+ Ions by Pumping into the YIG Absorption Band," *J. Appl. Phys.* **47**, 2483 (1976).

47. P.B. Klein and R.K. Chang, "Comparison of Second-Order Raman Scattering Measurements with a Phonon Density-of-States Calculation in GaSb," *Phys. Rev.* **B 14**, 2498 (1976).

48. R.G. Stafford, R.K. Chang, and P.J. Kindlmann, "Laser-Raman Monitoring of Ambient Sulfate Aerosols," in *Methods and Standards for Environmental Measurement, Proceedings of the 8th International Materials Research Symposium*, NBS Special Publication 464, W.H. Kirchoff, ed. (U.S. Government Printing Office, Washington, D.C., 1977), p. 659.

49. D.V. Murphy and R.K. Chang, "Pulse Stretching of Q-Switched Laser Emission by Intracavity Nonlinear Ellipse Rotation," *Opt. Commun.* **23**, 268 (1977).

50. R.L. Farrow, R.K. Chang, S. Mroczkowski, and F.H. Pollak, "Detection of Excess Crystalline As and Sb in III-V Oxide Interfaces by Raman Scattering," *Appl. Phys. Lett.* **31**, 768 (1977).

51. P.C. Black and R.K. Chang, "Laser-Raman Optical Multichannel Analyzer for Transient Gas Concentration Profile and Temperature Determination," *AIAA Journal* **16**, 295 (1978). National Technical Information Service No. N78-13405, Springfield, Va., 1978.

52. R.L. Farrow, P.C. Black, and R.K. Chang, "Resonant Two-Phonon Raman Scattering in GaSb near the E1 and E1 + Δ1 Gaps," in *Proceedings of the International Conference on Lattice Dynamics*, 1977, M. Balkanski, ed. (Flammarion Sciences, Paris, 1978), p. 182.

53. E.-H. Lee, R.E. Benner, J.B. Fenn, and R.K. Chang, "Angular Distribution of Fluorescence from Monodispersed Particles," *Appl. Opt.* **17**, 1980 (1978).

54. R.K. Chang and R.E. Benner, "Chemical Speciation and Concentration Monitoring of Sulfur Oxides by Laser-Raman Scattering," in *Environmental Protection Agency Proceedings on Primary Sulfate Emissions from Combustion Sources*, Vol. 1, EPA600/9-78-020a, August 1978, p. 97.

55. R.L. Farrow and R.K. Chang, "Interaction of E1 and E1 + Δ1 Exciton Recombination with LO Phonons in III-V Semiconductors," *Solid-State Electron.* **21**, 1347 (1978).

56. R.L. Farrow, R.K. Chang, and R.M. Martin, "Interference of E1 and E1 + Δ1 Hot Exciton Luminescence with LO Phonon Raman Scattering in III-V Semiconductors," in *Physics of Semiconductors*, 1978, B.L.H. Wilson, ed. (Institute of Physics, London, 1979), p. 485.

Yale University publications on morphology dependent resonances (MDRs)

57. R.E. Benner and R.K. Chang, "Utilization of Optical Fibers in Remote Inelastic Light Scattering Probes," in *Fiber Optics: Advances in Research and Development*, B. Bendow and S.S. Mitra, eds. (Plenum Press, New York, 1979), p. 625.

58. R.E. Benner, E.H. Lee, J.B. Fenn, and R.K. Chang, "Inelastic Emission from Monodispersed Particles," in *Proceedings of the Sixth International Conference on Raman Spectroscopy*, Vol. 2, E.D. Schmid, R.S. Krishnan, W. Kiefer, H.W. Schrotter, eds. (Heyden, 1978), p. 508.

59. E.-H. Lee, R.E. Benner, J.B. Fenn, and R.K. Chang, "Angular Distribution of Fluorescence from Liquids and Monodispersed Spheres by Evanescent Wave Excitation," *Appl. Opt.* **18**, 862 (1979).

60. M.B. Long, B.F. Webber, and R.K. Chang, "Instantaneous Two-Dimensional Concentration Measurements in a Jet Flow by Mie Scattering," *Appl. Phys. Lett.* **34**, 22 (1979).

61. B.F. Webber, M.B. Long, and R.K. Chang, "Two-Dimensional Average Concentration Measurements in a Jet Flow by Raman Scattering," *Appl. Phys. Lett.* **35**, 119 (1979).

62. D.V. Murphy, M.B. Long, R.K. Chang, and A.C. Eckbreth, "Spatially Resolved Coherent anti-Stokes Raman Spectroscopy from a Line across a CH_4 Jet," *Opt. Lett.* **4**, 167 (1979).

63. R.E. Benner, R. Dornhaus, M.B. Long, and R.K. Chang, "Inelastic Light Scattering from a Distribution of Microparticles," in *Microbeam Analysis 1979, Proceedings of the 14th Annual Conference of the Microbeam Analysis Society*, Dale E. Newbury, ed. (San Francisco Press, San Francisco, 1979), p. 191.

64. R. Dornhaus, R.L. Farrow, R.K. Chang, and R.M. Martin, "Interaction between Localized Carriers in the Accumulation Layer and Extended Bulk LO Phonons in InSb and GaSb: Raman Interference Lineshapes," in *Light Scattering in Solids*, J.L. Birman, H.Z. Cummins, and K.K. Rebane, eds. (Plenum Press, New York, 1979), p. 299.

65. R.E. Benner, R. Dornhaus, and R.K. Chang, "Angular Emission Profiles of Dye Molecules Excited by Surface Plasmon Waves at a Metal Surface," *Opt. Commun.* **30**, 145 (1979).

66. R.E. Benner, J.F. Owen, and R.K. Chang, "Radiation Patterns of Inelastic Reemission from Microparticles in Homogeneous Surroundings and near Dielectric or Metal Interfaces," *J. Phys. Chem.* **84**, 1602 (1980).

67. R. Dornhaus, M.B. Long, R.E. Benner, and R.K. Chang, "Time Development of SERS from Pyridine, Pyrimidine, Pyrazine, and Cyanide Adsorbed on Ag Electrodes during an Oxidation-Reduction Cycle," *Surf. Sci.* **93**, 240 (1980).

68. R. Dornhaus and R.K. Chang, "Comments on the 210–243 cm-1 Mode in Surface Enhanced Raman Scattering from the Pyridine-Ag System," *Solid State Commun.* **34**, 811 (1980).

69. R. Dornhaus, R.L. Farrow, and R.K. Chang, "Interactions between Carriers and LO Phonons in Bulk p-GaSb and p-InSb: Raman Interference Lineshapes," *Solid State Commun.* **35**, 123 (1980).

70. R. Dornhaus, R.E. Benner, R.K. Chang, and I. Chabay, "Surface Plasmon Contribution to SERS," *Surf. Sci.* **101**, 367 (1980).

71. R.E. Benner, R. Dornhaus, R.K. Chang, and B.L. Laube, "Correlations in the Raman Spectra of Cyanide Complexes Adsorbed on Ag Electrodes with Voltammograms," *Surf. Sci.* **101**, 341 (1980).

72. R.E. Benner, P.W. Barber, J.F. Owen, and R.K. Chang, "Observation of Structure Resonances in the Fluorescence Spectra from Microspheres," *Phys. Rev. Lett.* **44**, 475 (1980).

73. R.E. Benner, K.U. von Raben, R. Dornhaus, R.K. Chang, B.L. Laube, and F.A. Otter," Correlation of SERS with Cyclic Voltammetry for Cyanide Complexes Adsorbed on Cu Electrodes," *Surf. Sci.* **102**, 7 (1981).

74. R.K. Chang, J.F. Owen, P.W. Barber, B.J. Messinger, and R.E. Benner, "Inelastic Light Emission from Spherical Particles and Cylindrical Fibers," *J. Raman Spectrosc.* **10**, 178 (1981).

75. M.B. Long, B.T. Chu, and R.K. Chang, "Instantaneous Two-Dimensional Gas Concentration Measurements by Light Scattering," *AIAA Journal* **19**, 1151 (1981).

76. R.K. Chang, R.E. Benner, R. Dornhaus, K.U. von Raben, and B.L. Laube, "Enhanced Raman Scattering of Molecules Adsorbed on Ag, Cu, and Au Surfaces," in *Lasers and Applications*, W.O.N. Guimaraes, C.T. Lin, and A. Mooradian, eds. (Springer-Verlag, Berlin, 1981), p. 55.

77. B.J. Messinger, K.U. von Raben, R.K. Chang, and P.W. Barber, "Local Fields at the Surface of Noble Metal Microspheres," *Phys. Rev.* **B 24**, 649 (1981).

78. J.F. Owen, P.W. Barber, B.J. Messinger, and R.K. Chang, "Determination of Optical Fiber Diameter from Resonances in the Elastic Scattering Spectrum," *Opt. Lett.* **6**, 272 (1981).

79. J.F. Owen, R.K. Chang, and P.W. Barber, "Internal Electric Field Distributions of a Dielectric Cylinder at Resonance Wavelengths," *Opt. Lett.* **6**, 540 (1981).

80. J.F. Owen, P.W. Barber, P.B. Dorain, and R.K. Chang, "Enhancement of Fluorescence Induced by Microstructure Resonances of a Dielectric Fiber," *Phys. Rev. Lett.* **47**, 1075 (1981).

81. K.U. von Raben, R.K. Chang, and B.L. Laube, "Surface Enhanced Raman Scattering of Au $(CN)2^-$ Ions Adsorbed on Gold Colloids," *Chem. Phys. Lett.* **79**, 465 (1981).

82. D.V. Murphy and R.K. Chang, "Single-Pulse Broadband Rotational Coherent AntiStokes Raman-Scattering Thermometry of Cold N2 Gas," *Opt. Lett.* **6**, 233 (1981).

83. P.B. Dorain, K.U. von Raben, R.K. Chang, and B.L. Laube, "Catalytic Formation of $SO_{2,3}^-$ and $SO_{2,4}^-$ from SO_2 on Silver Observed by Surface Enhanced Raman Scattering," *Chem. Phys. Lett.* **84**, 405 (1981).

84. R.K. Chang and M.B. Long, "Optical Multichannel Detection," in *Light Scattering in Solids II, Topics in Applied Physics*, Vol. 50, M. Cardona and G. Güntherodt, eds. (Springer-Verlag, Heidelberg, 1982), p. 179.

85. K.U. von Raben, R.K. Chang, P.B. Dorain, and B.L. Laude, "SERS from Oxy-Anions and Oxides Adsorbed on Ag Powder," in *Raman Spectroscopy*, J. Lascombe and P.V. Huong, eds. (Wiley Heyden, Chichester, 1982), p. 81.

86. P.W. Barber, J.F. Owen, and R.K. Chang, "Resonant Scattering for Characterization of Axisymmetric Dielectric Objects," *IEEE Trans. Antennas Propag.* AP-30, 168 (1982).

87. D.V. Murphy, K.U. von Raben, R.K. Chang, and P.B. Dorain, "Surface Enhanced Hyper-Raman Scattering from $SO_{2,3}^-$ Adsorbed on Ag Powder," *Chem. Phys. Lett.* **85**, 43 (1982).

88. R.K. Chang and T.E. Furtak, eds., Surface Enhanced Raman Scattering (Plenum Press, New York, 1982).

89. T.T. Chen, J.F. Owen, R.K. Chang, and B.L. Laube, "Surface-Enhanced Raman Scattering of Water Adsorbed on Silver Electrodes," *Chem. Phys. Lett.* **89**, 356 (1982).

90. J.F. Owen, P.W. Barber, and R.K. Chang, "Morphology-Dependent Raman Spectra from Microparticles," in *Microbeam Analysis 1982*, K.F.J. Heinrich, ed. (San Francisco Press, 1982), p. 255.

91. J.F. Owen, R.K. Chang, and P.W. Barber, "Morphology-Dependent Resonances in Raman Scattering, Fluorescence Emission, and Elastic Scattering from Microparticles," *Aerosol Sci. Technol.* **1**, 293 (1982).

Yale University publications on surface enhanced Raman scattering (SERS)

92. T.T. Chen, K.U. von Raben, J.F. Owen, R.K. Chang, and B.L. Laube, "Laser Illumination Effects on the Surface Morphology, Cyclic Voltammetry, and Surface-Enhanced Raman Scattering of Ag Electrodes," *Chem. Phys. Lett.* **91**, 494 (1982).

93. D.V. Murphy, K.U. von Raben, T.T. Chen, J.F. Owen, R.K. Chang, and B.L. Laube, "Surface Enhanced Second-Harmonic Generation from Ag Electrodes and Adsorbates during Electrochemical Cycling," *Surf. Sci.* **124**, 529 (1983).

94. J.F. Owen, T.T. Chen, R.K. Chang, and B.L. Laube, "Competition among Adsorbed Species for Surface Sites on Ag Electrodes as Determined by SERS," *Surf. Sci.* **125**, 679 (1983).

95. R.K. Chang, M.B. Long, D.V. Murphy, and B.T. Chu, "Laser Scattering for Concentration and Temperature Determination in Turbulence and Combustion Studies," in *Proceedings of an International Conference on Lasers*, May, 1980 (Wiley, New York, 1983), p. 203.

96. J.F. Owen, T.T. Chen, R.K. Chang, and B.L. Laube, "SERS of H20, Pyridine, and Halides Adsorbed on Ag Electrodes during Electrochemical Cycling," *J. Electroanal. Chem.* **150**, 389 (1983).

97. P.W. Barber, R.K. Chang, and H. Massoudi, "Surface-Enhanced Electric Intensities on Large Silver Spheroids," *Phys. Rev. Lett.* **50**, 997 (1983).

98. R.K. Chang, "Surface Enhancement Mechanisms," in *Proceedings of the International Conference on Time-Resolved Vibrational Spectroscopy*, 1982 (Academic Press, New York, 1983), p. 369.

99. R.E. Benner, K.U. von Raben, K.C. Lee, J.F. Owen, R.K. Chang, and B.L. Laube, "Raman Scattering from Pt $(CN)_{2,4}^{-}$ Adsorbed on Pt Colloids," *Chem. Phys. Lett.* **96**, 65 (1983).

100. K.U. von Raben, P.B. Dorain, T.T. Chen, and R.K. Chang, "NO_2^- and NO_3^- Formation on Oxygenated Ag Surfaces Exposed to Nitrogen Oxide Gases: A SERS Study," *Chem. Phys. Lett.* **95**, 269 (1983).

101. J.F. Owen, T.T. Chen, R.K. Chang, and B.L. Laube, "Irreversible Loss of Adatoms on Ag Electrodes during Potential Cycling Determined from Surface Enhanced Raman Intensities," *Surf. Sci.* **131**, 195 (1983).

102. P.W. Barber, R.K. Chang, and H. Massoudi, "Electrodynamic Calculations of the Surface-Enhanced Electric Intensities on Large Ag Spheroids," *Phys. Rev.* **B 27**, 7251 (1983).

103. J.B. Snow, J.-B. Zheng, and R.K. Chang, "Spatially and Spectrally Resolved Multi-point Coherent anti-Stokes Raman Scattering from N2 and O2 Flows," *Opt. Lett.* **8**, 599 (1983).

104. J.B. Zheng, A. Leipertz, J.B. Snow, and R.K. Chang, "Simultaneous Observation of Rotational Coherent Stokes Raman Scattering and Coherent anti-Stokes Raman Scattering in Air and Nitrogen," *Opt. Lett.* **8**, 350 (1983).

105. R.K. Chang, "Surface Enhanced Raman Scattering: An Overview of Experiments," *J. de Phys. Colloq.* **44**, C10-283 (1983).

106. K.U. von Raben, K.C. Lee, R.K. Chang, and R.E. Benner, "Enhanced Raman Scattering from Cesium Suboxides in Photocathodes: Ag-O-Cs (S-1) and Bi-Ag-O-Cs (S-10)," *J. Appl. Phys.* **55**, 3907 (1984).

107. R.K. Chang and B.L. Laube, "Surface Enhanced Raman Scattering and Nonlinear Optics Applied to Electrochemistry," *CRC Critical Rev. Solid State Materials Sci.* **12**, 1 (1984).

108. T.T. Chen, K.U. von Raben, D.V. Murphy, R.K. Chang, and B.L. Laube, "Surface Enhanced Raman Scattering and Second-Harmonic Generation from CN^- Complexes and $SO_{2,4}^{-}$ on Ag Electrodes during Oxidation-Reduction Cycles," *Surf. Sci.* **143**, 369 (1984).

109. J.F. Owen and R.K. Chang, "Increased Enhancement in Surface Enhanced Raman Scattering from Ag Electrodes with the Addition of Pyridine to the Electrolyte," *Chem. Phys. Lett.* **104**, 59 (1984).

110. T.T. Chen, K.E. Smith, J.F. Owen, and R.K. Chang, "The Metal Cation Effect on the SERS of Interfacial D20 and H20," *Chem. Phys. Lett.* **108**, 32 (1984).

111. J.F. Owen and R.K. Chang, "SERS of H20 Adsorbed on a Ag Electrode: The Role of High Salt Concentration," *Chem. Phys. Lett.* **104**, 510 (1984).

112. J.-B. Zheng, J.B. Snow, D.V. Murphy, A. Leipertz, R.K. Chang, and R.L. Farrow, "Experimental Comparison of Broadband Rotational Coherent anti-Stokes Raman Scattering (CARS) and Broadband Vibrational CARS in a Flame," *Opt. Lett.* **9**, 341 (1984).

113. H.-M. Tzeng, K.F. Wall, M.B. Long, and R.K. Chang, "Evaporation and Condensation Rates of Liquid Droplets Deduced from Structure Resonances in the Fluorescence Spectra," *Opt. Lett.* **9**, 273 (1984).

114. K.U. von Raben, R.K. Chang, B.L. Laube, and P.W. Barber, "Wavelength Dependence of Surface Enhanced Raman Scattering from Ag Colloids with Adsorbed CN^- Complexes, $SO_{2,3}^-$ and Pyridine," *J. Phys. Chem.* **88**, 5290 (1984).

115. T.T. Chen, R.K. Chang, and B.L. Laube, "Detection of Interfacial Hydroxyl Ions (OH- and OD^-) on Ag Electrodes by Surface Enhanced Raman Scattering," *Chem. Phys. Lett.* **108**, 39 (1984).

116. P.B. Dorain, K.U. von Raben, and R.K. Chang, "Voltage Dependence of the pH at the Interface of a Ag Electrode: A SERS Measurement of a Phosphate-Phosphoric Acid System," *Surf. Sci.* **148**, 439 (1984).

117. P.B. Dorain, T.T. Chen, and R.K. Chang, "Probing Molecular Species at the Electrode-Electrolyte Interface with Surface Enhanced Raman Scattering," in *Proceedings of the SPIE Conference on Applications of Laser Chemistry and Diagnostics*, Vol. 482 (SPIE, Bellingham, Washington, 1984), p. 116.

118. T.T. Chen, J.F. Owen, and R.K. Chang, "Cation-Induced Superequivalent Adsorption of C1- Ions: Temporal and Potential Evolution of SERS," in *Dynamics on Surfaces*, B. Pullman, J. Jortner, A. Nitzan, and B. Gerber, eds. (D. Reidel Publishing Co., Dordrecht, 1984), p. 365.

Yale University publications on morphology dependent resonances (MDRs)

119. H.-M. Tzeng, K.F. Wall, M.B. Long, and R.K. Chang, "Laser Emission from Individual Droplets at Wavelengths Corresponding to Morphology-Dependent Resonances," *Opt. Lett.* **9**, 499 (1984).

120. J.B. Snow, S.-X. Qian, and R.K. Chang, "Stimulated Raman Scattering from Individual Water and Ethanol Droplets at Morphology-Dependent Resonances," *Opt. Lett.* **10**, 37 (1985).

121. T.T. Chen and R.K. Chang, "Surface Enhanced Raman Scattering of Interfacial DOD, HOD, and OD^-," *Surf. Sci.* **158**, 325 (1985).

122. H.-M. Tzeng, M.B. Long, R.K. Chang, and P.W. Barber, "Laser-Induced Shape Distortions of Flowing Droplets Deduced from Morphology-Dependent Resonances in Fluorescence Spectra," *Opt. Lett.* **10**, 209 (1985).

123. R.K. Chang, "SERS from Adsorbates on Electrodes during Oxidation-Reduction Cycles," *Surf. Sci.* **158**, 405 (1985).

124. S.-X. Qian, J.B. Snow, and R.K. Chang, "Coherent Raman Mixing and Coherent Anti-Stokes Raman Scattering from Individual Micrometer-Size Droplets," *Opt. Lett.* **10**, 499 (1985).

125. K.U. von Raben, R.K. Chang, P.B. Dorain, and B.L. Laube, "Adsorbed Sulfur and Nitrogen Oxides on Silver Powders Detected by Surface Enhanced Raman Scattering," in *Lasers as Reactants and Probes in Chemistry*, W.M. Jackson and A.B. Harvey, eds. (Howard University Press, Washington, DC, 1985), p. 249.

126. S.-X. Qian, J.B. Snow, and R.K. Chang, "Nonlinear Optical Processes in Micron-Size Droplets," in *Laser Spectroscopy VII*, T.W. Hänsch and Y.R. Shen, eds. (Springer-Verlag, Berlin, 1985), p. 204.

127. H.-M. Tzeng, M.B. Long, R.K. Chang, and P.W. Barber, "Size and Shape Variations of Liquid Droplets Deduced from Morphology-Dependent Resonances in Fluorescence Spectra," in *Proceedings of the SPIE Particle Sizing and Spray Analysis Conference*, Vol. 573 (SPIE, Bellingham, Washington, 1985), p. 80.

128. M.P. Cline, P.W. Barber, and R.K. Chang, "Surface Enhanced Electric Intensities on Transition- and Noble-Metal Spheroids," *J. Opt. Soc. Am.* **B 3**, 15 (1986).

129. J.B. Snow, J.-B. Zheng, and R.K. Chang, "Increased Sensitivity of a Vidicon Optical Multichannel Analyzer with a Detachable Electrostatic Image Intensifier," *Appl. Opt.* **25**, 172 (1986).

130. S.-X. Qian, J.B. Snow, H.-M. Tzeng, and R.K. Chang, "Lasing Droplets: Highlighting the Liquid-Air Interface by Laser Emission," *Science* **231**, 486 (1986).

131. S.-X. Qian and R.K. Chang, "Multiorder Stokes Emission from Micrometer-Size Droplets," *Phys. Rev. Lett.* **56**, 926 (1986).

132. J.B. Snow, S.-X. Qian, and R.K. Chang, "Nonlinear Optics with a Micrometer-Size Droplet," *Opt. News* **12**(5), 5 (1986).

133. R.K. Chang, S.-X. Qian, and J. Eickmans, "Stimulated Raman Scattering, Phase Modulation, and Coherent Anti-Stokes Raman Scattering from Single Micrometer-Size Liquid Droplets," in *Methods of Laser Spectroscopy*, Y. Prior, A. Ben-Reuven, and M. Rosenbluh, eds. (Plenum Press, New York, 1986), p. 249.

134. S.-X. Qian and R.K. Chang, "Phase-Modulation-Broadened Line Shapes from Micrometer-Size CS2 Droplets," *Opt. Lett.* **11**, 371 (1986).

135. K.F. Wall and R.K. Chang, "I2-Vapor Notch Filter with Optical Multichannel Detection of Low-Frequency-Shift Inelastic Scattering from Surface-Enhanced Raman-Scattering Active Electrodes," *Opt. Lett.* **11**, 493 (1986).

136. K.F. Wall and R.K. Chang, "Separation of the Low-Frequency Mode from the Inelastic Continuum Scattering of a SERS Active Electrode," *Chem. Phys. Lett.* **129**, 144 (1986).

137. K.F. Wall, M.L.A. Temperini, and R.K. Chang, "Potential-Dependent Measurements of the Low-Frequency Mode, the Inelastic Continuum, Elastic Scattering, and SERS from a Ag Electrode," *Chem. Phys. Lett.* **129**, 253 (1986).

138. J.H. Eickmans and R.K. Chang, "Stimulated Raman Scattering Spectra from Single Water Droplets," in *Proceedings of the Tenth International Conference on Raman Spectroscopy*, W.L. Peticolas and B. Hudson, eds. (University of Oregon, Eugene, 1986), p. 8-7.

139. W.-F. Hsieh, H.-M. Tzeng, and R.K. Chang, "High Intensity Laser Interactions with Micrometer-Size Dye Droplets," in *Special Issue of the Annual Report of the Institute of Physics*, Academia Sinica (Taiwan), Vol. 16, 1986, p. 1.

140. R.K. Chang, "Micrometer-Size Droplets as Optical Cavities: Lasing and Other Non-linear Effects," in *Advances in Laser Science-II*, M. Lapp, W.C. Stwalley, and G.A. Kenney-Wallace, eds. (American Institute of Physics, New York, 1987), p. 509.

141. J.H. Eickmans, S.-X. Qian, and R.K. Chang, "Detection of Water Droplet Size and Anion Species by Nonlinear Optical Scattering," *Part. Charact.* **4**, 85 (1987).

142. J.H. Eickmans, W.-F. Hsieh, and R.K. Chang, "Laser-Induced Explosion of H20 Droplets: Spatially Resolved Spectra," *Opt. Lett.* **12**, 22 (1987).

143. W.P. Acker, B. Schlicht, R.K. Chang, and P.W. Barber, "Quadrupole Mode in the Extinction and Elastic-Scattering Spectra of Gold Microstructures," *Opt. Lett.* **12**, 465 (1987).

144. D.S. Benincasa, P.W. Barber, J.-Z. Zhang, W.-F. Hsieh, and R.K. Chang, "Spatial Distribution of the Internal and Near-Field Intensities of Large Cylindrical and Spherical Scatterers," *Appl. Opt.* **26**, 1348 (1987).

145. R.K. Chang, "Surface Enhanced Raman Scattering at Electrodes: A Status Report," *Ber. Bunsenges. Phys. Chem.* **91**, 296 (1987).

146. W.-F. Hsieh, J.-B. Zheng, C.F. Wood, B.T. Chu, and R.K. Chang, "Propagation Velocity of Laser-Induced Plasma inside and outside a Transparent Droplet," *Opt. Lett.* **12**, 576 (1987).

147. B. Schlicht, K.F. Wall, R.K. Chang, and P.W. Barber, "Light Scattering by Two Parallel Glass Fibers," *J. Opt. Soc. Am.* **A 4**, 800 (1987).

148. W.-F. Hsieh, J.H. Eickmans, and R.K. Chang, "Internal and External Laser-Induced Avalanche Breakdown of Single Droplets in an Argon Atmosphere," *J. Opt. Soc. Am.* **B 4**, 1816 (1987).

149. J.H. Eickmans, W.-F. Hsieh, and R.K. Chang, "Plasma Spectroscopy of H, Li, and Na in Plumes Resulting from Laser-Induced Droplet Explosion," *Appl. Opt.* **26**, 3721 (1987).

150. J.-Z. Zhang, J.K. Lam, C.F. Wood, B.T. Chu, and R.K. Chang, "Explosive Vaporization of a Large Transparent Droplet Irradiated by a High Intensity Laser," *Appl. Opt.* **26**, 4731 (1987).

151. R.K. Chang, "Nonlinear Optical Phenomena in Single Micron-Size Droplets," in *Laser Optics of Condensed Matter*, J.L. Birman, H.Z. Cummins, and A.A. Kaplyanskii, eds. (Plenum Press, New York, 1988), p. 193.

152. C.F. Wood, D.H. Leach, J.-Z. Zhang, R.K. Chang, and P.W. Barber, "Time-Resolved Shadowgraphs of Large Individual Water and Ethanol Droplets Vaporized by a Pulsed CO2 Laser," *Appl. Opt.* **27**, 2279 (1988).

153. J.-Z. Zhang, D.H. Leach, and R.K. Chang, "Photon Lifetime within a Droplet: Temporal Determination of Elastic and Stimulated Raman Scattering," *Opt. Lett.* **13**, 270 (1988).

154. R.K. Chang, J.H. Eickmans, W.-F. Hsieh, C.F. Wood, J.-Z. Zhang, and J.-B. Zheng, "Laser-Induced Breakdown in Large Transparent Water Droplets," *Appl. Opt.* **27**, 2377 (1988).

155. W.-F. Hsieh, J.-B. Zheng, and R.K. Chang, "Time Dependence of Multiorder Stimulated Raman Scattering from Single Droplets," *Opt. Lett.* **13**, 497 (1988).

156. R.K. Chang, D.H. Leach, and J.-Z. Zhang, "The Q-Factor of Micrometer-Size Droplets as Optical Cavities," in *Proceedings of the U.S.-Mexico Workshop on Electrodynamics of Interfaces and Composite Systems*, Taxco, Mexico, August 10–14, 1987, R.G. Barrera and W.L. Mochan, eds. (World Scientific, Singapore, 1988), p. 373.

157. W.P. Acker, B. Yip, D.H. Leach, and R.K. Chang, "The Use of a Charge-Coupled Device and Position Sensitive Resistive Anode Detector for Multiorder Spontaneous Raman Spectroscopy from Silicon," *J. Appl. Phys.* **64**, 2263 (1988).

158. J.-B. Zheng, W.-F. Hsieh, S.-C. Chen, and R.K. Chang, "Temporally and Spatially Resolved Spectroscopy of Laser-Induced Plasma from a Droplet," *Opt. Lett.* **13**, 559 (1988).

159. J.-Z. Zhang and R.K. Chang, "Shape Distortion of a Single Water Droplet by Laser-Induced Electrostriction," *Opt. Lett.* **13**, 916 (1988).

160. P.W. Barber and R.K. Chang, eds., Optical Effects Associated with Small Particles (World Scientific, Singapore, 1988).

161. J.-Z. Zhang and R.K. Chang, "Generation and Suppression of Stimulated Brillouin Scattering in Single Liquid Droplets," *J. Opt. Soc. Am.* **B 6**, 151 (1989).

162. W.P. Acker, D.H. Leach, and R.K. Chang, "Third-Order Optical Sum Frequency Generation in Micrometer-Sized Liquid Droplets," *Opt. Lett.* **14**, 402 (1989).

163. W.P. Acker, D.H. Leach, and R.K. Chang, "Stokes and Anti-Stokes Hyper-Raman Scattering from Benzene, Deuterated Benzene, and Carbon Tetrachloride," *Chem. Phys. Lett.* **155**, 491 (1989).

164. W.-F. Hsieh, J.-B. Zheng, and R.K. Chang, "Transmission through Plasma Created by Laser-Induced Breakdown of Water Droplets," *Opt. Lett.* **14**, 1014 (1989).

165. J.-B. Zheng, W.-F. Hsieh, S.-C. Chen, and R.K. Chang, "Growth, Decay, and Quenching of Stimulated Raman Scattering in Transparent Liquid Droplets," in *Laser Materials and Laser Spectroscopy*, Z. Wang and Z. Zhang, eds. (World Scientific, Singapore, 1989), p. 259.

166. J.-Z. Zhang, G. Chen, and R.K. Chang, "Pumping of Stimulated Raman Scattering by Stimulated Brillouin Scattering within a Single Droplet: Input Laser Linewidth Effects," *J. Opt. Soc. Am.* **B 7**, 108 (1990).

167. R.K. Chang, "Raman Spectroscopic Techniques in Interfacial Electrochemistry," in *Spectroscopic and Diffraction Techniques in Interfacial Electrochemistry*, C. Gutiérrez and C. Melendres, eds. (Kluwer Academic Publishers, Dordrecht, 1990), p. 155.

168. R.K. Chang and A.S. Kwok, "High Intensity Laser Beam Interactions with Single Droplets," in *Proceedings of the AGARD Conference on Atmospheric Propagation in the UV, Visible, IR and MM-Wave Region and Related Systems Aspects*, Copenhagen, October 9–13, 1989, AGARD CP-454 (NATO, Neuilly sur Seine, France, 1990), p. 20-1.

169. A.S. Kwok, C.F. Wood, and R.K. Chang, "Fluorescence Imaging of CO2 Laser-Heated Droplets," *Opt. Lett.* **15**, 664 (1990).

170. W.P. Acker, A. Serpengüzel, R.K. Chang, and S.C. Hill, "Stimulated Raman Scattering of Fuel Droplets: Chemical Concentration and Size Determination," *Appl. Phys.* **B 51**, 9 (1990).

171. D.H. Leach, W.P. Acker, and R.K. Chang, "Effect of the Phase Velocity and Spatial Overlap of Spherical Resonances on Sum-Frequency Generation in Droplets," *Opt. Lett.* **15**, 895 (1990).

172. R.K. Chang, "Some of Bloembergen's Nonlinear Optical Effects Revisited in Single Micrometer-Sized Droplets," in *Resonances*, M.D. Levenson, E. Mazur, P.S. Pershan, and Y.R. Shen, eds. (World Scientific, Singapore, 1990), p. 200.

173. G. Chen, W.P. Acker, R.K. Chang, and S.C. Hill, "Fine Structures in the Angular Distribution of Stimulated Raman Scattering from Single Droplets," *Opt. Lett.* **16**, 117 (1991).

174. J.-B. Zheng, W.-F. Hsieh, S.-C. Chen, and R.K. Chang, "Laser-Induced Breakout and Detonation Waves in Droplets: I. Experiments," *J. Opt. Soc. Am.* **B 8**, 319 (1991).

175. G. Chen, R.K. Chang, S.C. Hill, and P.W. Barber, "Frequency Splitting of Degenerate Spherical Cavity Mode: Stimulated Raman Scattering Spectrum of Deformed Droplets," *Opt. Lett.* **16**, 1269 (1991).

176. R.K. Chang, G. Chen, S.C. Hill, and P.W. Barber, "Nonlinear Optical Processes in

Droplets with Single-Mode Laser Excitation", in *Proceedings of the SPIE Conference on Nonlinear Optics and Materials*, Vol. 1497 (SPIE, Bellingham, Washington, 1991), p. 2.

177. A. Serpengüzel, G. Chen, R.K. Chang, "Stimulated Raman Scattering of Aqueous Droplets Containing Ions: Concentration and Size Determination," *Particulate Sc. and Tech.* **8**, 179 (1990).

178. A. Serpengüzel, J.C. Swindal, R.K. Chang, and W.P. Acker, "Two-Dimensional Imaging of Sprays with Fluorescence, Lasing, and Stimulated Raman Scattering," *Appl. Opt.* **31**, 3543 (1992).

179. A. Serpengüzel, G. Chen, R.K. Chang, and W.-F. Hsieh, "Heuristic Model for the Growth and Coupling of Nonlinear Processes in Droplets," *J. Opt. Soc. Am.* **B 9**, 871 (1992).

180. D.H. Leach, R.K. Chang, and W.P. Acker, "Stimulated Anti-Stokes Raman Scattering in Microdroplets," *Opt. Lett.* **17**, 387 (1992).

181. R.K. Chang and Ali Serpengüzel, "Characteristics and Applications of Stimulated Raman Scattering in Microdroplets," in *Thirteenth International Conference on Raman Spectroscopy*, 1992, Würzburg, Germany, W. Kiefer, M. Cardona, G. Schaak, F. W. Schneider, and H. W. Schrötter, eds. (John Wiley and Sons, 1992) p. 33.

182. David H. Leach, Richard K. Chang, William P. Acker, and Steven C. Hill, "Third Order Sum Frequency Generation in Droplets: Experimental Results," *J. Opt. Soc. Am.* **B 10**, 34 (1993).

183. Steven C. Hill, David H. Leach, and Richard K. Chang, "Third Order Sum Frequency Generation in Droplets: Model with Numerical Results for Third Harmonic Generation," *J. Opt. Soc. Am.* **B 10**, 16 (1993).

184. Frederick H. Wirth, Kimberly A. Juvan, David H. Leach, J. Christian Swindal, Richard K. Chang, and P. T. Leung, "Phonon-Retention Effects on Stimulated Brillouin Scattering from Micrometer Droplets Illuminated with Multiple-Short Laser Pulses," *Opt. Lett.* **17**, 1334 (1992).

185. Alfred S. Kwok and Richard K. Chang, "Fluorescence Seeding of Weaker-Gain Raman Modes in Microdroplets: Enhancement of Stimulated Raman Scattering," *Opt. Lett.* **17**, 1262 (1992).

186. J. Christian Swindal, David H. Leach, Richard K. Chang, and Kenneth Young, "Precession of Morphology-Dependent Resonances in Non-Spherical Droplets," *Opt. Lett.* **18**, 191 (1993).

187. A.S. Kwok, A. Serpengüzel, W.-F. Hsieh, R.K. Chang, and J.B. Gillespie, "Two-photon Pumped Lasing in Microdroplets," *Opt. Lett.* **17**, 1435 (1992).

188. Gang Chen, J. Christian Swindal, and Richard K. Chang, "Frequency Splitting and Precession of Cavity Modes of a Droplet Deformed by Inertial Forces," in *Proceedings of Shanghai International Symposium on Quantum Optics*, Vol. 1726 (SPIE, Bellingham, Washington, 1992) p. 292.

189. Gang Chen, Dipakbin Q. Chowdhury, Richard K. Chang, and Wen-Feng Hsieh, "Laser-Induced Radiation Leakage from Microdroplets," *J. Opt. Soc. Am.* **B 10**, 620 (1993).

190. Gang Chen, Ali Serpengüzel, Richard K. Chang, and William P. Acker, "Relative Evaporation Rates of Droplets in a Segmented Stream Determined by Cavity Droplet Fluorescence Peak Shifts," in *Proceedings of the SPIE Conference on Laser Applications in Combustion and Combustion Diagnostics*, Vol. 1862 (SPIE, Bellingham, Washington, 1993) p. 200.

191. Steven C. Hill and Richard K. Chang, "On Modeling Nonlinear Optical Mixing Processes in Droplets," in *Proceedings of the SPIE Conference on Laser Applications in Combustion and Combustion Diagnostics*, Vol. 1862 (SPIE, Bellingham, Washington, 1993) p. 309.

192. Ali Serpengüzel, Richard K. Chang, William P. Acker, and Rodney L. Sung, "Laser Diagnostic Techniques for Characterizing Droplet Size, Composition, and Differential Evaporation in Fuel Sprays," in *XXIV FISITA Congress*, 7–11 June, 1992 London, (Mechanical Engineering Publications Ltd, London, 1992). C389/417, paper number 925030, p. 107.

193. D. H. Leach, D. Q. Chowdhury, and R. K. Chang, "The Effect of Goos-Hanchen Shift on Geometrical Optics Model for Spherical Cavity Mode Spacing," *J. Opt. Soc. Am. A.* **11**, 1110 (1994).

194. Thomas J. Mach, Rodney L. Sung, Peter M. Liiva, William P. Acker, J. Christian Swindal, and Richard K. Chang, "Experimental Determination of Fuel Additive Effects on Leidenfrost Temperature and Deposit Formation," *SAE International Congress and Exposition*, Detroit, MI, March 1–5, 1993, SAE Technical paper number 930774.

195. G. Chen, Md. M. Mazumder, Y. R. Chemla, A. Serpengüzel, R. K. Chang, and S. C. Hill, "Wavelength Variation of Laser Emission Along the Entire Rim of Slightly Deformed Droplets," *Opt. Lett.* **18**, 1993 (1993).

196. Janice L. Cheung, Alfred S. Kwok, Kimberly A. Juvan, David H. Leach, and Richard K. Chang, "Stimulated Low-Frequency Emission from Anisotropic Molecules in Microdroplets," *Chem. Phys. Lett.* **213**, 309 (1993).

197. Alfred S. Kwok and Richard K. Chang, "Stimulated Resonance Raman Scattering of Rhodamine 6G," *Opt. Lett.* **18**, 1703 (1993).

198. Alfred S. Kwok and Richard K. Chang, "Suppression of Lasing by Stimulated Raman Scattering in Microdroplets," *Opt. Lett.* **18**, 1597 (1993).

199. Alfred S. Kwok and Richard K. Chang, "Detection of Minority Species in Microdroplets: Enhancement of Stimulated Raman Scattering," *Optics and Photonics News* **4**, 34 (1993).

200. J. Christian Swindal, G. Chen, A. Serpengüzel, R. K. Chang, and W. P. Acker, "Spray Diagnostics with Lasing and Stimulated Raman Scattering," *Progress in Astronautics*

and Aeronautics Vol. 171, "Recent Advances in Spray Combustion: Spray Combustion Measurements and Model Simulation", Vol. II edited by K. K. Kuo, (American Institute of Aeronautics and Astronautics, Washington, DC 1996) p 63.

201. Steven C. Hill and Richard K. Chang, "Nonlinear Optics in Droplets," in *Studies in Classical and Quantum Nonlinear Optics*, O. Keller, ed., Nova Science Publishers, New York (1995), p 171.

202. Gang Chen, Md. Mohiuddin Mazumder, J. Christian Swindal, Karl Schaschek, and Richard K. Chang, "Nonlinear Optical Emission and Scattering in Micrometer-sized Droplets," in *Proceedings of the AGARD Symposium on High Power Microwaves (HPM)*, Ottawa, Canada, May 2–5, 1994, AGARD CP-564 (NATO, Neuilly Sur Sein, France) 1995, p. 24-1.

203. Md. Mohiuddin Mazumder, Steven C. Hill, Dipakbin Q. Chowdhury, and Richard K. Chang, "Dispersive Optical Bistability in a Dielectric Sphere," *J. Opt. Soc. Am.* **B 12**, 297 (1995).

204. J. U. Nöckel, A. D. Stone and R. K. Chang, "Q-spoiling and Directionality in Deformed Ring Cavities," *Opt. Lett.* **19**, 1693 (1994).

205. Janice L. Cheung, Justin M. Hartings and Richard K. Chang, "Different Temporal Behavior for the Forward- and Backward-Circulating Radiation within a Microdroplet," *Opt. Lett.* **20**, 1089 (1995).

206. Md. Mohiuddin Mazumder, Gang Chen, Richard K. Chang, James B. Gillespie, "Wavelength Shifts of Dye Lasing in Microdroplets: Effect of Absorption Change," *Opt. Lett.* **20**, 287 (1995).

207. Md. Mohiuddin Mazumder, Karl Schaschek, Richard K. Chang and James B. Gillespie, "Efficient Pumping of Minority Species Simulated Raman Scattering (SRS) by Majority Species SRS in a Microdroplet of a Binary Mixture," *Chem. Phys. Lett.* **239**, 361 (1995).

208. Steven C. Hill, Ronald G. Pinnick, Paul Nachman, Gang Chen, Richard K. Chang, Michael W. Mayo, and Gilbert L. Fernandez, "Aerosol-Fluorescence Spectrum Analyzer: Real-time Measurement of Emission Spectra of Individual Airborne Particles," *Appl. Opt.* **34**, 7149 (1995).

209. Md. Mohiuddin Mazumder, Gang Chen, Peter J. Kindlmann, R. K. Chang, and James B. Gillespie, "Temperature-Dependent Wavelength Shifts of Dye Lasing in Microdroplets with a Thermochromic Additive," *Opt. Lett.* **20**, 1668 (1995).

210. Paul Nachman, Gang Chen, R. G. Pinnick, Steven C. Hill, Richard K. Chang, Michael W. Mayo, and Gilbert L. Fernandez, "Conditional-Sampling Spectrograph Detection System for Fluorescence Measurements of Individual Airborne Biological Particles," *Appl. Opt.* **35**, 1069 (1996).

211. A. Mekis, J. U. Nöckel, G. Chen, A. D. Stone and R. K. Chang, "Chaos and Q-spoiling in Lasing Droplets," *Phys. Rev. Lett.* **75**, 2682 (1995).

212. Richard K. Chang, Gang Chen and Md. Mohiuddin Mazumder, "Nonlinear Optics in Micro-Meter Sized Droplets," in *Proceedings of the NATO ASI Series E*, 314, Les Houches, France, 1995, Kluwer Academic Publishers, The Netherlands, 1996, pp. 75–99.

213. Gang Chen, Md. Mohiuddin Mazumder, Richard K. Chang, J. Christian Swindal and William P. Acker, "Laser Diagnostics for Droplet Characterization," *Progress in Energy & Combustion Science* **22**, 163 (1996).

214. J. L. Cheung, J. M. Hartings, and R. K. Chang, "Nonlinear Optics of Microdroplets Illuminated by Picosecond Laser Pulses," in *Handbook of Optical Properties*, Vol. II, Optics of Small Particles, Interfaces and Surfaces, CRC Press, p. 223 (1996).

215. Md. Mohiuddin Mazumder, Dipakbin Q. Chowdhury, Steven C. Hill, and Richard K. Chang, "Optical Resonances of a Spherical Dielectric Microcavity: Effects of Perturbations," in *Optical Processes in Microcavities*, edited by Richard K. Chang and Anthony J. Campillo, (World Scientific Publishers), p. 209 (1996).

216. J. Christian Swindal, Gang Chen, Karl Schaschek, and Richard K. Chang, and Thomas Jackson, "Measurement of the Evaporation Rates of Closely Spaced Flowing Droplets by Optical Cavity Resonances," *Journal of Atomization and Sprays* **6**, 331 (1996).

217. Md. Mohiuddin Mazumder, Mitchell H. Fields, Justin M. Hartings, Xiaoyun Pu, Alfred S. Kwok, Karl Schaschek and Richard K. Chang, "Enhancing Stimulated Raman Scattering of Weaker Gain Raman Modes in Microdroplets by Seeding and Efficient Pumping," in *Nonlinear Frequency Generation and Conversion*, Vol. 2700 (SPIE, Bellingham, Washington, 1996) p. 352.

218. Gang Chen, Paul Nachman, Ronald G. Pinnick, Steven C. Hill, and Richard K. Chang, "Conditional-firing Aerosol-Fluorescence Spectrum Analyzer for Individual Airborne Particles with Pulsed 266-nm Laser Excitation," *Opt. Lett.* **21**, 1307 (1996).

219. Jens U. Nöckel, A. Douglas Stone, Gang Chen, Helen L. Grossman and Richard K. Chang, "Directional Emission from Asymmetric Resonant Cavities," *Opt. Lett.* **21**, 1609 (1996).

220. Mitchell H. Fields, Jürgen Popp, and Richard K. Chang, "External Seeding of Stimulated Raman Scattering in Microdroplets," *Opt. Lett.* **21**, 1457 (1996).

221. Justin M. Hartings, Janice L. Cheung, and Richard K. Chang, "Temporal beating of nondegenerate azimuthal modes in nonspherical microdroplets: Technique to determine the distortion amplitude," *Appl. Opt.* **37**, (1998).

222. Jürgen Popp, Mitchell H. Fields, Richard K. Chang, "Injection Seeding of Lasing in Microdroplets," *Opt. Lett.* **22**, 139 (1997).

223. Justin Hartings, Xiaoyun Pu, Janice L. Cheung, Richard K. Chang, "Laser Induced Distortion for Increased Input Coupling of Light to Droplet-Cavity Modes," *JOSA* **B 14**, 1842 (1997).

224. Jürgen Popp, Mitchell H. Fields, and Richard K. Chang, "Q-Switching by Saturable Absorption in Microdroplets: Elastic Scattering and Laser Emission," *Opt. Lett.* **22**, 1296 (1997).

225. Justin Hartings, Andrew Poon, Xiaoyun Pu, Richard K. Chang and Thomas Leslie, "Second Harmonic Generation and Fluorescence Images from Surfactants on Hanging Droplets," *Chem. Phys. Lett.* **281**, 389 (1998).

226. R.K. Chang and Anthony J. Campillo, eds., Optical Processes in Microcavities (World Scientific Publishers, Singapore, 1996).

227. Ronald G. Pinnick, Steven C. Hill, Paul Nachman, Gorden Videen, Gang Chen and R.K. Chang "Aerosol Fluorescence Spectrum Analyzer for Measurement of Single Micron-sized Airborne Biological Particles," *Aerosol Science and Technology* **28**, 95 (1998).

228. Andrew W. Poon, James A. Lock and Richard K. Chang, "Spiral morphology-dependent resonances in an optical fiber: Effects of fiber tilt and focused Gaussian beam illumination," *Opt. Lett.* **23**, 1105 (1998).

229. Stephen Holler, Yongle Pan, Richard K. Chang, Jerold Bottiger, Steven C. Hill, David B. Hillis, "Two-dimensional Angular Light Scattering for the Characterization of Airborne Micro-particles," *Opt. Lett.* **23**, 1489 (1998).

230. Richard K. Chang, Mitchell H. Fields and Jürgen Popp, "Nonlinear Spectroscopy with Microspheres: Mechanisms and Applications" in *Proceedings of the XIII International Conference on Laser Spectroscopy*, Hangzhou, China, June 2–7, 1997, p. 339.

231. Yongle Pan, Stephen Holler, Richard K. Chang, Steven Hill, Ronald Pinnick, Stanley Niles, and Jerold Bottiger, "Single-Shot Fluorescence Spectra of Individual Micron-Sized Bio-Aerosols Illuminated by a 351 nm or 266 nm UV-Laser," *Opt. Lett.* **24**, 116 (1999).

232. Stephen Holler, Yongle Pan, Jerold Bottiger, Steven Hill, David B. Hillis and Richard K. Chang, "Two-dimensional Angular Scattering Measurements of Single Airborne Micro-Particles", in *Air Monitoring and Detection of Chemical and Biological Agents*, Joseph Leonelli, Mark L. Althouse, Editors, Proceedings of SPIE 3533, 64 (1999).

233. Nicholas Fell, Ronald Pinnick, Steven Hill, Gorden Videen, Stanley Niles, Richard K. Chang, Stephen Holler, Yongle Pan, Jerold Bottiger and Burt Bronk, "Concentration, Size, and Excitation Power Effects on Fluorescence from Microdroplets and Microparticles Containing Tryptophan and Bacteria", in *Air Monitoring and Detection of Chemical and Biological Agents*, Joseph Leonelli, Mark L. Althouse, Editors, Proceedings of SPIE 3533, 52 (1999).

234. Seongsik Chang and Richard K. Chang, "Optics Within a Deformed Cavity" *Proceedings of the Sixteenth International Conference on Raman Spectroscopy*, Cape Town, South Africa, September 6–11, (1998), A. M. Heynes, ed., John Wiley & Sons Publishers, p. 26.

235. Seongsik Chang, Nathan B. Rex and Richard K. Chang, "Chemical Lasing in Pendant Droplets: Lasing Spectra, Emission Pattern and Cavity Lifetime Measurements", *JOSA* **B 16**, 1224 (1999).

236. Seongsik Chang, Nathan B. Rex, Richard K. Chang, Gabel Chong and Louis B. Guido, "Stimulated emission and lasing in whispering-gallery modes of GaN microdisk cavities", *Appl. Phys. Lett.* **75**, 166 (1999).

237. Andrew W. Poon and Richard K. Chang "Simultaneous measurement of relative outer diameter along fiber axis: nm resolution in diameter and mm coverage in length" *Technical Digest Symposium on Optical Fiber Measurements*, sponsored by NIST, Boulder, CO Sept. 15–17, 1998.

238. Stephen Holler, Mario Surbek, Richard K. Chang and Yong-le Pan, "Two-dimensional Angular Optical Scattering Patterns as Droplets Evolve into Clusters", *Opt. Lett.* **24**, 1185 (1999).

239. Steven Hill, Ronald Pinnick, Yong-le Pan, Steve Holler, Richard K. Chang, Jerold Bottigter, Bean Chen, Chun-Sing Orr, and Greg Feather, "Real-time Measurement of Fluorescence Spectra from Single Airborne Biological Particles," *Field Analytical Chemistry and Technology* **3**, 221 (1999).

240. Mitchell Fields, Jürgen Popp, Richard K. Chang, "Nonlinear Optics in Microspheres", *Progress in Optics*, 41, edited by Emil Wolf (Elsevier Science Publishers), 1, (2000).

241. Yong-Le Pan, Stephen Holler, Richard K. Chang, Steven C. Hill, Ronald G. Pinnick, Stanley Niles, Jerold R. Bottiger, and Burt V. Bronk, "Real-time detection and characterization of individual flowing airborne biological particles: fluorescence spectra and elastic scattering measurements" in P*roceeding of SPIE Conference on Air Monitoring and Detection of Chemical and Biological Agents II*, Vol. 3855 (SPIE Bellingham, Washington), (2000), p. 117.

242. Stephen Holler, Mario Surbek, Yong-le Pan, Richard K. Chang, "Two-dimensional angular optical scattering (TAOS) measurements of single airborne microparticles: Transition to clusters and dual wavelength absorption studies" in *Proceedings of SPIE Conference on Environmental Monitoring and Remediation Technologies II*, Vol. 3853 (SPIE Bellingham, Washington), (1999), p. 132.

243. Nathan B. Rex, Richard K. Chang, Louis J. Guido, "Threshold Minimization and Directional Laser Emission from GaN Microdisks" *Proceedings of SPIE/Laser Resonators III, Conference*, Vol. 3930 (SPIE Bellingham, Washington, (2000), p.163.

244. Yong-Le Pan, Ronald G. Pinnick, Steven C. Hill, Stanley Niles, Stephen Holler, Jerold R. Bottiger, Jean-Pierre Wolf, and Richard K. Chang, "Dynamics of Photon-induced Degradation and Fluorescence in Riboflavin Microparticles," *Appl. Phys.* **B 72**, 449 (2001).

245. S. Chang, J. Nöckel, R.K. Chang, A. D. Stone, "Observation of Emission from Chaotic Lasing Modes in Deformed Microspheres: displacement by the stable orbit modes," *JOSA* **B 17**, 1 (2000).

246. Steven C. Hill, Yong-le Pan, Stephen Holler, Richard K. Chang, Veronique Boutou, Jin Yu, Stephane Ramstein, and Jean-Pierre Wolf, "Enhanced Backward-Directed Multi-Photon-Excited Fluorescence from Dielectric Microcavities," *Phys. Rev. Lett.* **85**, 54 (2000).

247. Nathan Rex, Seongsik Chang, Richard K. Chang, Gabel Chong, Louis Guido "Threshold Lowering in GaN Lasers by Means of Spatially Selective Optical Pumping," *IEEE Photonics Technology Letters* **13**, 1 (2001).

248. Stephen Holler, Jean-Claude Auger, Brian Stout, Yongle Pan, Jerold R. Bottiger, Richard K. Chang, Gorden Videen, "Observations and Calculations of Light Scattering from Clusters of Spheres", *Appl. Opt.* **39**, 6873 (2000).

249. Andrew W. Poon, Steven L. Kan, Paul A. Tick, Daniel A. Nolan, Richard K. Chang, "Preferential input and output coupling of oval-shaped optical microcavities: Fluorescent emission from oval-shaped dye-doped glass fibers," *Proceedings of SPIE Photonics Taiwan 2000 Conference*, Taipei, Taiwan (July, 2000).

250. Jean-Pierre Wolf, Yong-le Pan, Gordon M. Turner, Matthew C. Beard, Charles A. Schmuttenmaer, Stephen Holler, Richard K. Chang, "Ballistic Trajectories of Optical Wavepackets within Microcavities", *Phys. Rev.* **A 64**, 023808-1 (2000).

251. Yong-le Pan, Patrick Cobler, Alexander Potter, Tim Chou, Richard K. Chang, Ronald Pinnick, Steven Hill, Jean-Pierre Wolf, "High-speed, High-sensitivity Aerosol Fluorescence Spectrum Detection Using a 32-anode PMT Detector", *Review of Scientific Instruments* **72**, 1831 (2001).

252. A. Poon, F. Courvoisier, R.K. Chang, "Multi-mode resonances in square-shaped optical micro-cavities", *Opt. Lett.* **16**, 632 (2001).

253. Steven C. Hill, Ronald G. Pinnick, Stanley Niles, Nicholas F. Fell, Jr., Yongle Pan, Jerold Bottiger, Stephen Holler, and R.K. Chang, "Fluorescence from airborne microparticles: dependence on size, concentration of fluorophores and illumination intensity," *Appl. Opt.* **40**, 3005 (2001).

254. Andrew W. Poon, Richard K. Chang and Dipak Q. Chowdhury, "Measurement of fiber-cladding diameter uniformity by use of whispering-gallery modes: nanometer resolution in diameter variations along millimeter - centimeter length," *Opt. Lett.* **26**, 1867 (2001).

255. Andrew W. Poon, Francois Courvoisier and Richard K. Chang, "Evanescent and refractive side-coupled high-Q resonances of non-circular 2-D micropillars", *Proceedings of Laser Resonators, SPIE/High-Power Lasers and Applications Symposium*, 4270, San Jose, CA. Jan. 20–26, 2001, p131.

256. Yong-le Pan and Richard K. Chang, "A Multi-Anode Photomultiplier: Many Detectors in One," *Laser Focus World*, May 2001, p. 235.

257. N. Rex, H. Türeci, H. Schwefel, R.K. Chang and A. Douglas Stone, "Fresnel filtering in lasing emission from scarred modes of wave-chaotic optical resonators," *Phys. Rev. Lett.* **88**, 094102 (2002).

258. Yong-Le Pan, Steven C. Hill, Jean-Pierre Wolf, Stephen Holler, Richard K. Chang and Jerold Bottiger, "Backward-Enhanced Fluorescence from Clusters of Microspheres and Particles of Tryptophan," *Appl. Opt.* **41**, 2994 (2002).

259. Catherine Favre, Veronique Boutou, Steven C. Hill, Wiebke Zimmer, Marcel Krenz, Hendrik Lambrecht, Jin Yu, Richard K. Chang, Ludger Woeste, Jean-Pierre Wolf, "White-light Nanosource with Directional Emission," *Phys. Rev. Lett.* **89**, 035002-1 (2002).

260. Chern, G.D., A.W. Poon, and R.K. Chang. "Directional Laser Emission from Square, Spiral and Mismatched Semi-Circular Dye-Doped Polymer Based Micropillar Cavities," in *Summaries of Papers Presented at the Quantum Electronics and Laser Science Conference*, 25–26. Long Beach, CA, USA: Opt. Soc. America, 2002.

261. Yong-Le Pan, Justin Hartings, Ronald G. Pinnick, Steven C. Hill, Justin Halverson, and Richard K. Chang, "Single-Particle Fluorescence Spectrometer for Ambient Aerosols," *Aerosol and Science and Technology* **37**, 628 (2003).

262. Yong-Le Pan and Richard K. Chang, "Highly Efficient Prism Coupling to Whispering Gallery Modes of a Square-Cavity," *Appl. Phys. Lett.* **82**, 487 (2003).

263. Veronique Boutou, Catherine Favre, Steve Hill, Yong-Le Pan, Richard K, Chang, and Jean-Pierre Wolf, "Backward Enhanced Emission from Multiphoton Processes in Aerosols," *Appl. Phys.* **B 75**, 145 (2002).

264. Jens U. Nöckel and Richard K. Chang, "2D Microcavities: Theory and Experiments," *Experimental Methods in the Physical Sciences* **40**, edited by Roger D. van Zee and Patrick Looney, Elsevier Science B.V. 2002, p 185.

265. Harald Schwefel, Nathan B. Rex, Hakan E. Türeci, Richard K. Chang, A. Douglas Stone, Tahar Ben-Messaoud, and Joseph Zyss, "Dramatic Shape Sensitivity of Directional Emission Patterns from Similarly Deformed Cylindrical Polymer Lasers," *JOSA* **B**, 923 (2004).

266. Yong-Le Pan, Kevin B. Aptowicz, Matt Hart, Jay D. Eversole, and Richard K. Chang, "Characterizing and Monitoring Respiratory Aerosols by Light Scattering," *Opt. Lett.* **28**, 589 (2003).

267. Ali Serpengüzel, Serkan Küçükşenel and Richard K. Chang, "Microdroplet Identification and Size Measurement in Sprays with Lasing Images," *Opt. Express* **10**, 1118 (2002).

268. S. Holler, S. Zomer, G. Crosta, Y. Pan, R.K. Chang, and J. Bottiger, "Multivariate analysis of two-dimensional angular optical scattering (TAOS) patterns from aggregates," *Appl. Opt.* **43**, 6198 (2004).

269. G. D. Chern, H. E. Türeci, A. Douglas Stone, R.K. Chang, M. Kneissl and N. M. Johnson "Uni-Directional Lasing from InGaN Multiple Quantum-Well Spiral-Shaped Micropillars," *Appl. Phys. Lett.* **83**, 1710 (2003).

270. Yong-Le Pan, Veronique Boutou, Richard K. Chang, İlker Özden, Kristina Davitt and Arto Nurmikko, "Application of Light Emitting Diodes for Bioaerosol Fluorescence Detection," *Opt. Lett.* **28**, 1707 (2003).

271. Ronald G. Pinnick, Steven C. Hill, Yong-Le Pan, and Richard K. Chang, "Fluorescence Spectra of Atmospheric Aerosol at Adelphi, Maryland, USA: Measurement and Classification of Single-Particles Containing Organic Carbon," *Atmospheric Environment* **38**, 1657 (2004).

272. G. D. Chern, A. W. Poon, R.K. Chang, T. Ben-Messaoud, O. Alloschery, E. Toussaere, J. Zyss, and S.-Y. Kou, "Direct Evidence of Open Ray Orbits in a Square 2-d Resonator of Dye-Doped Polymers," *Opt. Lett.* **29**, 1674 (2004).

273. Yong-Le Pan, Veronique Boutou, Jerold R. Bottiger, Samuel S. Zhang, Jean-Pierre Wolf, and Richard K. Chang, "A Puff of Air Sorts Bioaerosols for Pathogen Identification," *Aerosol Science and Technology* **38**, 598 (2004).

274. Francois Courvoisier, Veronique Boutou, Jean-Pierre Wolf, Richard K. Chang, and Joseph Zyss, "Deciphering Output Coupling Mechanism in Spiral Microcavities with Femtosecond Light Bullets," *Opt. Lett.* **30**, 1 (2005).

275. M. Kneissl, M. Teepe, N. Miyashia, N.M. Johnson, G.D. Chern, and R.K. Chang, "Current-Injection Spiral-Shaped Microcavity Disk Laser Diodes with Unidirectional Emission," *Appl. Phys. Lett.* **84**, 2485 (2004).

276. H.G.L. Schwefel, H.E. Türeci, A. Douglas Stone and R.K. Chang, "Chapter 9: Progress in Asymmetric Resonant Cavities: Using Shape as A Design Parameter in Dielectric Microcavity Lasers," *Optical Microcavities*, Edited by Kerry Vahala, Advanced Series in Applied Physics 5, World Scientific Publishing 2004, p. 415.

277. Kevin B. Aptowicz, Yong-Le Pan, Richard K. Chang, Ronald G. Pinnick, Steven C. Hill, Richard L. Tober, Anish Goyal, Thomas Jeys, and Burt Bronk, "Two-Dimensional Angular Optical Scattering Patterns of Microdroplets in the Mid Infrared with Strong and Weak Absorption," *Opt. Lett.* **29**, 1965 (2004).

278. Kevin B. Aptowicz and Richard K. Chang, "Angularly-Resolved Elastic Scatter from Single Particles Collected Over A Large Solid Angle and with High Resolution," *Journal of Physics: Conference Series* **6**, 90–96 (2005).

279. Kristina Davitt, Yoon-Kyu Song, William M. Patterson III, Arto V. Nurmikko, Maria Gherasimova, Jung Han, Yong-Le Pan and Richard K. Chang, "290 and 340 nm UV LED Arrays for Single Particle Fluorescence Detection," *Optics Express* **13**, No. 23, 9548–9555 (2005).

280. Y.L. Pan, P.J. Cobler, S.A. Rhodes, J. Halverson, and R.K. Chang, "Separating hazardous aerosols from ambient aerosols: role of fluorescence-spectral determination, aerodynamic deflector and pulse aerodynamic localizer (PAL)," *Proc. of SPIE* Vol. 5990 59900Y (2005).

281. Matt Frain, David P. Schmidt, Yong-Le Pan, and Richard K. Chang, "Selective Deflection and Localization of Flowing Aerosols Onto A Substrate," *J. of Aerosol Science and Technology* **40**: 218–225 (2006).

282. Gustavo E. Fernandes, Yong-Le Pan, Richard K. Chang, Kevin B. Aptowicz, and Ronald G. Pinnick, "Simultaneous Forward- and Backward-Hemisphere Elastic-Light-Scattering patterns of Respirable-Size Aerosols," *Opt. Lett.* **31**, 1–3 (2006).

283. Richard K. Chang, Gustavo E. Fernandes, and Michael Kneissl, "The Quest for Uni-Directionality with WGMs in μ-Lasers: Coupled Oscillators and Amplifiers," *8th International Conference on Transparent Optical Networks ICTON 2006*, Nottingham, UK, June 18–22, 2006, Paper Mo. D1.1, Conference Proceedings 1, pp. 47–51.

<u>Yale University publications on light scattering from aerosols</u>

284. K.B. Aptowicz, R.G. Pinnick, S.C. Hill, Y.L. Pan, and R.K. Chang, "Optical scattering patterns from single urban aerosol particles at Adelphi, Maryland, USA: A classification relating to particle morphologies," *Journal of Geophysical Research* **III**, D12212 (2006).

285. O.I. Sindoni, R. Saija, M.A. Iati, F. Borghese, P. Denti, G.E. Fernandes, Y.L. Pan, and R.K. Chang, "Optical scattering by biological aerosols: experimental and computational results on spore simulants," *Opt. Express* **14**, No. 15, 6942 (2006).

286. Gustavo E. Fernandes, Grace D. Chern, Qinghai Song, Lei Xu, Michael Kneissl, Noble M. Johnson, and Richard K. Chang. "Varying the Overlap of Direct-Coupling between Spiral and Semicircle Semiconductor Microdisk Lasers," in *2007 9th International Conference on Transparent Optical Networks*, 212–15. Rome, Italy: IEEE, 2007.

287. Yong-Le Pan, Jay D. Eversole, Paul H. Kaye, Virginia Foot, Ronald G. Pinnick, Steven C. Hill, Michael W. Mayo, Jerold Bottiger, Alan Huston, Vasanthi Sivaprakasam, Richard K. Chang, "Bio-Aerosol Fluorescence: Detecting and characterizing bio-aerosols via UV light-induced fluorescence spectroscopy," *Optics of Biological Particles*, edited by A. Hoekstra *et al.*, Springer 2007, pp. 63–164. [Book chapter]

288. Kristina Davitt, Yoon-Kyu Song, William R. Patterson III, Arto V. Nurmikko, Yong-Le Pan, Richard K. Chang, Jung Han, Maria Gherasimova, Patrick J. Cobler, Paul D. Butler, and Vincent Palermo, "Spectroscopic Sorting of Aerosols by a Compact Sensor Employing UV LEDs," *Aerosol Science and Technology*, Volume: 40 Issue: 12 Pages: 1047–1051 (DEC 2006).

289. G.D. Chern, G.E. Fernandes, and R.K. Chang, Q. Song and L. Xu, M. Kneissl and N.M. Johnson, "High-Q-preserving coupling between a spiral and a semicircle mu-cavity," *Opt. Lett.*, Volume: 32 Issue: 9 Pages: 1093–1095 (May 1, 2007).

290. Jean-Claude Auger, Kevin B. Aptowicz, Ronald G. Pinnick, Yong-Le Pan, and Richard K. Chang, " Angularly resolved light scattering from aerosolized spores: observations and calculations," *Opt. Lett.*, Volume: 32 Issue: 22 Pages: 3358–3360 (Nov 15, 2007).

291. Yong-Le Pan, Ronald G. Pinnick, Steven C. Hill, James M. Rosen, and Richard K. Chang, "Single-particle laser-induced-fluorescence spectra of biological and other organic-carbon aerosols in the atmosphere: Measurements at New Haven Connecticut, and Las Cruces, New Mexico," *Journal Of Geophysical Research-Atmospheres*, Volume: 112 Issue: D24 Article Number: D24S19 (Dec 28, 2007).

292. Fernandes, Gustavo E., Laurent Guyot, Grace D. Chern, Michael Kneissl, Noble M. Johnson, Qinghai Song, Lei Xu, and Richard K. Chang. "Wavelength and Intensity Switching in Directly Coupled Semiconductor Microdisk Lasers." *Opt. Lett.* **33**, no. 6 (March 15, 2008): 605.

293. H. Huang, Y. L. Pan, S. C. Hill, R. K. Chang, "Deposition of Selected Airborne Particles into a Microfluidic Flow Cytometer for Bioanalysis". ARL Technical Report (ARL-TR-4446, 2008).

294. Y. L. Pan, R. K. Chang, S. C. Hill, and R. G. Pinnick, "Using single-particle fluorescence to detect bioaerosols," *Optics & Photonics News* **19**(9), 29–33 (2008).

295. Y. L. Pan, R. G. Pinnick, S. C. Hill, H. Huang, R. K. Chang, "Dual-wavelength-excitation single-particle fluorescence spectrometer/particle sorter for real-time measurement of organic carbon and biological aerosols," *Optically Based Biological and Chemical Detection for Defense IV, Proc. SPIE*, Vol. 7116, 71160J-1-8 (2008).

296. Richard K. Chang and Yong-Le Pan, "Linear and non-linear spectroscopy of microparticles: Basic principles, new techniques and promising applications," *Conference on Spectroscopy and Dynamics of Microparticles*, University of Bristol, Bristol, England, UK, July 02–04, 2007, Faraday Discussions, Volume: 137 Pages: 9–36 (2008).

297. Gustavo E. Fernandes, Laurent Guyot, Grace D. Chern, Michael Kneissl, Noble M. Johnson, Qinghai Song, Lei Xu, and Richard K. Chang, "Wavelength and intensity switching in directly coupled semiconductor microdisk lasers," *Opt. Lett.*, Volume: 33 Issue: 6 Pages: 605–607 (Mar 15, 2008).

298. J. Kasparian, V. Boutou, J.-P. Wolf, Y.-L. Pan, R.K. Chang, "Angular distribution of non-linear optical emission from spheroidal microparticles," *Applied Physics B-Lasers and Optics*, Volume: 91 Issue: 1 Pages: 167–171 (Apr 2008).

299. Hermes C. Huang, Yong-Le Pan, Steven C. Hill, Ronald G. Pinnick, and Richard K. Chang, "Real-time measurement of dual-wavelength laser-induced fluorescence spectra of individual aerosol particles," *Optics Express*, Volume: 16 Issue: 21 Pages: 16523–16528 (Oct 13, 2008).

300. Hill, Steven C., Michael W. Mayo, and Richard K. Chang. "Fluorescence of Bacteria, Pollens, and Naturally Occurring Airborne Particles: Excitation/Emission Spectra:" ARL Technical Report, (ARL-TR-4722, February 2009).

301. Yong-Le Pan, Ronald G. Pinnick, Steven C. Hill, and Richard K. Chang, "Particle-Fluorescence Spectrometer for Real-Time Single-Particle Measurements of Atmospheric Organic Carbon and Biological Aerosol," *Environmental Science & Technology*, Volume: 43 Issue: 2 Pages: 429–434 (Jan 15, 2009).

302. Y. L. Pan, J. Bowersett, S. C. Hill, R. G. Pinnick, R. K. Chang, "The development of an aerodynamic focusing nozzle for bioaerosol detection", ARL Technical Report (ARL-TR-5026, Oct, 2009).

303. R. G. Pinnick, Y. L. Pan, S. C. Hill, K. Aptowicz, and R. K. Chang, "Laser-induced fluorescence spectra and angular elastic scattering patterns of single atmospheric aerosol particles," in *Fundamentals and Applications in Aerosol Spectroscopy*, eds., R. Signorell and J.P. Reid (Taylor-Francis, December, 2010), pp. 297–320.

304. J.-C. Auger, G.E. Fernandes, K.B. Aptowicz, Yong-Le Pan, R.K. Chang, "Influence of surface roughness on the elastic-light scattering patterns of micron-sized aerosol particles," *Applied Physics B-Lasers and Optics*, Volume: 99 Issue: 1-2 Pages: 229–234 (Apr 2010).

305. Yong-Le Pan, Steven C. Hill, Ronald G. Pinnick, Hermes Huang, Jerold R. Bottiger, Richard K. Chang, "Fluorescence spectra of atmospheric aerosol particles measured using one or two excitation wavelengths: Comparison of classification schemes employing different emission and scattering results," *Optics Express*, Volume: 18 Issue: 12 Pages: 12436–12457 (Jun 7, 2010).

306. R. G. Pinnick, Y. L. Pan, S. C. Hill, K. Aptowicz, and R. K. Chang, "Laser-induced fluorescence spectra and angular elastic scattering patterns of single atmospheric aerosol particles," in *Fundamentals and Applications in Aerosol Spectroscopy*, eds., R. Signorell and J.P. Reid (TaylorFrancis, December, 2010), pp. 297–320.

307. Y. L. Pan, S. C. Hill, R. G. Pinnick, J. M. House, R. C. Flagan, R. K. Chang, "Dual-excitation-wavelength fluorescence spectra and elastic scattering for differentiation of single airborne pollen and fungal particles", *Atmospheric Environments* **45**(8), 1555–1563 (2011).

308. Yong-Le Pan, Matthew J. Berg, Samuel Shao-Ming Zhang, Heeso Noh, Hui Cao, Richard K. Chang, Gorden Videen, "Measurement and Autocorrelation Analysis of Two-Dimensional Light-Scattering Patterns from Living Cells for Label-Free Classification," *Cytometry*, Part A Volume: 79A Issue: 4 Pages: 284–292 (Apr 2011).

309. G. F. Crosta, Y. L. Pan, R. K. Chang, "Automated classification of single airborne particles from two-dimension, angle-resolved optical scattering (TAOS) patterns": *Sensing Technologies for Global Health, Military Medicine, Disaster Response, and Environmental Monitoring and biometric Technology for Human Identification VIII, SPIE Proc.* 8029, 80290Y (2011).

310. Yong-Le Pan, Hermes Huang, Richard K. Chang, "Clustered and integrated fluorescence spectra from single atmospheric aerosol particles excited by a 263-and 351-nm laser at

New Haven, CT, and Adelphi, MD," *Journal of Quantitative Spectroscopy & Radiative Transfer*, Volume: 113 Issue: 17 Pages: 2213–2221 (Nov 2012).

311. Giovanni Franco Crosta, Yong-Le Pan, Kevin B. Aptowicz, Caterina Casati, Ronald G. Pinnick, Richard K. Chang, Gorden W. Videen, "Automated classification of single airborne particles from two-dimensional angle-resolved optical scattering (TAOS) patterns by non-linear filtering", *Journal of Quantitative Spectroscopy & Radiative Transfer*, Volume: 131 Special Issue: SI Pages: 215–233 (Dec 2013).

312. K.B. Aptowicz, Yong-Le Pan, Sean D. Martin, Elena Fernandez, Richard K. Chang, R.G. Pinnick, "Decomposition of atmospheric aerosol phase function by particle size and asphericity from measurements of single particle optical scattering patterns," *Journal of Quantitative Spectroscopy & Radiative Transfer*, Volume: 131 Special Issue: SI Pages: 13–23 (Dec 2013).

313. G. F. Crosta, Y. L. Pan, G. Videen, K. B. Aptowicz, R. K. Chang, "Discrimination of airborne material particles from light scattering (TAOS) patterns", *Proc. SPIE 8723, Sensing Technologies for Global Health, Military Medicine, and Environmental Monitoring III*, 872318, May 29, 2013 (Invited Paper).

Scientists Working with Richard at Yale University from 1966 to 2008

RK Chang Lab

	PhD Students	Post-Doctoral Fellows	Visiting Professors/ Research Scientists	MS Students
1960's				
	James Ralston '70			Lee Galbraith
	Bernell Argyle '71			
	Joan L. Lewis '71			
	Fred Parsons '71			
	Ronald Wadsack '71			
1970's				
	Daniel Fouche '73	Matthias Selders		Edward Yi Chen
	Jin-Joo Song '74	Stanley Sussman		Bruce Weber
	Paul Klein '75	Hiro Masui		James Moore
	Philip Black '77	Richard Stafford		Gary Ready
	El-Hang Lee (Fenn) '78	Rolf Dornhaus		
	Roger Farrow '79	Takao Kobayashi		
	Marshall Long '80			
	Daniel Murphy '81			
1980's				
	Ulrich von Raben '82	Mohid Bansai	Peter Barber	Kimberly Joven
	Huey-Ming Tzeng (Long) '85	Robert Benner	Paul Dorain	
	Kevin Wall '86	James Owen	Tom T. Chen	
	Wen-Feng Hsieh '88	Johannes Eickmans	Kwan Ching Lee	
	William Acker '89	Burghard Schlicht	Jia-Biao Zheng	
	David Leach '90	Carol Wood	Alfred Leipertz	
	Jian-Zhi Zhang '90	Judith Snow	Yu-Fen Li	
			Shi-Xiong Qian	
			Maria Temperini	
			Andreas Otto	
			Shu-Chi Chen	
			Frederik Wirth	

	<u>PhD Students</u>	<u>Post-Doctoral Fellows</u>	<u>Visiting Professors/ Research Scientists</u>	<u>MS Students</u>
<u>1990's</u>				
	Ali Serpengüzel '92	William Acker	Gershon Kurizki	Thomas Heldt
	Alfred Kwok '93	Robert Benner	Kenneth Young	Kang-Bin Chua
	Paul Gang Chen '94	Dipakbin Chowdhury	Xiaoyun Pu	
	J Christian Swindal '94	Karl Schaschek	Yongle Pan	
	Janice Cheung '95	Jurgen Popp	Hiroshi Taniguchi	
	Mohiuddin Mazumder '96		Jean-Pierre Wolf	
	Justin Hartings '97			
	Mitchell Fields '97			
	Seongsik Chang '98			
	Steven Holler '99			
<u>2000's</u>				
	Andrew Wing On Poon '01		Yongle Pan	<u>Senior Projects:</u>
	Nathan B. Rex '01		Veronique Boutou	Yam R. Chemla
	Kevin Aptowicz '05			Gina LaRossa
	Grace Chern Metcalfe '05		<u>Exchange Students:</u>	Jeff Moses
	Gustavo Fernandes '08		Manfred Rothdach	Alex Potter
	Hermes Huang '09		Mario Surbek	Steven Kan
			François Courvoisier	Vanessa Wood
			Qinghai Song	

Memorial Article

Journal of Quantitative Spectroscopy and Radiative Transfer

Journal of Quantitative Spectroscopy & Radiative Transfer 255 (2020) 107273

Contents lists available at ScienceDirect

Journal of Quantitative Spectroscopy & Radiative Transfer

ELSEVIER

journal homepage: www.elsevier.com/locate/jqsrt

Richard K. Chang: In memoriam

Yong-Le Pan[a,*], Gorden Videen[a], Pang-Mei Natasha Chang[b], Sung-Wen Pu Chang[c]

[a] CCDC-Army Research Laboratory, 2800 Powder Mill Road, Adelphi, MD 20783, USA
[b] 69 Fifth Ave., New York City, NY 10003, USA
[c] 1084 Bird Bay Way, Venice, FL 34285, USA

ARTICLE INFO

Article history:
Received 14 August 2020
Accepted 20 August 2020
Available online 21 August 2020

ABSTRACT

We sketch the life and scientific legacy of Prof. Richard K. (Kounai) Chang (张国鼐), 1940–2020, the Henry Ford II Professor of Applied Physics, Professor of Physics and Electrical Engineering at Yale University. We briefly outline his research from linear and non-linear optics; light interactions with solids and liquids in various surface, micro-shapes and nano-structures; optical effects and processes in microresonators; and later to laser spectroscopic characterization of sprays, combustion, and environmental pollutants. Prof. Chang has many colleagues, collaborators, students, and friends among the JQSRT community.

Published by Elsevier Ltd.

1. Introduction

Richard Kounai Chang (张国鼐) was born to a large prominent Shanghainese family in 1940 in Hong Kong. The first character in his Chinese name, 国, is his generation name, which derives from a couplet composed by his great-grandfather for the descendants of the Chang family – "fine kingdom, bright sovereignty" (嘉国邦明) – with each generation taking a character in turn and repeating after four generations. 鼐, the second character in Richard's Chinese name means the highest level incense burner used by the emperor, and is a homonym for "patience," to reflect that Richard was born in a China at war, first against the Japanese, then in a civil war between the Nationalists and Communists.

Because of the country's unrest, Richard attended elementary schools in Chungking (重庆, the wartime capital), Shanghai, Hong Kong and Tokyo, where the family emigrated to in 1949 as target of the rising communists in China. When the family moved again to Sao Paulo, Brazil, Richard began at a British high school, then moved on his own to the United States in 1956 to complete high school at the Stony Brook School, Long Island, New York. He obtained his Bachelor of Science (B.S.) degree from Massachusetts Institute of Technology (MIT) in 1961, his Master of Arts (M.A.) and Doctor of Philosophy degree (Ph.D.) from Harvard University in 1962 and 1965, respectively. He did one year of post-doctoral work at Harvard (1965–1966).

While an undergraduate, Richard met Sung-Wen Pu ("Sunny") through family friends. Sunny studied Biology at Brandeis University, and later pursued her Ph.D. there. The couple married in 1961, and had three children, two born in Boston, and the youngest born in New Haven, Connecticut when the family moved there in July, 1966. Richard began his academic career at Yale University as an Assistant Professor in Engineering and Applied Science. He retired from Yale in 2008 as the Henry Ford II Professor of Applied Physics and Professor of Physics and Electrical Engineering, having contributed 42 years of his life to Yale. Fig. 1 showed one of his portraits taken in 2008.

At Yale, Prof. Chang taught and trained numerous students at the very highest standards. Generations of undergraduate students, graduate students, young colleagues, and collaborators around the world have benefited from his wisdom, advice, and mentoring. He is a model of generosity, integrity and professionalism, and an admired and beloved true gentleman. Most will remember his calm, friendly, patient and polite demeanor. His presence has impacted his former students, many of whom are leading teams at their current universities, institutes, companies, or other organizations, many of whom continued collaborating with Richard for many years after leaving Yale.

In this paper, we will sketch Prof. Chang's scientific activities, which mainly focused on his contributions to our broadly and deeply fundamental understanding of various linear and nonlinear optical phenomena, and to the development of new laser spectroscopic technologies. Richard was one of the pioneers in the study of morphology dependent resonances (MDRs), or whispering-gallery modes (WGM). These studies led to research on laser emission from microdroplets and solid microspheres that has expanded with numerous applications. In a seemingly unrelated field, he made remarkable contributions in the field of second-harmonic generation, surface-enhanced Raman scattering, and the development of detection and characterization of airborne aerosol

* Corresponding author.
E-mail address: yongle.pan.civ@mail.mil (Y.-L. Pan).

https://doi.org/10.1016/j.jqsrt.2020.107273
0022-4073/Published by Elsevier Ltd.

(used with permission)

Fig. 1. Richard K. Chang, Henry Ford II Professor of Applied Physics, Professor of Electrical Engineering and Physics, when he retired from Yale University in 2008.

Fig. 2. Prof. Chang in 2012 with his wife Sunny, Gorden Videen and Yong-Le Pan (right to left) in front of Yale Mory's, where Prof. Chang treated many guests and celebrated the graduations of his students. It is decorated with photos of Yale alumni like George W. Bush, Bill Clinton.

particles. His distinguishable work has been demonstrated with the more than 300 peer-reviewed journal articles, 7 books and book chapters, 11 awarded US patents, and about 13,000 peer citations. The long list of co-authors from Americas, Asia, and Europe reflects the influence and wide collaborations he has had (Fig. 2).

2. Scientific career and activities

Richard Chang started his research career at Harvard University with his advisor, the Nobel Laureate Nicolaas Bloembergen, who was recognized for his work in developing the driving principles behind nonlinear optics for laser spectroscopy. In 1966, Prof. Chang was hired at Yale University, where his research group eventually occupied more than 20 research labs and offices of the Becton Center. Long after his retirement, you can see the sticker RKC on lab equipment and components throughout the center where he trained numerous undergraduate students, graduate students, post-doctoral researchers, and visiting professors and scientists. Prof. Chang spent more than 40 years with his students and colleagues there except for a few monthly sabbatical visits to Europe and Asian. In the below sections we will outline his key scientific research.

2.1. Surface second-harmonic generation and nonlinear spectroscopy

The laser and semiconductor were milestone scientific discoveries in the 1960s. Richard Chang was in the thick of things, performing pioneering research in second-harmonic generation (SHG) and nonlinear spectroscopic measurements from semiconductors and metals, using and expanding upon the new technologies [1,2]. The very first experimental method and apparatus used to detect the SHG from surfaces showed how SHG depends on the laser illuminating polarization and incident angle. The derived theoretical expression described how SHG from a surface relies on laser intensity, polarity, field distribution, as well as material symmetry and susceptibility [1]. The nonlinear susceptibility and frequency-dependent dispersion measurements from a wide range of materials such as GaAs, InAs, InSb, and ZnTe supplied systemic understanding for SHG, third harmonic generation (THG), and other linear or non-linear phenomena [2,3]. The observations of nonlinear processes and the corresponding spectroscopy were extended to microparticle systems, especially in droplets [4–6]. When YP cleaned Prof. Chang's labs during 2007–2008, he found hundreds of different semiconductor samples used for linear and nonlinear measurements in the drawers in room BC014A, as well as several Berglund-Liu vibrating-orifice-aerosol generators (VOAG) used for producing droplets in room BC016A-B, BC017, and BC018. This initial research established the foundation for obtaining SHG and nonlinear spectroscopy from the boundary and surface between solid-air, liquid-air, solid-liquid, and even micron- and nano-sized particles.

2.2. Optical effects and process in microcavities and small particles

Prof. Chang helped lay the foundations in the exploration and use of morphology dependent resonances (MDR), or whispering-gallery modes (WGM), associated with microcavities or microresonators. YP remembers the story about how Richard observed and interpreted this phenomena for the first time. Prof. Chang and his colleagues were measuring fluorescence spectra from dye-doped polystyrene microspheres and noticed sharp peaks superposed on the broad spectra which were usually observed from bulk material. Initially, they thought these sharp peaks were noise, but soon after, they realized these were resonant peaks corresponding to the natural modes of the microsphere. They are formed from the constructive interference of the light wave traveling around the surface of the microsphere trapped by total internal reflection. The enhanced effects can be formed from inelastic (fluorescence and Raman) scattering, elastic scattering, or the illuminating light source [7–10]. This study has dramatically expanded into numerous research fields, such as quantum electrodynamics (QED), optical add/drop filter in communication, bacterial and viral optical sensors, optoelectronic oscillators and laser stabilizers, generation of microlasers and optical combs [4–25]. The early fundamental studies of how the resonances occur rapidly expanded to explain how the illumination position and coupling method of the light source affect the resonance, what the electronic field distribution is within the microcavity, and what the near-field and far-field spatial distribution of the emission looks like. While the most fundamental cavities are the spherical droplets in the early experiments, it was found that the MDRs can occur in many different shapes of microcavities, including spheres, disks, rings, oblate/prolate spheroids, ovals, triangles, rectangles, squares, polygons, or any of their deformed /disturbed shapes [7–14]. By measuring the shift of the resonant peaks from a very high-Q microcavity, a sensor can even detect a single virus or single molecule [15,16]; Using a combination of multiple particular resonant frequencies from microresonators, these adding/dropping filters have wide applications in optical communication [17–19]; From the early observation of laser emission from droplets [11], microlasers on solid chips have been realized with directional emission [20–23] and even by electronic pumping other than through an optical pump, towards a applicable laser device [23]; The invention of the optical frequency comb (OFC), which is a new laser source consisting of a series of discrete, equally spaced frequency lines, was awarded the Nobel Prize in Physics to John L. Hall and Theodor W. Hänsch in 2005. The OFC initially was produced by a four-wave mixing process from expensive, clumsy mode-locked laser systems. Recently, using a particular nonlinear-material-formed micro-resonators, the OFC was able to be generated on a chip covering a wide frequency range from UV, blue, Green, red, and to infra-red (IR) (e.g. [24,25]). It is amazing that the MDR phenomena has expanded into so many very exciting fundamental and application research fields.

2.3. Resonant-enhanced, stimulated, and surface-enhanced Raman scattering

Prof. Chang's pioneering research in Raman scattering started along many fronts, from investigations of Raman scattering from semiconductor [26,27] and gas [28], resonant-enhanced Raman scattering [26,28], MDR enhanced Raman scattering [10,29], surface enhanced Raman scattering (SERS, [30–36]), to stimulated Raman scattering [37–41]. About a quarter of his publications related to Raman scattering. Raman spectroscopy remains one of the main chemical identification methods, providing the molecular rovibrational energy distribution from inelastic scattering of photons by matter. This analytical method has been used widely in research and industry. However, scientists struggled with extremely weak signals from its very small cross section about 10^{-6}, resulting in low signal-to-noise ratios. Gaining a stronger Raman scattering signal has been a big challenge since its discovery, and Prof. Chang's research has been focused on demonstrating how to obtain large Raman scattering using various enhancement techniques.

In the Raman-scattering process, molecules absorb the laser photons and are excited to a virtual electronic energy level, then transit to the ground state by emitting the Raman scattering photons. If the exciting laser wavelength is selected close to an electronic transition of the molecules, this photon absorption resonance can greatly enhance the Raman scattering intensity by a factor of thousands. This Resonant-Raman scattering does not change the spectral profiles, but the intensity varies with the detuning of the incident laser wavelength. Prof. Chang and his colleagues first demonstrated that they could obtain 3 orders of magnitude enhancement in the Stokes and the first-overtone intensities from resonance Raman scattering to resonance fluorescence in I_2 gas [28]. Perhaps the most well-known Raman-enhanced technology today is surface-enhanced Raman scattering (SERS), in which molecules adsorbed onto rough metal surfaces or nanostructures can enhance the Raman signal as much as 10^{12}, making single-molecule detection possible. There are two primary mechanisms in explaining such SERS phenomena. The electromagnetic theory attributes the excitation of localized surface plasmons, while the chemical theory proposes the formation of charge-transfer complexes. Similar to resonant-Raman, SERS does not change the spectral peak positions, and would not be affected by the tuning of the excitation wavelength, but it can display different relative peak intensities from spontaneous Raman spectra. For instance, some bands appear only in the bulk-phase Raman spectrum or only in the SERS spectrum. This suggests that the selection rules in the two processes are different [31]. Enormous enhancement factors have been measured with 10^6 to 10^8 times in certain metal surfaces resulting from submonolayer coverage of molecular adsorbates at the interface [31]. Surface plasmon excitation can increase SERS intensities by at least 10 times while metal islands of Ag can give much larger SERS intensities [32]. Meanwhile, experimental and theoretical calculations indicated that the surfaces of silver spheroids can lead to field enhancement, this effect dropped dramatically as the spheroids size increased, in addition to shifting the resonances to longer wavelengths and broadening them. The spheroids can also produce a new set of higher-order resonances appearing at shorter wavelengths [34,35].

The stimulated Raman scattering (SRS) process occurs when an optical feedback at the Stokes-shifted frequency adds to the spontaneous process and gets amplified. This feedback can be supplied by a strong laser or enhancement from a microcavity through MDR effects. It depends on the spontaneous Raman cross section and linewidth, the path length of the medium, the input intensity, and the optical feedback at the right frequency [36]. Dominant SRS peaks occurring droplets were found to consist of a series of equally spaced peaks which were attributed to the MDRs of the droplet. The incident laser and SRS intensities inside the droplets were high enough to produce non-linear optical effects, like third-order sum frequency generation and two-photon absorption [36–41]. Obviously, SRS does modify the spectral profiles from spontaneous, resonant, and surface enhanced Raman scattering. Furthermore, if the Raman scattering process occurs through the presence of a microcavity, the emissions would be enhanced by the MDR effects at the resonance frequencies determined by the dimension and structures of the microresonator or nanostructures. These enhanced peaks may coincidently overlap with the spontaneous peaks, but mostly they do not. The enhancement factor strongly depends on the high Q factor of the resonator. These peak positions also can be either in the TE or TM modes. In all such studies with droplet resonators, it is important to remember that the resonances depend on the droplet size, and these resonant modes will shift as the liquid droplet evaporates. The information resulting from Prof. Chang's investigations greatly advanced our understanding of the various Raman scattering processes, hastening the development of science and technology in this field [26–41].

2.4. Development of microlaser and micro-opto-electronic devices

Since the invention of the laser in 1960, scientists have continually worked to improve its quantum efficiency, size, power, and wavelength range. Printing lasers and other optical and electronic components with integrated circuits (IC) on a board had been a dream for the ideal optic-electronic systems. Starting from the discovery of a lasing droplet [11], Prof. Chang's fundamental research advanced this process [20,21,42–48]. By studying the effects of deforming the microcavity into a disk, ellipse, quadrupole and hexadecapole, large differences in the far-field emission patterns were observed. The differing geometries of the unstable manifolds of periodic orbits were the decisive factors in determining the far-field pattern. Surprisingly, strongly chaotic ray dynamics was compatible with highly directional emission in the far field [42,43], lasers emit in multiple or all directions for a microcavity-based lasing system. It was found by introducing special cavity shapes, unidirectional laser emission could also be achieved [20,21,44,45]. Using designed spatially selective optical pumping or electronic injection, the lasing threshold was greatly reduced, and the quantum efficiency was significantly increased [23,46]. Using a combination of resonators with slightly different size and shapes, add/drop filters with large free spectral range, selective mode, high-coupling efficiency device can be produced [19,47,48]. Other devices such

(used with permission)

Fig. 3. The American Association for Aerosol Research (AAAR) awarded the David Sinclair Award to Prof. Richard K. Chang in 2005.

as a three-element directly coupled microdisk device consisting of one spiral microdisk with two semicircular microdisks was able to control individual microdisk lasing while achieving optimal transfer of optical power between adjacent microdisks. By controlling the transparency of the center element via an injection current, the edge elements were able to exchange their amplified spontaneous emission. In this manner, on-off-on switching of the output intensity, as well as discontinuous shifts in the output wavelength, was achieved as a function of increasing injection current [48]. All these studies opened new visions for the world in developing micro-opto-electronic devices.

2.5. Laser spectroscopic technologies for the detection and characterization of airborne aerosol particles

The Yale Center for Laser Diagnostics was established by Prof. Chang and his colleagues in early 1980's, and it is currently directed by Prof. Marshall Long who is Prof. Chang's former graduate student. The center gathers experts from the Departments of Applied Physics, Electrical Engineering, and Mechanical Engineering, and use their strengths in laser physics, spectroscopy, surface science, detector technology, and signal processing to study turbulence, combustion, molecule-surface interactions, and airborne aerosol particles. Prof. Chang spent tremendous energy managing the labs, training young researchers, and generating great scientific results in this center. The American Association for Aerosol Research (AAAR) David Sinclair Award in 2005 was awarded to Prof. Chang for his extensive and continuous work in the aerosol field for more than 30 years. It is "inventive, seminal, and very often beautiful and proper adjectives to describe his work." (Fig. 3). The award was in recognition of Prof. Chang's pioneering research in developing laser spectroscopic technologies for the detection and characterization of airborne aerosol particles.

In the 1990s, detecting harmful aerosol particles became increasingly important not only from an environmental standpoint, but also as a national security concern over the potential threat that biowarfare agents could pose. This was a driver of Prof. Chang's latter research, in which he worked to develop methodologies for real-time, in-*situ* detection and characterization of aerosol particles. His goal was to focus on the laser spectroscopic methods he had worked to develop without relying on reagents. To realize these goals, Prof. Chang and YP with various graduate students at Yale collaborated with researchers at the Army Research Laboratory (ARL), primarily led by Dr. Ronald Pinnick, Dr. Steven Hill, and GV. Together, we drove the field in developing fluorescence and elastic scattering-based biosensors [49–69]. We were the first to measure fluorescence from single aerosol particles and discovered that the biomolecules excited by UV lasers can be used to distinguish between bioaerosols and non-bioaerosols. The single-particle fluorescence from *Bacillus anthracis* was far larger than that from common mineral dust (kaolin) and other particles [49]. By detecting and synchronizing the arrival of flowing particles with a conditionally fired UV laser pulse, single particle fluorescence detection was made practical [50]. YP remembers that the collaboration formed during a lunch break at the CLEO conference in Baltimore of May, 1998, where we discussed how to improve the system detection ability. Returning from the conference, YL began the experimental work of the bioaerosol detection project with then Ph.D. student Stephan Holler, who is currently a professor at Fordham University.

With three key improvements, we soon obtained high SNR spectra from *B. subtilis* (a simulant for *anthrax*) at 1 μm size excited by a single laser pulse at 266 nm or 351 nm in June, 1998 [51]. In September, Prof. Chang, Stephan Holler and YP drove to Adelphi to duplicate the system at ARL, where scientists from the Navy Research Laboratory (NRL), Air Force Research Laboratory (AFRL), Edgewood Chemical and Biological Center (ECBC), and the National Institute for Occupational Safety and Health (NIOSH) brought their interesting bioaerosol simulants, interference, and background aerosols to the lab to measure single-particle fluorescence spectra [52]. We demonstrated the system and recorded high SNR single particle fluorescence spectra. To address the low frame rates, Prof. Chang put forward an idea to use a multiple-anode photomultiplier tube (PMT) for spectral detection. Hamamatsu introduced such units to the CLEO community, but they were not available on the market. We approached the Vtech Engineering Corp. at Boston to develop special electronics for the PMT, and the high-speed, high-sensitivity single-particle fluorescence spectrometer (SPFS) were created. This system was capable of obtaining two dispersed fluorescence spectra (280–700 nm) and the scattering size from individual single flowing aerosol particles using two wavelength excitation at 263 and 351 nm, it could detect and analyze particles at a maximum rate of 20,000 particles per second, and could even sort and collect suspect particles for further examination based on the real-time analyses [53–57]. Prof. Chang's group also led the field in using light-emitting diodes (LED) and LED arrays as the miniaturized excitation light source for bioaerosol detection. This revolutionary proof-of-concept demonstration pointed the way to making smaller and less expensive deployable biosensors [58,59].

While fluorescence and Raman scattering is primarily dependent on aerosol composition, the distribution of elastically scattered light is primarily dependent on aerosol size and shape. To harness this information and gain more aerosol signatures, we built an optical system that can measure the two-dimensional angle-resolved optical scattering patterns (TAOS) of individual aerosol particles using one CCD detector. The device can record the scattering patterns over a very large solid angle ($>2\pi$ steradians) for an aerosol particle illuminated by a single laser pulse. Thousands of scattering patterns from laboratory-generated biothreat simulants and atmospheric aerosol particles have been recorded and analyzed, and it was shown that the TAOS patterns do provide additional discriminating information to distinguish some classes of bioaerosol particles from interferents and background aerosol particles [60–62]. One of the interesting outcomes of the SPFS system was a better understanding of natural and anthropogenic aerosols, as numerous measurement campaigns were carried out to determine the natural background from different regions in different seasons [62–64]. Other studies to determine the best detector and light source and excitation wavelength, as well as how fluorescence emission depends on particle size, concentration of fluorophores, illumination intensity, collecting direction, and photodegradation [65–68] also have been investigated. All these efforts have been widely recognized by the aerosol research society, such as Dr. Richard DeFreez commented "The ARL/Yale team has generated numerous key developments in the relevant technology and science" in his invited review paper "LIF Bio-Aerosol Threat Triggers: Then and Now" [69].

3. Scattered stories about Prof. Chang's family life

Many of Prof. Chang's students were surprised to learn that he had not been a stellar student in his youth. Nor had his academic path been straightforward or assured. At his various international elementary schools, he did just enough work to get by and was happy to receive the letter grades "C" for average work, saying proudly that C "was the best" as it was the first letter of his surname. While he loved riding his bicycle and playing ball with his friends, his antics included climbing to the open Jeeps of visiting officials and sometimes sending a car rolling accidentally. When he was fourteen, his family moved to Sao Paulo, Brazil, where he attended a British school and took up every dare his equally prank-obsessed friends challenged him to, invariably landing all of them in school detention every Saturday morning. There, Richard was proud when he figured out a way to complete quickly the assignment of 1,000 lines of "I will not " by writing with four pens at the same time, and even rotating them so the teacher would detect no pressure pattern.

One Saturday morning in January 1956, Richard's mother was having tea at a friend's home in Sao Paulo and noticed that her friend's son, a classmate of Richard's, was at home. Is your son sick and had to be absent from school? Richard's mother asked. Her friend informed her that there was no school on Saturdays. Richard's mother proudly responded that her son went to school every Saturday. "Only juvenile delinquents go to school on Saturdays!" came her friend's response. Shocked, embarrassed and angry, Richard's mother confronted him when she returned home and learned the truth. His mother, who had received her own Ph.D. from the London School of Economics declared, "No more fooling around in Brazil! I am sending you to America to study!"

This is how Richard found himself alone on a ship bound to America the next month. Not being familiar with the American school system, Richard's mother was unsure what grade to enroll Richard. Nor did she remember the Northern hemisphere was well in the middle of a school year (not summer) and cold winter. She did know, however, that Richard needed a change in environment if he wished to fulfill whatever potential he had. Her message must have gotten through to Richard as he vowed to himself on the ship that he would become a new serious person.

It was with this determination that Richard began his education at the Stony Brook School, a Christian preparatory school in Long Island, New York, which had the school motto of "Character before Career." Richard arrived in the middle of his junior year. As a scholarship student, he worked several hours a week with his roommate, another scholarship student, to thoroughly clean and mop the school's kitchen and cafeteria floors after the last meal was served on Saturday evenings. To make pocket money, he sometimes did cleaning during the weekends for staff living on campus.

He spent his first summer in America working on a dairy farm in upstate New York. The first Saturday he tried to sleep in, the farmer's wife woke him at 5:30 a.m. and said that the cows did not take the weekends off. Richard said he realized then that education was his only escape from hard labor. When sent to cut saplings in the field, he cut ten trees and studied ten SAT words. The following summer, he was an assistant to Jacques-Andre Istel, an alumnus of Stony Brook known as the "father of modern skydiving." Part of Richard's job as an assistant was to swim daily in the pool with Jacques Istel who was fully dressed in heavy parachuting gear.

Although Richard's half year as a junior went smoothly, the first semester of his senior year presented considerable challenges when he had to apply to college. His facility in English was passable for a foreign student and his math was good, so he could only apply to engineering colleges. However, every engineering college required that the prospective applicant had taken high school Chemistry which Richard had not. The Chemistry teacher at Stony Brook told Richard that he would give Richard the Chemistry textbook and his syllabus for the course. Richard could study Chemistry on his own for the next two and half months, and the teacher would answer any questions Richard did not understand. If Richard passed the school's final Chemistry exam by December, he would give Richard a whole year's credit for chemistry, thus making him eligible to apply to an engineering college. Much to his delight and surprise, Richard passed the Chemistry test. He applied to and was accepted at MIT, with a major in Electrical Engineering.

Richard's acceptance to MIT had a profound effect on him. First, he realized the influence that a single teacher, in this case, his chemistry teacher, Mr. Goldberg - who considered him individually and came up with a creative solution for him (as well as wrote him a glowing recommendation letter) - could have. Secondly, he felt quite certain that, in any other country, without consistent good marks, he would not have been given this opportunity. For him, America was indeed the land of opportunity. These two realizations, that a single person could make a big difference, and that there are many ways to judge a person's potential, helped to shape his future thoughts and actions towards his own students.

In his freshman year at MIT, Richard had a job as a busboy in the cafeteria. He was occasionally teased by some Hong Kong classmates for having to work, as they knew how prosperous his family had been before the Chinese Communist Party took power. Not letting the teasing get to him, Richard would describe the taunts as "water falling down a duck's back." In his sophomore year, Richard was accepted to the MIT Coop Program where he could work at IBM for one semester and attend MIT classes the next semester. Summer jobs at IBM were guaranteed. The only catch was that a student had to study and pass certain courses on their own in order to fulfill the MIT course requirements.

Richard worked at IBM in Poughkeepsie, Kingston, Rio de Janeiro, and the Thomas J. Watson Lab in Yorktown Heights. He loved working at IBM, and it was his intention to work there after graduation. While working a semester at the Kingston IBM Lab in 1959, Richard went to visit his cousin for Thanksgiving. She was staying with a host family in Queens, New York. The family had two sons and three daughters, the second of which, Sunny, was studying at Brandeis University in Waltham, Massachusetts, not too far from MIT. Before Richard left the house, he promised Sunny's father, who had liked him immediately, that he would look Sunny up once he returned to MIT next semester. After their first date, Richard decided that Sunny was the woman he wanted to marry. It took her a while to reach that same conclusion. However, they did get married in 1961 and were happily married for almost 59 years!

During Richard's senior year at MIT, he and Sunny often discussed his working for IBM after graduation. It was during one of those discussions, that he mentioned how wonderfully precise IBM was in wanting their employees to be there at 7:13 in the morning. Sunny could not believe that a company would want their employees to be there at 7:13. Why not 7:10 or 7:15 am? She asked, why would anyone want to work for a company that wanted you there at 7:13? She suggested that perhaps he ought to apply to graduate school. Perhaps with a Master's degree, one need not be there at 7:13. Richard thought about her suggestion and agreed. As Sunny was still a junior in college, Richard wanted to be in the same area, so Harvard seemed like a good choice. The couple got the application, filled it out while sitting in the parking lot behind Gordon McKay Lab, and submitted it.

Miraculously, Richard was accepted at Harvard for a Master's degree program. A couple of decades later, however, one of Richard's colleagues at Yale (Prof. Bob Narendra), told him what had happened when his application had come before the Harvard Admissions Committee. The committee asked: Is this guy for real? Why is he applying only for a Master's degree, not a Ph.D.? The committee called Richard's reference, Professor Epstein at MIT, who explained that Richard was thinking about getting married and needing a job soon, so he wanted to get just a Master's degree. Professor Epstein then told Harvard to accept Richard as he had a lot of potential and would be a good Ph.D. candidate. At MIT, Richard had taken Professor Epstein's electrical engineering course. The mid-term exam average was at a low 29%, so Professor Epstein worked out the solutions for the exam problems on the board before returning the blue examination books back to the students. When all the blue books had been returned to the students, Richard did not receive his. He went to see Professor Epstein who said that all the books had been returned, and he did not know why Richard's exam book was missing. He told Richard that since the solution was now revealed, he would give Richard whatever grade Richard thought he deserved. Richard hesitated, and finally said, "96%." Professor Epstein got very angry and declared that under no circumstances would he give Richard a 96. How dare Richard take advantage of the situation! Their meeting did not go well. Some three weeks later, Professor Epstein found Richard's misplaced blue exam book and indeed Richard had gotten around 96%. Professor Epstein then told Richard that if he ever needed a recommendation for graduate school or work, he would help him. This was how Richard got accepted to Harvard for a "nonexistent" Master's degree program.

The 5 years at Harvard went by quickly. Sometimes life was rather hectic. Sunny and Richard got married, many exciting scientific experiments were performed, many new friendships were formed and two new family members (son Kevin and daughter Natasha) were added with a third on the way! In the beginning Sunny worked full time in the Harvard University Biological Laboratories. In 1964, she returned to her alma mater Brandeis University to work on her Ph.D. in Biology which became interrupted by Richard's acceptance of a position at Yale in 1966.

Richard began his life at Yale on July 1, 1966 when the whole family moved to their house on Paradise Avenue in Hamden, Connecticut where their daughter Nicole was born. All three children were raised in the rich musical environment surrounding Yale, attended the Hamden public schools, and received their undergraduate education at Harvard University before they pursued professional education. All three have their own families of two children each (Fig. 4).

Both Richard and Sunny were involved in each other's professional life. Sunny accompanied Richard to numerous conferences and partook in many scientific discussions; they were inseparable. Sunny became coordinator of a New Haven high school gifted program after receiving a Connecticut Teaching Certification from the Yale Master of Arts Teaching Program in 1968. Richard accepted more than ten of Sunny's gifted inner-city high school students to work in his lab throughout his Yale career. In the 1970's, Sunny and a colleague of hers, Frances Bennett (wife of Yale Physics Professor William Bennett) approached Professor Robert Wheeler, then chairman of Yale's Department of Engineering and Applied Physics, about starting a Saturday morning program to introduce interesting engineering and physics topics to local high school students. There would be scientific lectures, demonstrations and lab visits sponsored by different professors in the department. Professor Wheeler agreed in principle as long as a Yale faculty member would be willing to be in charge. Richard agreed to be in charge of it and the Saturday morning "Frontiers of Applied Science" for high school students became a reality. Many high school students, both in and out of state, participated in the program.

Richard also supported Sunny in completing a low-residency Ph.D. in Education from Columbia University's Teachers College in 1995. Richard's support included all the household chores and more importantly, a deal with Sunny, that she would do at least one full hour of physical exercise with him daily. Although Sunny was initially annoyed at the contingency requirement, she credits the daily exercise (June 1993–June 1995) as crucial to her being able to work full time while pursuing her degree.

In early 1996, at the age of 55, Richard was diagnosed with Parkinson's disease. Physical exercise played a pivotal role in Richard's daily life, and he knew his body well enough to complain of something amiss for more than a year before his diagnosis. He underwent numerous tests, and neurologist after neurologist told him that he did not have enough symptoms for a diagnosis. Studies have shown that there is a probable link between exposure to insecticides and Parkinson's disease. As none of his relatives ever had Parkinson's disease, Richard and Sunny surmised that possibly Richard's Parkinson's was the result of his continual exposure to DDT while he was living in Brazil. During the two years he lived in Brazil, Richard slept with a DDT spray can next to his bed. While in bed, he would hold the DDT can and aim for any mosquito that flew above him and marvel at the way the mosquitos would dive down after just one spray of DDT. He had such fun! Sunny often told Richard that he was lucky to only have to contend with Parkinson's, not cancer also, considering how much DDT he had breathed in during his teens.

As the disease progressed, Richard had difficulty walking and talking. These problems did not place any limitations on their lifestyle. GV remembers inviting Richard to attend a NATO workshop held in Novosibirsk, Russia in 2005. Despite his Parkinson's and the logistics involved in getting to the meeting, Richard said he would be there. He and Sunny showed up and even had a mini family reunion in the Siberian forest with their son-in-law. Less than one month after his retirement in June 2008, Richard had a

Fig. 4. Richard and Sunny with their family gathered at a luncheon at Singapore's Tower Club in 2016 Easter.

drowning episode in a swimming pool near their retirement house in Brattleboro, Vermont. He gradually recovered after two days in a coma. This episode significantly hurt his retirement life; however, we never heard any complaint from him. He always was smiling, patient and generous, while on a slow, challenging recovery filled with physical therapy and exercise to regain his health.

Despite severe health setbacks, Richard and Sunny always were positive and happy. It cannot be doubted that the source of their happiness was their strong faith in God. They accepted challenges in stride and celebrated whatever gifts of joy came into their lives in any form, always finding a beam of sunshine no matter how cloudy the skies. Most importantly they spread their joy to everyone they met. While they used science to try to unravel some of the magic of God's creation, they accepted His mysteries through their faith. Richard was a visionary, always looking at new technologies and how they could improve people's lives and probing with friends and colleagues how they could be used to advance the human condition. He and Sunny always hoped that tomorrow would be better than today.

Richard Chang will be remembered by many as a true gentleman, kind-hearted, with a thoughtful and loving nature, and always wearing a smile. He has been a model for us to learn how to enjoy and celebrate life. He always considered and worried for the students and always put others before himself. His demeanor calmed those whom he met and he strongly affected our lives, as well as many others. After Richard's 60th birthday celebration, organized by Professor Marshall Long, Richard received a handwritten note from D. Allan Bromley, dean of Yale Engineerings and Applied Physics (also science advisor to President George W.H. Bush), saying that he had attended many such Yale events but had never been to one where a professor was so beloved by his students.

Richard and Sunny settled in Singapore (in the building next to their son and his family) in 2014 to escape the New England winters and get much-needed domestic help. In this final chapter, Richard enjoyed great food, travelling in Asia, swimming, musical performances, numerous visits from his daughters and their families, visits from former students or colleagues, and much time with his son's family, especially his two young granddaughters. We are extremely saddened for the passing of Prof. Chang on February 16, 2020 in Singapore. It is a huge loss to his family, to us, to his friends, and to the whole scientific community.

Declaration of Competing Interest

The authors (Yong-Le Pan, Gorden Videen, Pang-Mei Natasha Chang, and Sung-Wen Pu Chang) declare that they have no known competing financial interests or personal relationships that could have appeared to influence the work reported in this paper.

Supplementary materials

Supplementary material associated with this article can be found, in the online version, at doi:10.1016/j.jqsrt.2020.107273.

References

[1] Bloembergen N, Chang RK, JHA SS, et al. Optical second-harmonic generation in reflection from media with inversion symmetry. Phys Rev 1968;174(3):813–22.

[2] Parsons FG, Chen EY, Chang RK. Dispersion of nonlinear optical susceptibilities in hexagonal II-VI semiconductors. Phys Rev Lett 1971;27(21):1436–9.

[3] Parsons FG, Chang RK. Measurement of the nonlinear susceptibility dispersion by dye lasers. Opt Commun 1971;3(3):173–8.

[4] Fields MH, Popp J, Chang RK. Nonlinear optics in microspheres. Prog Opt 2000;41:1–95.

[5] Chang RK. Some of Bloembergen's nonlinear optical effects revisited in single micrometer-sized droplets. In: Resonances. Singapore: World Scientific Publishing Co; 1990. p. 200–11.

[6] Leach DH, Chang RK, Acker WP, Hill SC. Third-order sum-frequency generation in droplets: experimental results. J Opt Soc Am B 1993;10(1):34–45.

[7] Benner RE, Barber PW, Owen JF, Chang RK. Observation of structure resonances in the fluorescence spectra from microspheres. Phys Rev Lett 1980;44:475–8.

[8] Barber PW, Chang edited RK. Optical effects associated with small particles. Singapore: World Scientific Publishing Co; 1988.

[9] Chang RK, Campillo A J. Optical processes in microcavities edited. Singapore: World Scientific Publishing Co; 1996.

[10] Owen JF, Chang RK, Barber PW. Morphology-dependent resonances in Raman scattering, fluorescence emission, and elastic scattering from microparticles. Aerosol Sci Technol 1982;1(3):293–302.

[11] Tzeng H-M, Wall KF, Long MB, Chang RK. Laser emission from individual droplets at wavelengths corresponding to morphology-dependent resonances. Opt Lett 1984;9(11):499–501.

[12] Messinger BJ, Ulrich von Raben K, Chang RK, Barber PW. Local fields at the surface of noble metal microspheres. Phys Rev B 1981;24(2):649–57.

[13] Wolf JP, Pan YL, Turner GM, et al. Ballistic trajectories of optical wave packets within microcavities. Phys Rev A 2001;64(2):023808.
[14] Hill SC, Boutou V, Yu J, Ramstein S, Wolf JP, Holler S, et al. Enhanced backward-directed multiphoton-excited fluorescence from dielectric microcavities. Phys Rev Lett 2000;85(1):54–7.
[15] Dantham VR, Holler S, Kolchenko V, et al. Taking whispering gallery-mode single virus detection and sizing to the limit. Appl Phys Lett 2012;101(4):043704.
[16] Armani AM, Kulkarni RP, Fraser SE, Flagan RC, Vahala KJ. Label-free, single-molecule detection with optical microcavities. Science 2007;317(5839):783–7.
[17] Little BE, Chu ST, Haus HA, et al. Microring resonator channel dropping filters. J Lightwave Technol 1997;15(6):998–1005.
[18] Little BE, Foresi JS, Steinmeyer G, et al. Ultra-compact Si-SiO_2 microring resonator optical channel dropping filters. IEEE Photon Technol Lett 1998;10(4):549–51.
[19] Poon AW, Courvoisier F, Chang RK. Multimode resonances in square-shaped optical microcavities. Opt Lett 2001;26(9):632–4.
[20] Nöckel JU, Stone AD, Chang RK. Q spoiling and directionality in deformed ring cavities. Opt Lett 1994;19(21):1693–5.
[21] Nockel JU, Stone AD, Chen G, et al. Directional emission from asymmetric resonant cavities. Opt Lett 1996;21(19):1609–11.
[22] Gmachl C, Capasso F, Narimanov EE, et al. High-power directional emission from microlasers with chaotic resonators. Science 1998;280(5369):1556–64.
[23] Kneissl M, Teepe M, Miyashita N, et al. Current-injection spiral-shaped microcavity disk laser diodes with unidirectional emission. Appl Phys Lett 2004;84(14):2485–7.
[24] Guo X, Zou CL, Jung H, et al. Efficient generation of a near-visible frequency comb via cherenkov-like radiation from a Kerr Microcomb. Phys Rev Appl 2018;10(1):014012.
[25] Jung H, Stoll R, Guo X, Fischer D, Tang HX. Green, red, and IR frequency comb line generation from single IR pump in AlN microring resonator. Optica 2014;1(6):396–9.
[26] Ralston JM, Wadsack RL, Chang RK. Resonant cancelation of raman scattering from CdS and Si. Phys Rev Lett 1970;25(12):814–18.
[27] Farrow RL, Chang RK, Mroczkowski S. Detection of excess crystalline As and Sb in III-V oxide interfaces by Raman scattering. Appl Phys Lett 1977;31(11):768–70.
[28] Fouche DG, Chang RK. Observation of resonance raman scattering below the dissociation limit in I2 vapor. Phys Rev Lett 1972;29:536–9.
[29] Chen TT, Von Raben KU, Owen JF, Chang RK. Laser illumination effects on the surface morphology, cyclic voltammetry, and surface-enhanced Raman scattering of Ag electrodes. Chem Phys Lett 1982;91(6):494–500.
[30] Dornhaus R, Long MB, Benner RE, Chang RK. Time development of SERS from pyridine, pyrimidine, pyrazine, and cyanide adsorbed on ag electrodes during an oxidation-reduction cycle. Surf Sci 1980;93(1):240–62.
[31] Chang RK, Laube BL. Surface-enhanced Raman scattering and nonlinear optics applied to electrochemistry. Crit Rev Solid State Mater Sci 1984;12(1):1–73.
[32] Dornhaus R, Benner RE, Chang RK, Chabay I. Surface plasmon contribution to SERS. Surf Sci 1980;101(1–3):367–73.
[33] Vonraben KV, Chang RK, Laube BL. "Surface enhanced Raman scattering of Au(CN)2– ions adsorbed on gold colloids. Chem Phys Lett 1981;79(3):465–9.
[34] Barber PW, Chang RK, Massoudi H. Surface-enhanced electric intensities on large silver spheroids. Phys Rev Lett 1983;50(13):97–100.
[35] Barber PW, Chang RK, Massoudi H. Electrodynamic calculations of the surface-enhanced electric intensities on large Ag spheroids. Phys Rev B 1983;27(12):7251–61.
[36] Fields M H, Popp J, Chang R K. External seeding of stimulated Raman scattering in droplets. Opt Lett 1996;2118:1456–9.
[37] Snow JB, Qian SX, Chang RK. Stimulated Raman scattering from individual water and ethanol droplets at morphology-dependent resonances. Opt Lett 1985;10(1):37–9.
[38] Zhang JZ, Leach DH, KChang R. Photon lifetime within a droplet: temporal determination of elastic and stimulated Raman scattering. Opt Lett 1988;13(4):270–2.
[39] Chen G, Chang RK, Hill SC, Barber PW. Frequency splitting of degenerate spherical cavity modes: stimulated Raman scattering spectrum of deformed droplets. Opt Lett 1991;16(16):1269–71.
[40] Qian SX, Snow JB, Chang RK. Coherent Raman mixing and coherent anti-Stokes Raman scattering from individual micrometer-size droplets. Opt Lett 1985;10(10):499–501.
[41] Qian SX, Chang RK. Multi-order Stokes emission from micrometer-size droplets. Phys Rev Lett 1986;56(9):926–9.
[42] Schwefel HGL, Rex NB, Tureci HE, et al. Dramatic shape sensitivity of directional emission patterns from similarly deformed cylindrical polymer lasers. J Opt Soc Am B 2004;21(5):923–34.
[43] Chang S, Chang RK, Stone AD, et al. Observation of emission from chaotic lasing modes in deformed microspheres: displacement by the stable-orbit modes. J Opt Soc Am B 2000;17(11):1828–34.
[44] Chern GD, Tureci HE, Stone AD, et al. Unidirectional lasing from InGaN multiple-quantum-well spiral-shaped micropillars. Appl Phys Lett 2003;83(9):1710–12.
[45] Rex NB, Tureci HE, Schwefel HGL, et al. Fresnel filtering in lasing emission from scarred modes of wave-chaotic optical resonators. Phys Rev Lett 2002;88(9):094102.
[46] Rex NB, Chang RK, Guido LJ. Threshold lowering in GaN micropillar lasers by means of spatially selective optical pumping. IEEE Photon Technol Lett 2001;13(1):1–3.
[47] Pan YL, Chang RK. Highly efficient prism coupling to whispering gallery modes of a square mu cavity. Appl Phys Lett 2000;82(4):487–9.
[48] Fernandes GE, Guyot L, Chern GD, et al. Wavelength and intensity switching in directly coupled semiconductor lmicrodisk lasers. Opt Lett 2008;33(6):605–7.
[49] Hill SC, Pinnick RG, Nachman P, et al. Aerosol-fluorescence spectrum analyzer: real-time measurement of emission spectra of airborne biological particles. Appl Opt 1995;34(30):7149–55.
[50] Chen G, Nachman P, Pinnick RG, et al. Conditional-firing aerosol-fluorescence spectrum analyzer for individual airborne particles with pulsed 266-nm laser excitation. Opt Lett 1996;21(16):1307–9.
[51] Pan YL, Holler S, Chang RK, Hill S, Pinnick RG, Niles S, et al. Single-shot fluorescence spectra of individual micron-sized bio-aerosols illuminated by 351 nm and 266 nm laser. Opt Lett 1999;24(2):116–18.
[52] Hill SC, Pinnick RG, Niles S, Pan YL, Holler S, Chang RK, et al. Real-time measurement of fluorescence spectra from single airborne biological particles. Field Anal Chem Technol 1999;3:221–39.
[53] Pan YL, Cobler P, Potter A, Chou T, Holler S, Chang RK, et al. High-speed, high--sensitivity aerosol fluorescence spectrum detection by using a 32-anode PMT detector. Rev Sci Instrum 2001;72(3):1831–6.
[54] Pan YL, Hartings J, Pinnick RG, Hill SC, Halverson J, Chang RK. Single-particle fluorescence spectrometer for ambient aerosols. Aerosol Sci Technol 2003;37(8):628–39.
[55] Huang HC, Pan YL, Hill SC, Pinnick RG, Chang RK. Real-time measurement of dual-wavelength laser-induced fluorescence spectra of individual aerosol particles. Opt Express 2008;16(21):16523–8.
[56] Pan YL, Hill SC, Pinnick RG, Huang H, Bottiger JR, Chang RK. Fluorescence spectra of atmospheric aerosol particles measured using one or two excitation wavelengths: comparison of classification schemes employing different emission and scattering results. Opt Express 2010;18(12):12436–57.
[57] Pan YL, Boutou V, Bottiger JR, Zhang SS, Wolf JP, Chang RK. A puff of air sorts bioaerosols for pathogen identification. Aerosol Sci Technol 2004;38:598–602.
[58] Pan YL, Boutou V, Chang RK, Ozden I, Davitt K, Nurmikko AV. Application of light-emitting diodes for aerosol fluorescence detection. Opt Lett 2003;28(18):1707–9.
[59] Davitt K, Song YK, Patterson WR, Nurmikko AV, Gherasimova M, Han J, et al. 290 and 340 nm UV LED arrays for fluorescence detection from single airborne particles. Opt Express 2005;13(23):9548–55.
[60] Pan YL, Aptowicz KB, Hart M, Eversole JD, Chang RK. Characterizing and monitoring respiratory aerosols by light scattering. Opt Lett 2003;28(8):589–91.
[61] Aptowicz KB, Pan YL, Chang RK, Pinnick RG, Hill SC, Tober RL, et al. Two-dimensional angular optical scattering patterns of microdroplets in the mid infrared with strong and weak absorption. Opt Lett 2004;29:1965–7.
[62] Aptowicz KB, Pinnick RG, Hill SC, Pan YL, Chang RK. Optical scattering patterns from single urban aerosol particles at Adelphi, Maryland, USA: a classification relating to particle morphologies. J Geophys Res 2006;111:D12212.
[63] Pinnick RG, Hill SC, Pan YL, Chang RK. Fluorescence spectra of atmospheric aerosol at Adelphi, Maryland, USA: measurement and classification of single particles containing organic carbon. Atmos Environ 2004;38:1657–72.
[64] Pan YL, Pinnick RG, Hill SC, Rosen JM, Chang RK. Single-particle laser-induced-fluorescence spectra of biological and other organic-carbon aerosols in the atmosphere: measurements at New Haven, CT and Las Cruces, NM, USA. J Geophys Res 2007;112:D24S19.
[65] Hill SC, Pinnick RG, Niles S, Fell NF Jr., Pan YL, Bottiger JR, et al. Fluorescence from airborne microparticles: dependence on size, concentration of fluorophores, and illumination intensity. Appl Opt 2001;40(18):3005–13.
[66] Pan YL, Holler S, Pinnick RG, Hill SC, Niles S, Bottiger JR, et al. Dynamics of photon-induced degradation and fluorescence spectra in riboflavin microparticles. Appl Phys B 2001;72:449–54.
[67] Hill SC, Boutou V, Yu J, Ramstein S, Wolf JP, Pan YL, et al. Enhanced backward-directed multi-photon-excited fluorescence from dielectric microcavity. Phys Rev Lett 2000;85(1):54–7.
[68] Hill SC, Pinnick RG, Niles S, Fell NF Jr., Pan YL, Bottiger JR, Bronk BV, Holler S, Chang RK. Fluorescence from airborne microparticles: dependence on size, concentration of fluorophores, and illumination intensity. Appl. Opt 2002;40(18):3005–13.
[69] DeFreez R. LIF bio-aerosol threat triggers: then and now. SPIE proceedings, optically based biological and chemical detection for defence V, 7484; 2009. 74840H.

Index

www.ingramcontent.com/pod-product-compliance
Ingram Content Group UK Ltd.
Pitfield, Milton Keynes, MK11 3LW, UK
UKHW051834120926
8588IPUK00006B/52

9 789819 803248